Laboratory Investigations

A Manual for General Biology
Fourth Edition

Michael B. Clark

Suspended Animations, Publisher
www.suspendedanimations.net

Laboratory Investigations Lab Book

A Manual for General Biology

Fourth Edition

Many Illustrations by: Ron Leishman

ISBN 978-1-885380-02-9

Published in the United States of America by:

Suspended Animations

20275 Deerhorn Valley Road

Jamul, CA 91935

877-468-4777

www.suspendedanimations.net

January 2016

Table of Contents

Notes to the Student

Welcome to *Laboratory Investigations*

This book is written for you—the non-major biology student. ***Laboratory Investigations*** uses an approach that is easy to understand and allows you to build your knowledge of science one step at a time no matter what your previous background. A lab class is quite different from a lecture class. Mostly, it is more "doing" than listening and more "looking" than taking notes. Some students are apprehensive because of these differences, and they worry about their chances of success. The comments and suggestions below should answer some of your questions about how to succeed in biology lab class.

- Come early to the first day of lab class. Check out the other students. One (or more) of them is going to be your lab partner. Some students are not as interested in learning biology. Find someone who is serious about success. Be a good partner for them. Look for several other students who also are good lab partners, and form a study group for tests.
- Be prepared. Briefly read the lab activities before you come to class. If you and your lab partner are prepared, you will learn more during the lab and will finish the work before the end of class.
- Review the topics in this lab book that pertain to your upcoming lecture test. You will discover that many of the lab chapters are helpful with understanding material presented in lecture class. This can be especially helpful if your lecture class is ahead of the lab schedule. For example, if you are having a lecture test covering protein synthesis and your class hasn't done that lab yet, then review what is covered in this book under that topic. It will help you for the lecture exam.
- We use words sparingly, constructing each sentence as a clue to something important about the topic. Read carefully, and work step-by-step through the instructions and explanations. Pay particular attention to the ***bold-italics*** terms and concepts. These designate the central themes and definitions. They are important ideas, and many will be found on your lab tests.
- Use the blank space in the left margin of the page for lab notes and detailed answers to the questions.
- Take time to review your work. Most students leave as soon as they finish the lab. Later, they are surprised when their answers are incorrect or incomplete. Take advantage of the last half-hour of each lab. It is an excellent opportunity for checking your understanding of the topic with your instructor and other students.
- Always practice safety in the lab classroom. Make sure that you understand the instructions for properly using the lab equipment, lab materials, and chemical solutions. Ask your instructor whenever you have a question or any confusion about safe lab procedures.
- Finally, have fun! Take an active interest in the lab activities and the time will pass quickly and your chances for success will increase. Good luck to you, and let us know how ***Laboratory Investigations*** has helped you. We welcome your comments and suggestions.

Foreword

Laboratory Investigations covers all the topics you expect to find in a traditional biology lab book for non-majors. In addition, there are human oriented and special-topic labs, and we have added tutorial information to many of the labs that is specifically designed to assist students with difficult lecture concepts. All labs have been classroom tested by more than 100,000 students. The changes suggested by those students, their teachers, and the lab technicians have been incorporated into the fourth edition of ***Laboratory Investigations***. Our goal is to provide basic labs that are self-contained and totally dependable. Other lab manuals usually require that most of the organizational work be handled by the teacher and the lab technician. They provide a skeletal set of instructions and guiding questions leaving most of the technical problems and student questions unaddressed. The teacher and the lab technician are in service to these typical lab books. In our opinion, that idea is exactly backwards.

A good lab book should serve everyone: the students, the lab technicians, and the teachers who adopt it. ***Laboratory Investigations*** and its supplementary materials were created with that in mind. Because our labs are easy to understand, students become self-directed. The open-ended format of each lab activity allows you to expand a favorite topic or add your own experiments to the lab without disrupting the overall organization. The lab technician will be pleased to discover that our labs are easy to set up and maintain during the week, and experiments are written with supply budgets and safety in mind. Every lab contains reminders to the students about proper lab procedures. However, those statements are not meant to replace your thorough discussions of safe lab behavior and proper equipment handling that are a necessary part of every lab class.

Be sure to consult our Instructor Guide and Lab Technician Guide. They are full of helpful hints and are especially useful for any new instructors or tutors in your biology program. Visit our website www.suspendedanimations.net and search for all of the available resources that are intended for your use. Also, if you email us with particular interests or needs, we can supply you with even more resources. Finally, ***Laboratory Investigations*** owes its success to many people. We would like to thank the teachers and students and lab technicians at the more than 100 schools who have used our lab books. And we offer a very special thank you to those people whose comments and suggestions have helped us to make ***Laboratory Investigations*** better every year. We do not wait for the next edition to incorporate your suggestions. We incorporate all important edits into each year's new printing. Thank you from all of us at Suspended Animations Publishing.

LAB

1

Measurement

Some people insist that measuring things is the only way to get a fair deal from another person. But others say that measuring things is at the heart of all mistrust between people. Whatever the correct judgment may be, historians tell us that measurement systems are based on political and economic needs and can be found wherever people have a network of dealing with each other.

Originally, measurement systems depended on the type of material being exchanged. For example, a farmer selling apples would price them by the cart-load, not by the bucket. But when selling milk, a cart-load of liquid would have been ridiculous and a bucket more appropriate. These kinds of systems developed and changed as they were used. Eventually, ***standard*** "cart-loads" and "buckets" were established, and these became the basis for the systems of measurement we use today. As science developed, the need for scientific measurement began to overlap the measurement systems used for trade. But there are problems with adapting trade measurement systems for use in science. In grade school children are taught the English System which uses inches, feet, yards; cups, pints, quarts, gallons; ounces, pounds, and tons. However, a quart in Canada is equal to 1.136 liters. A quart in the United States is equal to 0.946 liters. And in Mexico there are no quarts, only liters. For trade purposes, these inequities can be worked out, but in science or industry where precision is important, the English System simply will not do.

One hundred years ago the International Metric System was devised to standardize measurement around the world. This system provides exact precision in ***powers of 10***, and there is a standard reference unit used in each measurement category. On either side of the reference unit, the units increase by 10, 100, and 1000 or decrease by $\frac{1}{10}$, $\frac{1}{100}$, and $\frac{1}{1000}$. The universal standard unit for weight is the ***gram*** (g). The standard for volume is the ***liter*** (l). The standard unit for length is the ***meter*** (m). Science embraced the Metric System and so did many of the countries in the world. It was anticipated that the United States would be converted to the Metric System by now. Because we have not yet converted, it remains necessary for us to be able to convert quarts to liters, pounds to grams, and inches to meters. In today's lab you will be introduced to the Metric System and learn to use simple measuring equipment that involve different units of the Metric System.

Exercise #1	**Some Basics of the Metric System**
Exercise #2	**Conversions: Metric ←→ English**
Exercise #3	**Personal List of Metric Reference**
Exercise #4	**A Test of Your Measuring Skills**
Exercise #5	**A Simple Scientific Experiment**

Exercise #1
Some Basics of the Metric System

Fractions

A *fraction* represents a part of the whole (One), and there are two methods of expressing it:

- First—***Simple Fractions***. Examples are $\frac{1}{4}$, $\frac{1}{2}$, $\frac{9}{25}$, etc.
- Second—***Decimal Fractions***. Examples are 0.25, 0.50, 0.36, etc.

Rule 1: If you want to convert a simple fraction into a decimal fraction, then divide the top number of the simple fraction by the bottom number. Therefore, the simple fraction $\frac{9}{25}$ is converted into decimal form by:

$$25\overline{)\,9.00\,}\quad = 0.36$$

Rule 2: If you must convert a decimal fraction into a simple fraction, then put the decimal fraction over 1.00 (this represents the whole), and reduce the top and bottom numbers to the simplest fraction. Therefore, the decimal fraction 0.36 converts into:

$\frac{0.36}{1.00}$ or $\frac{36}{100}$ or $\frac{9}{25}$ (expressed as a simple fraction)

Show your work in figuring out the following problems.

? Question

1. $\frac{3}{4}$ = __________ (decimal fraction)

2. $\frac{9}{20}$ = __________ (decimal fraction)

3. 0.4 = __________ (simple fraction)

4. 0.15 = __________ (simple fraction)

Percents %

In the previous discussion we saw that fractions are parts of the number One. ***Percents*** (%) are parts of the number 100, and in order to calculate them, you must start with a decimal fraction. In order to convert the decimal fraction into a percent, multiply the decimal fraction by 100. The decimal fraction 0.23 is equal to 23%. When multiplying by 100, move the decimal point two positions to the right. (See the "Factors of 10 — Multiplication Rule" for more help with this.)

Convert the following decimal fractions into percents.

? Question

1. 0.79 = ______%
2. 0.07 = ______%
3. 0.18 = ______%
4. 0.001 = ______%
5. 0.809 = ______%

Factors of 10

Because the Metric System is based on units that differ from each other by factors of 10, we need to review how the decimal position moves when converting within metric units. This rule allows you to convert large metric units like meters into small metric units like millimeters.

Multiplication Rule: When multiplying a number by 10, 100, 1000, etc., move the decimal position ***to the right*** by the number of 0's (zeros) in the multiplier. For example:

By the way, you add more 0's as you move the decimal point to an empty space.

25 x 10 = 25.0 = 250

25 x 100 = 25.00 = 2500

25 x 1000 = __________
2.5 x 100 = __________
0.25 x 1000 = __________
0.002507 x 100,000 = __________

Try these for practice.

Division Rule: When dividing a number by 10, 100, 1000, etc., move the decimal position ***to the left*** by the number of 0's (zeros) in the divisor. For example:

25 ÷ 10 = 2.5. = 2.5

25 ÷ 100 = .2.5. = 0.25

25 ÷ 1000 = 0.2.5. = 0.025

Again, you add more 0's as you move the decimal point to an empty space.

0.25 ÷ 10 = __________

2.5 ÷ 10 = __________

2.5 ÷ 1000 = __________

Try these for practice.

This rule allows you to convert small metric units like millimeters into large metric units like meters. In science, the rule about decimal fractions like .025 is that you always put a "0" (zero) in front of the decimal point so that there is no confusion about the number being smaller than the number one. Therefore, we write the above example as 0.025.

Metric Prefixes

The Metric System is a standardized method for defining length, volume, and weight. The standard metric units are ***meter*** (m), ***liter*** (l), and ***gram*** (g). There are over a dozen prefixes used with each of these standard units, and they define "portions" of the units that differ from each other by factors of 10. In this lab you will use only three of the metric prefixes. Memorize them now.

Table 1.1. Metric Prefixes Used in Biology Lab. The prefixes milli, centi, and kilo are commonly used.

Prefix	Abbreviation	Compared to Standard	Examples
Milli	m	$\frac{1}{1000}$	mm, ml, mg
Centi	c	$\frac{1}{100}$	cm, *, *
Kilo	k	1000X	km, *, kg

* cl, cg, and kl are not used in this lab.

? Question

1. If you convert from a centi metric unit to a milli metric unit, are you going to a smaller unit or to a larger unit in size?

2. How much difference is there between a centi unit and a milli unit?

3. How many decimal positions would you move for a conversion from centi to milli?

4. Converting from centi to milli, would you be moving the decimal position to the left or to the right? __________ You would be using the _______________ rule.

5. Let's review your answers with an example.

   24.5 centimeters = ______ millimeters

6. Now, let's try it in reverse.

   53 millimeters = _____ centimeters

The Metric unit for length is the meter.

Length

The standard reference for length in the Metric System is the ***meter*** (abbreviated as **m**). You will be using three prefixes with this standard unit: milli $\frac{1}{1000}$, centi $\frac{1}{100}$, and kilo (1,000). Remember to use the rules pertaining to the movement of the decimal point.

? Question

1. The abbreviation for millimeter is mm. How many millimeters are in a meter? _____

2. The abbreviation for centimeter is cm. How many centimeters are in a meter? _____

3. The abbreviation for kilometer is km. How many kilometers are in a meter? _____ (Did you get tricked by this question?)

4. How many meters are in a kilometer? _____

5. How many mm are in 1 cm? _____

6. How many cm are in a mm? _____

7. 40 cm = _____ m.

8. 40 cm is what fraction of a meter? _____

9. 40 cm is what % of a meter? _____

10. $\frac{32m}{10mm}$ = _____ (Hint: You must have the same units on the top and bottom before doing the division.)

11. 1.2 m x 30 = _____ m = __________ cm.

The Metric unit for volume is the liter.

Volume

The standard reference for volume in the Metric System is the ***liter*** (abbreviated as **l**). The prefixes used for metric length units also apply to volume units. However, milli is the only prefix that we will use during this lab. The milliliter (ml) is a very common unit for the scientific measurement of small volumes of liquid.

? Question

1. How many milliliters are in a liter?

2. How many liters are in a milliliter?

3. 355 ml = _____ l (This is a familiar volume for a canned soda.)

4. 750 ml = _____ l (This is a familiar volume for wine bottles.)

5. 15 ml = _____ l (This volume is one tablespoon.)

6. What % of a liter is 15 ml? _____

7. $\frac{1}{2}$ liter = _____ ml

Weight

The standard reference for weight in the Metric System is the ***gram*** (abbreviated as **g**). The metric prefixes that we will use during this lab are milli and kilo.

? Question

1. How many grams are in a kilogram? _____

2. How many kilograms are in a gram? _____

3. How many milligrams are in a gram? _____

4. How many milligrams are in a kilogram? _____

5. 454 g = _____ kg (This is the weight of a small loaf of bread.)

6. What % of a kilogram is 454 g? _____

7. 2.265 kg is the weight for a small bag of sugar. You are baking cupcakes for a school fund drive. It takes 100 g of sugar to make one batch of cupcakes. How many batches of cupcakes can you make with one small bag of sugar? (Show your work.)

The Metric unit for weight is the gram.

Exercise #2
Conversions: Metric ↔ English

Metric conversions – How hard can it be?

You are familiar with the English System of measurement. In this Exercise we will review some conversions between the English and Metric Systems.

Length

Materials

✓ a combination meterstick/yardstick

Procedure

- Make a line on a piece of paper exactly 10 inches long.
- Measure that same line in centimeters. 10 inches = _____ cm. Use this relationship to answer the following questions.

? Question

1. How many centimeters are in one inch? _____ cm = 1 inch
2. How many centimeters are in one foot? _____ cm = 1 foot
3. How many centimeters are in one yard? _____ cm = 1 yard

(Check your answers with the measuring stick.)

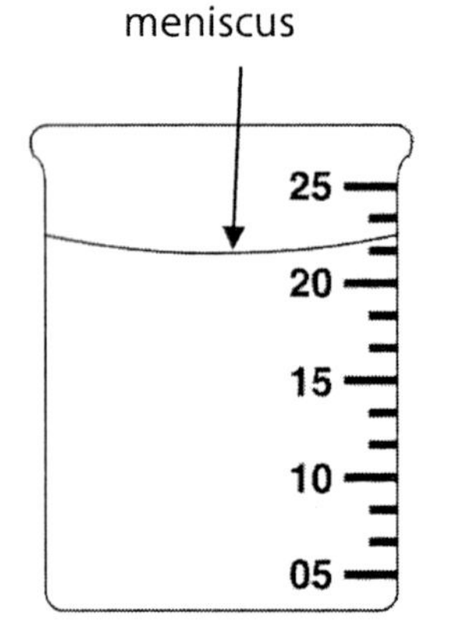

The curve of the water line is called the meniscus. This is where you measure.

Volume

Materials

✓ 1-liter graduated cylinder
✓ 10-milliliter graduated cylinder
✓ 1-quart graduated cylinder
✓ eye-dropper
✓ teaspoon

Procedure

- Fill the container marked "1 quart" with water. (There may be a painted line indicating the exact 1-quart amount.)

- Pour the 1 quart of water into a graduated cylinder for measuring liters.

  How many ml are there in a quart? _____ ml = 1 quart.
  Which is the greater volume? 4 Liters or 1 Gallon

- Count the number of drops of water it takes to fill the small graduated cylinder to the 1-milliliter mark.

  1 ml = _____ drops

- Fill the 10-milliliter cylinder to the 5-milliliter mark. Pour that amount into the teaspoon.

  5 ml = _____ teaspoon

Weight

Try this weight problem. Home Depot bought sacks of cement from a company in Mexico. The sign above the cement display reads: 100 lb/$4.95. The 100 lb cement sacks were also marked "45 kg."

? Question

1. How many pounds are in one kilogram?
2. How much do you weigh in kilograms?

Temperature

The English measurement of temperature is in degrees Fahrenheit (°F). Using this scale, water freezes at 32°F and boils at 212°F. There is a scientific temperature scale that is more like the metric scale. It uses Celsius (°C). On this scale water freezes at 0°C and boils at 100°C.

Procedure

- Use the dual scale thermometer illustration to answer the next questions.

? Question

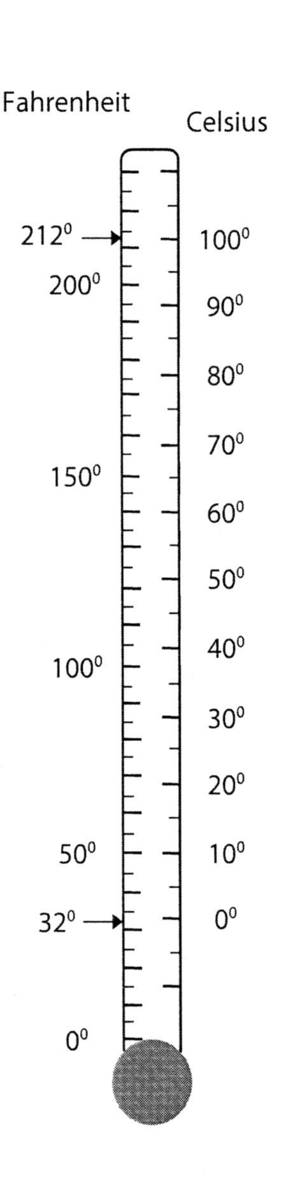

1. How many °F are there between the 0°C mark and the 100°C mark? _____

   Therefore, how many °F are there in one °C? _____

2. Water freezes at what temperature in °F? _____
   At what temperature in °C? _____

3. Water boils at what temperature in °F? _____
   At what temperature in °C? _____

4. What is your favorite air temperature in °F? _____
   What would that be in °C? _____

5. What is your idea of a "hot day" in °F? _____
   What temperature is that in °C?

6. Your normal body temperature is 98.6°F. Your child's forehead seems to be hot. You grab a Celsius thermometer by mistake and take her temperature. It reads 37°. Should you rush her to the hospital?

7. The formula for converting °C into °F is: **°F = °C x 1.8 + 32**. Why must the number 32 must be added?

8. Your cookbook says that roast beef is rare at 140°, medium at 160°, and well done at 170°. You like your beef cooked between medium and rare and only have a meat thermometer in °C. What temperature will the thermometer reach for the roast to be done the way you like it?

9. The water temperature gauge on your new Ford Truck reads 85°C. Are you overheating your engine?

Exercise #3
Personal List of Metric References

Think of something easy to remember that you can associate with each of the metric units below. Perhaps a centimeter might be the width of one of your fingernails. Be specific. Whatever you choose as a reference, make sure it's something you won't forget.

My Personal Metric Reference

Metric Unit

Length:

mm ____

cm ____

km ____

Volume:

ml __________________________ How many drops?___

l ____

Weight

mg ____

g ____

kg __________________________ How many pounds?___

Exercise #4
A Test of Your Measuring Skills

How to Weigh an Object

Weighing balances are used to measure the weight of an object. It is important that the object to be weighed is put inside a weighing container so that the material is not spilled on the balance pan.

- **Step 1:** "Zero" the scale. (Your instructor will show you how to do this.)
- **Step 2:** Then, weigh the weighing container. Why is it important to weigh the container first?
- **Step 3:** Put the substance or object into the weighing container, and then weigh them together.
- **Step 4:** Determine the difference between the weights recorded for Step 2 and Step 3.

Materials

✓ table salt

Procedure

- Your instructor will give you specific directions for using each type of weighing scale. Make sure that you "zero" the scales before weighing.
- Weigh 2.7 g of table salt, and put that amount onto a piece of paper. This is the recommended daily intake of salt in your diet.
- Weigh 11.6 g of table salt, and put that amount next to the other pile. This is the typical daily salt intake by people in our country.

Use a weighing paper or tray for the salt.

Exercise #5
A Simple Scientific Experiment

The scientific method begins with observation. There is something that we want to understand, and this is followed by a possible explanation based on our observations. That explanation is called the ***hypothesis***. Although the hypothesis represents a question, it is written as a declarative statement related to an experiment designed to test whether the hypothesis is true or not. For example, the question, "Are pennies before 1982 the same weight as pennies after 1982?" can be changed into an experimental hypothesis: "Pennies before and after 1982 are the same weight."

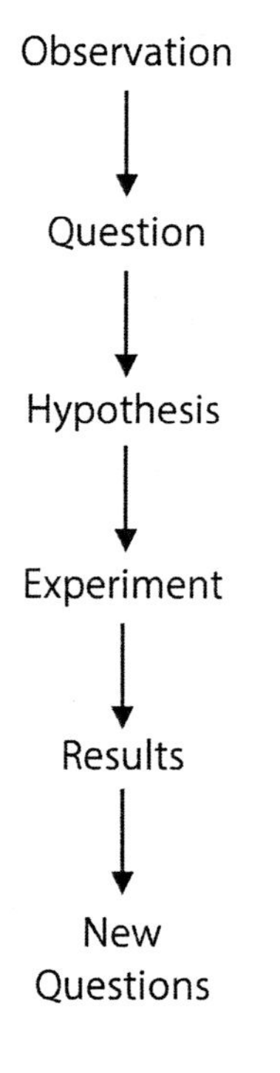

Procedure

- There are pennies in each date category on the lab table.
- Weigh a group of 10 coins from each date category to get a more accurate estimate of the weight of a single coin (divide the group weight by 10).
- Record the results.

Before 1982 penny: ______ g
After 1982 penny: ______ g
So, did the experiment tell us the hypothesis is true or false?

This experiment leads to more questions. You discovered that older pennies (pre-1982) seem to be a different weight from more recent pennies (post-1982). You wonder why. Is the weight difference because the pennies are not the same size, or because they are not made of the same metals? Now, it's your job to test the next hypothesis.

Procedure

- Determine how you would estimate the size of the coins by measuring the volume of water displaced by the coins.
- Use the graduated cylinder on the lab table and 10 coins together to get a better estimate of water displacement and the individual coin size. Do this for both pre-1982 and post-1982 pennies.
- One way to see if pennies are made of different metals is to cut them in half. Examine cut pennies at the demonstration table.
- Complete the practice lab report for this experiment.

Lab Report

Question:

Hypothesis:

Experimental Design:

Results:

Conclusions:

LAB

2

Experimentation

Experiments are designed to test the hypothesis.

This week's lab covers an introduction to experimental design. Experimentation is the primary tool of science. In fact, modern science only considers evidence that is based on direct empirical observation during rigorously controlled experiments (those that can be repeated by others). Even though health professionals and others in society must understand and prepare scientific reports, the scientific methodology seems daunting and even a bit boring. We hope this lab will convince you that the challenge is more interesting and understandable than many scientists display in overly sophisticated books and periodicals. There is an art to the scientific process. After all, most decisions in life do not have solid scientific explanations telling us exactly what is best to do.

We start with a discussion of the hypothesis from an insider's view. Most likely this is not a view that was presented in your previous science classes where you memorized the steps of the scientific method. You will learn some of the inherent bias and error "built into" the scientific method. It is a bit of a philosophy lesson. After a consideration of the hypothesis, you will learn some of the particular tools of experimentation including controls, blind and double-blind studies, and placebo effects. In addition, we offer a recommendation about when to stop experimentation and when to apply known knowledge (The 85% Rule). All of this is just a beginning consideration of how science is used and misused in the health professions.

Exercise #1	**The Hypothesis**
Exercise #2	**Experimental Design**
Exercise #3	**The 85% Rule**
Exercise #4	**Placebo Effects in Sport and Medicine**

Exercise #1
The Hypothesis

Should we try to prove the hypothesis correct, or try to prove it wrong?

The fundamental goal of science is to understand nature through observation and experimentation. The scientific method begins with observation. There is something in the world that we want to understand. Observation is followed by offering basic questions and possible explanations for what we are observing. "Do people with more good health behaviors live longer?" That question is the beginning of a ***hypothesis***. Although the hypothesis starts as a question, it is written as a declarative statement "People with more good health behaviors live longer." An experiment must then be designed to test whether the hypothesis is true or not. Often the hypothesis is stated in a very narrow and specific way because it is only a part of a much bigger explanation.

It turns out that the very specific way you state a hypothesis leads to a particular experimental design and also determines the kind of possible error that you might make in your study. Today, science generally agrees that all experiments should be designed to test what is called the ***Null Hypothesis***. The simplest statement of a null hypothesis is: **Sample A = Sample B**

When we like an idea, we have a natural bias to try to prove it correct. Statisticians found that the best way to test ideas is by trying to prove them wrong. People are naturally good at finding flaws in almost any idea. If we can't prove an idea wrong, then we feel more confident that the idea is probably correct. And we've avoided the natural bias of trying to prove our ideas correct. However, no matter how good science becomes at reducing mistakes and bias, there always remains some inherent error in its methods. Scientists accept that research errors do happen. Therefore, error is one of the first topics taught when scientists are trained to use the scientific method. There are two kinds of error that can be made when a researcher tests a null hypothesis.

Table 2.1. Error When Using the Scientific Method.

Two Possible Errors When Testing the Null Hypothesis	
Type I Error	The null hypothesis is actually true (Sample A = Sample B), but the experimenter mistakenly concludes it is false.
Type II Error	The null hypothesis is actually false (Sample A ≠ Sample B), but the experimenter mistakenly concludes it is true.

? Question

1. Is the hypothesis written as a question? Why not?

2. State the null hypothesis in your own words.

3. Is the correct approach of science to try to prove the hypothesis right or to try to prove it wrong?

4. The way that you state the null hypothesis determines the kind of _______ that you might make in your conclusions.

5. Define Type I error in your own words.

6. Define Type II error in your own words.

That theory is worthless. It isn't even wrong!

- Wolfgang Pauli (1900-1958) a pioneer of quantum mechanics

Avoiding the Catastrophes of Error

You have learned that science can make two kinds of mistakes—Type I and Type II. The next question is, "Can we avoid each of these errors?" It turns out that you can reduce Type I error to as low a risk as you choose. For example, if you want to reduce possible Type I error when using the Standard Error Test (a statistical test that you will use in the Data Analysis Lab), then you can require that the two sample means must differ by 3 S.E. units in order to conclude that they are different. (You will use a threshold of only 2 S.E. units in this class.) If you want to reduce Type I error even more, you could require 4 S.E. units as the threshold. This means that in order to avoid Type I error, I am willing to require more evidence, and then even stronger evidence. I could set the bar so high that no one can get enough evidence to prove the hypothesis wrong. I am now opening myself to Type II error. The Null hypothesis is actually wrong, but I'm forcing it to be considered correct. The dilemma of scientific testing is that when you reduce the chance of a Type I error, the chance of a Type II error automatically increases. This is because the factors creating these two kinds of error are linked. When either one of them is reduced, the other one tends to increase.

If you protect against Type I error, you automatically expose yourself to more Type II error.

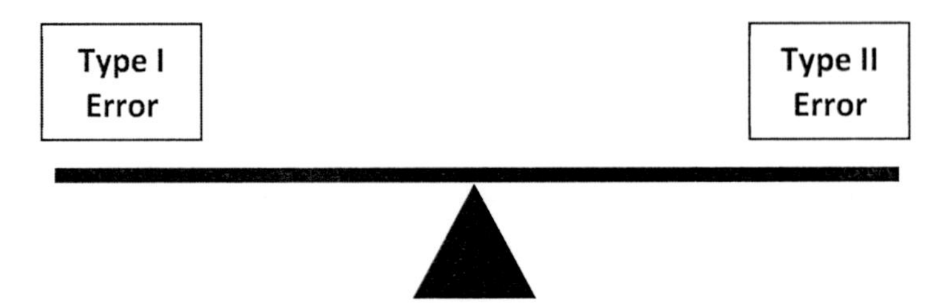

When scientists reduce Type I error, they must make adjustments in their experimental design (such as greatly increasing the sample size) to prevent Type II error from increasing out of their control. However, increasing the sample size will take more time and cost more money. Most investigators won't do it. So, if we are somewhat unwilling or unable to avoid error, how do we avoid a catastrophe when error does happen? One way of avoiding a catastrophe of Type I and Type II errors lies in the way that the null hypothesis is written. And there are two very important rules for stating the null hypothesis.

We love to prove an idea wrong.

- **Rule #1:** The null hypothesis is written in the equality form. (Heart rates on a test day are the same as heart rates on a normal day.)

- **Rule #2:** The null hypothesis should be stated in the "safest and most conservative way" to minimize danger. It must be stated in such a way that a Type II error won't result in someone dying or something equally serious.

The main reason that scientists use the equality form for stating the hypothesis is because this sets up a situation in which they can attempt to prove the hypothesis wrong. We excel at thinking up experiments designed to try to prove an idea wrong. It is much more difficult to devise ways of proving that an idea is correct, especially when not much is known about the phenomenon. So, scientists deliberately set the Type I error to a very high threshold (low risk), and then design experiments to try to prove the hypothesis wrong. However, when they set the Type I error to a low risk, this increases the chance of a Type II error. Remember, Type II error means that the null hypothesis (Sample A = Sample B) actually is false, but the experimenter mistakenly concludes it is true. Because science uses the approach of reducing Type I error to a minimum, it is automatically vulnerable to Type II error. Therefore, Rule #2 must be followed (safe and conservative hypothesis) to avoid the "catastrophe of error."

? Question

1. What are the two important rules for stating the null hypothesis?

2. What is the value of stating the hypothesis in an equality form? (Sample A = Sample B)

3. Does this help to reduce Type I error?

4. Why do scientists try to prove ideas wrong, rather than trying to prove them correct?

5. Does this help to reduce Type I error?

6. Can scientists set the Type I error to as low a risk as they want?

7. What happens as science reduces Type I error?

8. Why is it usually a challenge to greatly increase the sample size?

9. What kind of error is science most vulnerable to?

Type I or Type II

10. What is it about science that naturally leads to Type II error?

What is a Safe Null Hypothesis?

During the early years of World War II, London was being battered by daily bombing attacks. Many people had been killed, and the city was in chaos. It was a bleak time for the British people. Then a truly amazing man, the Prime Minister, took control of the situation. His name was Winston Churchill, and he inspired his people to persevere.

We will not cower to the scoundrels!

-Winston Churchill June, 1940

One of Churchill's brilliant inspirations occurred when he ordered a group of the best British scientists and mathematicians to meet with him. This stern and pugnacious man walked into the room of experts and stared at each of them without saying a word. After what must have seemed like an eternity, he bellowed, "We will not cower to the scoundrels! London must go on!" Then he proceeded to give these men and women an incredible challenge. They were to determine which areas of London would be safe from the bombing—safe for work, safe for transportation, safe for schools and the nation's children.

The experts were shocked. They had been ordered to assume an enormous responsibility. What if they were wrong? People would die! This was their state of mind as they worked feverishly to develop a method of testing various hypotheses so they could identify which areas would be safe from bombing. Soon they came to realize that there were two very different kinds of null hypotheses – the safe and the risky.

Scientists test a hypothesis by comparing data collected from two samples or by comparing a sample to an expected value. Remember, the simplest statement of the null hypothesis is Sample A = Sample B. But, how we define Sample A and B becomes crucial! In the bombing of London, the safest null hypothesis was:

"Zone A, an area they wondered if safe, is
the same as known bombing zones (Zone B)."
(Zone A is a bomb area.)

What if the hypothesis is false, but you think it's true?

With a "Safe" hypothesis, Nobody dies!

Why is this the safest way to state the null hypothesis? Remember, science tries to prove the hypothesis wrong. Science is at its best and most certain when proving a hypothesis wrong. Statistical comparisons were designed to test this hypothesis. The statistical threshold for making a Type I error was set at an extremely low risk. If England's scientists were successful at disproving this hypothesis ("Zone A is a bomb area"), then they could be reasonably certain that they had actually identified a place where bombs would not fall—a safe area for England's people.

Most important of all, the scientists were able to protect themselves against the danger of making Type II error. Remember, science is most vulnerable to Type II error. Type II error in a safely stated null hypothesis would have led them to mistakenly conclude that their hypothesis was true (that Zone A was a bomb area), and they would warn people away. It would be a mistake. But, no one would die because of it!

The Risky Null Hypothesis

"Zone A, an area they wondered if safe,
is the same as known unbombed zones (Zone B)."
(Zone A is a safe area.)

But, what if it's false, and you think it's true?

- People die!

This is the risky way to state the null hypothesis. It leads to catastrophe if a Type II error is made. A Type II error for this hypothesis would mistakenly conclude that Zone A is a safe area when it actually is a bomb area. People would die because of this mistake. It is a dramatic example that demonstrates why science should use only the "safe and conservative" way of investigating dangerous problems. A safe and conservative null hypothesis minimizes the danger coming from a Type II error. The "safe and conservative" null hypothesis became established in science. However, it is important to mention that not all researchers follow this guideline. Some even seem ignorant of it. By protecting against a Type II error, science makes it harder to prove their favorite idea correct. People don't like that! And there are bitter disagreements about what is a safe hypothesis.

? Question

1. Explain the importance of stating the null hypothesis in the "safest and most conservative way" to minimize danger.

2. Using the same approach as science, decide which hypothesis is the "safest and most conservative" null hypothesis. (Circle your choice.)

   Hypothesis A: Defendant is assumed to be innocent, and the trial tries to prove him guilty.

   Hypothesis B: Defendant is assumed to be guilty, and the trial tries to prove him innocent.

3. If you made a Type II error using Hypothesis **A**, what would be the result?

   Guilty man goes free   or   Innocent man goes to jail

4. If you made a Type II error using Hypothesis **B**, what would be the result?

   Guilty man goes free   or   Innocent man goes to jail

5. Any hypothesis must be considered in a context of possible bias and statistical error. For example, research into global warming can be approached from two directions.

   Hypothesis A: Area in question is an example of global warming.

   Hypothesis B: Area in question is an example of non-global warming.

   This is another example of a situation where we need to know which hypothesis would be the most dangerous. What is your opinion?

6. Another example of two different hypotheses involves the testing of drugs and treatments by the FDA.

   Hypothesis A: The test drug (or treatment) is safe.

   Hypothesis B: The test drug (or treatment) is unsafe.

The FDA requires the approach illustrated by Hypothesis **B** to be the acceptable method. Why have they chosen that approach?

7. Proponents of alternative medicine and nutritional supplements generally use the research approach represented by Hypothesis **A** below. They argue that this is the best way to be "open to" new treatments that are alternatives to traditional medicine. What is the difference between this approach and the one required by the FDA? What are your opinions?

   Hypothesis A: The test drug (or treatment) is effective.

   Hypothesis B: The test drug (or treatment) is ineffective.

It is easy to obtain confirmations,
or verifications, for nearly every theory –
if we look for confirmations.

- *Karl Popper (1902-1994),*
- *Philosopher, exponent of scientific falsification*

Exercise #2
Experimental Design

In science, we try to make sure that the work done during the day gets us a little closer to understanding what is really going on in the world. ***Experimental design*** is the planning of an experiment so that we get a yes or a no answer to the question we are asking.

The Experimental Control

The scientific method begins with a possible explanation of something we are observing. That explanation is a called the ***hypothesis.*** It generates the questions we are trying to answer in an experiment. For example, maybe adding more water to a lawn will make it grow faster. We could design an experiment to test that hypothesis. In the ***experimental design*** we might decide to water the lawn twice as much as normal rainfall. Let's suppose that this experiment is done for a month, and we observe that the lawn actually does grow quite a bit faster. Can we safely conclude that the hypothesis has been tested?

What if someone were to point out that the weather was warmer than normal during our experiment, and that is why our experimental lawn grew faster than normal? Or suppose that someone else said that our lawn really didn't grow any more than her lawn which was watered only by rainfall during the same month?

Because we can't answer these questions,
our "experimental design" has failed. Why?

Our critics will find our errors.

We did not control our experiment! An ***experimental control*** is a duplicate procedure that is set up exactly like the experiment except that the factor being tested (more water) is left out. So, in our experiment we should have monitored a nearby lawn that received only normal rainfall during the same warm month. Then we could compare the growth rate of that "control lawn" to our "experimental lawn" which was receiving extra watering. The results from observing the control lawn would have answered both questions from our critics.

All experiments need a control!

"Well, I guess we're the control group."

The blind study does not let the subjects know which treatment they are getting.

What are blind and double-blind studies

As we said previously, there are disagreements about which methods are required to be "real" science. Although there is agreement on most guidelines, not everyone follows them with equal fervor. Even if they did, bias would still remain because it seems to be a universal human trait. The most rigorous and objective experimentation and analysis are based on the "***principles of negation***" – constantly try to prove your ideas wrong. Testing of popular health remedies does not follow this guideline, but health professionals usually do. One of the research tools spawned from the principles of negation is the blind experiment.

The ***simple blind study*** does not let the subjects know which treatment they are getting (experiment or control). The purpose of this design is to reduce placebo effects. In ***double-blind experiments***, neither the researchers nor subjects know who is in the control or test groups. The purpose of this design is to reduce placebo effects and experimenter bias. A neutral third-party is the only one aware of the real grouping. All subjects are randomly assigned to be in either the test treatment or control group, and there can be an independent committee to supervise the study as a way of detecting any unexpected bias that may develop over time during the investigation. Sometimes an experimental treatment is so notably effective that everyone involved becomes aware that something is happening. Studies have been suspended in cases where tested drugs were so effective that it was unethical to withhold treatment to the control subjects and general public. Of course, there are disagreements in the "grey areas" of research findings, which is why the general principle of negation is valuable – "I should always try to disprove my hopeful drug."

? Question

1. What is the principle of negation?

2. Describe the control in a properly designed experiment.

3. Define simple blind study.

4. Define double-blind study.

Exercise #3
The 85% Rule

How many causes of an important health problem do you have to discover? When do you know enough to recommend treatments for an important health issue? Must you know everything that contributes to physical performance before saying anything really definite and helpful? We are swamped with a flood of new explanations that seem to come from everywhere – and so much competition of opinions. Do you say to yourself, "I thought there was a basic truth, but I don't know anymore"?

85% is enough. Get on with it!

We offer you the 85% Rule. This does not mean that you wait until you are 85% certain of something – we actually want you to be 95% certain of something. You'll learn about that in the Data Analysis Lab. Rather, we want you to focus on the central causes of 85% of a health problem. Determine what contributions to 85% of nutrition or performance improvements. In other words, when you have discovered the important variables that explain 85% of a particular situation, you can stop! If you don't stop, you will go on and on and lose sight of the important factors that would help patients or clients. Let us show you two situations where the 85% Rule can be used.

Case #1

What are you going to say when people come to you wanting to lose weight? Break into lab groups and take 5 minutes to come up with five things you would say to your patient (client).

1.

2.

Group Discussion

3.

4.

5.

Quickly check with other lab groups to see how many truly different suggestions were made by the entire class. How many? ______

Case #1 is on the short end of the 85% Rule. Physiologists have shown that only two variables are needed to explain body weight imbalance. Those two factors are energy input (Calories in food) and energy output (Calories used in metabolism). Of course, there are many individual factors affecting energy input and output, and each could be a tool for helping people. But in this case we have all that we need to begin our therapies. Consider what happens when we read one of many news articles where people don't stop with the already known explanations for weight imbalance.

News Flash!

"The Alabama group puts forth these ten explanations for obesity."

How many factors control weight imbalance?

1. Sleep debt
2. Pollution
3. Air conditioning
4. Decreased smoking
5. Medicine
6. Age and Ethnicity
7. Older moms
8. Ancestors' environment
9. Obesity linked to fertility
10. Unions of obese spouses

This news report has focused your attention on ten new explanations. Instead of doing that, stay focused on energy input and energy output, and use any "tricks" that will help you to help your patient. These tricks are many, and some even come from researchers who delve beyond the 85% Rule. None of this discussion is to disparage the findings of those who do not stop with the 85% Rule. But, this rule will keep you from getting lost in the forest of competing explanations.

Case #2

Most of the time you will discover that a handful of variables will explain 85% of the problem you face. However, this next example is an exception. It requires more than 15 causes to explain 85% of the deaths in 2011. (See Table 2.2.) The causes of death are listed in decreasing order of impact starting with Heart Failure (24.3% of deaths) and ending with Infections caused by aspirating food or liquids (0.6% of deaths).

Table 2.2. The Leading Causes of Death Listed in Cumulative Order. (Government Statistics, 2011)

Cause of Death	%	Cumulative %
Heart Failure	24.3	24.3
Cancer	23.3	47.6
Lower Respiratory	5.6	53.2
Cardiovascular	5.3	58.5
Accidents	5.0	63.5
Alzheimer's	3.4	66.9
Diabetes	2.8	69.7
Kidney Failure	2.1	71.8
Flu	2.0	73.8
Suicide	1.5	75.3
Blood Poisoning	1.4	76.7
Liver Failure	1.3	78.0
BP Kidney Failure	1.0	79.0
Parkinson's	0.9	79.9
Aspiration Infection	0.6	80.5

How many causes of death?

There is another variation of the 85% Rule to help in this case. Notice that the right-hand column of numbers is called Cumulative %. As you go down the list, all of the previous percent of deaths are being added together. These cumulative percents tell us how many more deaths are accounted for as we include more causes. By the time we get to #15 (Aspiration Infection) 80.5% of deaths have been accounted for.

The impact of each cause of death quickly decreases as we go down the list. It drops sharply at the third cause (Lower Respiratory), and another lessening happens with the sixth cause (Alzheimer's). Although the strict use of the 85% Rule doesn't help much in this case, we can see that nearly 50% of the death problem can be explained by causes #1 and #2. And, unless you want to chase after the entire list and beyond to get to 85%, you could stop with Cancer or perhaps Accidents. That would be using similar thinking to the 85% Rule. The focus is on the primary causes of a particular health problem. Your job now is to graph the Cumulative % of Deaths as More Causes of Death are included. Do that in Figure 2.1.

Graphing Activity

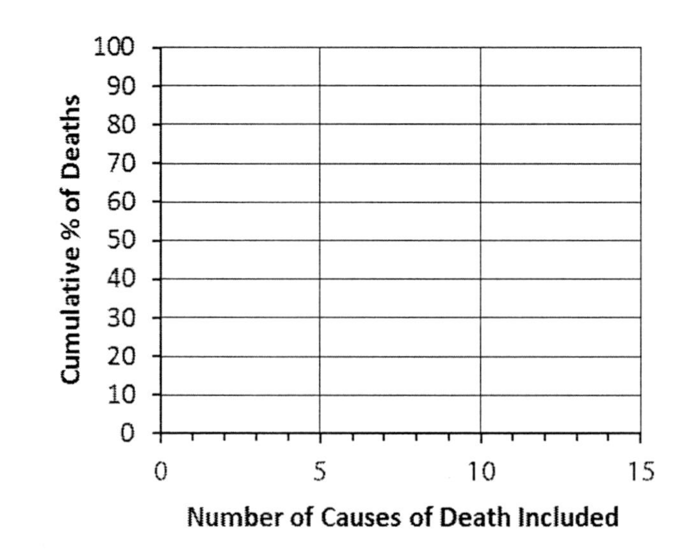

Figure 2.1. How many causes of death do we need to list? Plot the data from Table 2.2 into this graph.

? Question

1. State the 85% Rule in your own words.

2. What is the advantage of using the 85% Rule?

3. What would be your strongest objection to using the 85% Rule?

Exercise #4
Placebo Effects in Sport and Medicine

The placebo is a special control treatment sometimes used in human research.

Experiments in sport and medicine are usually designed to test whether a treatment will improve a health or performance problem. In setting up the experimental design there is a factor that we think may be affecting things. For example, the Coronary Drug Project (discussed next) tested whether niacin, aspirin, and a couple of other drugs benefit people with coronary heart disease. Those factors are called the ***experimental treatments*** in a scientific study. But in order to be more sure of the experimental treatment, we need to know what would happen without it. That situation (without the experimental factor) is called the ***control treatment***. You must control your experiment. The control treatment in the Coronary Drug Project included people exactly like the experimental groups except they did not receive an experimental drug.

Placebos are a kind of control treatment sometimes used in human research. The placebo is set up exactly like a control, except the subject being tested is told that he is getting the experimental treatment. The person is then given an inert substance or a known ineffective therapy. If the person responds in any way different from "no effect", then that effect on the subject is called the ***placebo effect***. It doesn't always happen, but when it does, what the heck is going on? And is it important in the health professions?

The placebo treatment showed a 50% reduction in mortality.

The Coronary Drug Project (started in 1965) is an example of a carefully designed study that tested the effects of various drugs on heart disease. It also revealed the importance of a placebo effect. The placebo group received a sugar pill (lactose) instead of any drug treatment. All groups were compared to each other and to a large control group of people who were not part of the project and received no treatments at all. Some of the treatments (niacin and aspirin) showed benefits of about a 10% lower mortality than the placebo group. But the placebo group surprised researchers with its own benefit of 50% the mortality of the control group.

We can explain the beneficial effects of certain drugs, but why should placebo treatment have benefit? This is the central question to all placebo effects. And three possible explanations are presented in the CDP's report:

- The placebo has a genuine physiological effect, or
- People who were willing to fully participate in a study (not drop out) are basically healthier people than those not willing to participate, or
- People who were willing to participate in the study were more diligent in other aspects of their lives, and those habits led to better health.

Research has shown that certain areas of the brain become more active during the placebo conditions. But this does not prove that these brain areas are the cause of the effect - only that the brain is responding. Some brain areas produce more dopamine which could be part of the placebo effect. Regardless of which theory is correct, the "good" effects of the placebo are typically reversed when the subject is told that the treatment is a placebo with no value. Placebos can cause intoxication, allergic reaction, greater weight-lifting ability, more endurance and speed, and many other interesting changes. Size and color of pills, "branding", high price, and even whether it is a capsule drug can matter. And it helps to see someone else who is already benefiting.

Is there any clue revealed by who responds to the placebo treatment?

Alzheimer's patients show no placebo response. Children show more response than adults. And some people just seem more susceptible than others. The placebo effect for strength-building is exaggerated when someone very fit recommends the placebo treatment. Men are very responsive when it involves body strength. It works better on those who think it will work. Maybe the person believes, and the belief helps. Or maybe she believes and starts doing something different that is actually the beneficial factor. In some clinical conditions such as irritable bowel syndrome, pain and anxiety, high blood pressure, and other disorders there can be quite long lasting benefits – sometimes weeks, months, or even a year or two. About 30% of people show some placebo response, although there is 0% in some conditions (infection) and as high as 80% in targeted lab studies that don't have particular clinical application.

So what can we conclude about placebos? The effects are significant but small in magnitude. Changes in physical strength are usually in the range of 2-5% above normal, and workout performance or event endurance rarely exceeds 10% benefit. There is evidence and strong opinion by some observers that the effect is too unreliable compared to "real" treatments. On the other hand, the placebo effect is so consistent that government regulation requires all drug testing to show results beyond normal control conditions and also beyond any previously demonstrated placebo effects. That is a standard indeed!

Men are very responsive to the placebo effect when it involves body strength.

Some people argue that the clinical use of placebo treatment is very limited. After all, patients can get better on their own, and maybe that's all that is happening. But when you consider that opiates don't relieve pain 30-50% of the time, and some patients can't tolerate them anyway, what can you do? Placebos do work for pain relief on 20-40% of patients, so that is something. And doctors can use placebos to see if their patients are malingering. But the World Medical Association states that placebos should be used in only two situations:

- When a placebo is necessary for the testing of important medical treatments, or

- When minor conditions are being investigated, and patients who receive the placebo are not subject to risk.

Placebos are effective pain relievers for 20-40% of patients.

What about the marketing and sales of health treatments, weight loss products, sports and training supplements, etc.? Most of these deliver results in the same range as the placebo effect. How would you know if a health treatment is beyond placebo? There are billions of marketing dollars spent on placebo treatments. What are you going to do?

? Question

1. What is the key difference between a normal experimental "control" and the placebo treatment?

2. How did the drug niacin compare with the placebo in the Coronary Drug Project?

3. How much better was the placebo group than the control group?

4. What were the three explanations for the placebo effect reported in the Coronary Drug Project?

5. One group of people show no placebo response, and this suggests that the thinking brain is involved. Which group is that?

6. The current government standard for judging the effectiveness of a drug or treatment is that its benefits must be greater than ________________.

7. Both niacin and caffeine have noticeable effects on all people. Niacin makes you feel "warm in the face", and caffeine "peps you up". Researchers have proven that if either or both are added to a supplement, that supplement will have a doubling of the placebo effect. Discuss how this can influence the health and exercise industries.

8. How could you effectively use a placebo treatment in the health professions? Be specific.

9. What health and exercise products use "branding" and "buff" pictures to promote their use? Why would these promoters spend large amounts of money to get endorsements of their products?

10. Is there any health or exercise product that can't be explained by the placebo effect? List them.

11. How could you counter the marketing of placebo products and treatments targeted at patients and clients of the health professions? Be specific.

Group Discussion

Take 10 minutes to share your answers to questions 7-11 with other lab groups. List the five best ideas of the class.

1.

2.

3.

4.

5.

LAB

3

The Microscope

Physical reality is easiest to comprehend when we can see it. But, we can't see the very small or very far away things. In past times people had to discover and explore the smaller world of atoms and the larger world of the universe by using indirect observations and imagination. It wasn't until the improvements of the microscope and the telescope in the 1600's that we could directly observe the worlds of the very small and the very large. Because of those inventions, biology and astronomy have become major arenas of scientific exploration.

Your unaided eye cannot see an object that is less than 0.1 mm in length. But with a compound or dissecting microscope, small objects seem huge. The microscope is an extension of your visual sense. During today's lab, you will learn to use this tool and study the appearance of organisms that ordinarily you cannot see.

Exercise #1 **Film on the Microscope**
Exercise #2 **A Pinhole Microscope**
Exercise #3 **Finding and Focusing**
Exercise #4 **Depth of Focus: Crossed Threads**
Exercise #5 **Estimating the Size of an Object**
Exercise #6 **Searching the Fly Wing**

Exercise #1
Film on the Microscope

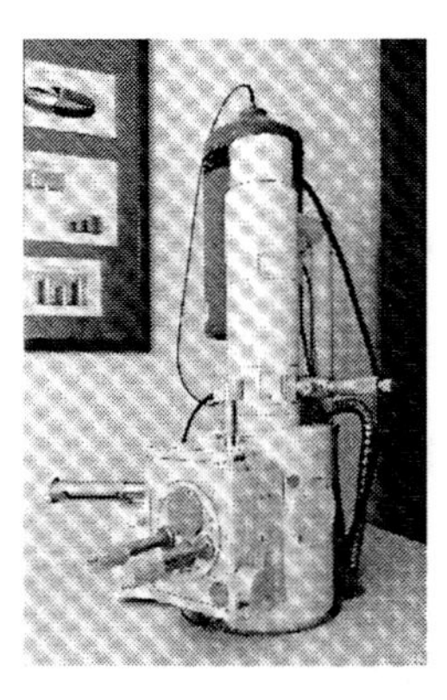

Electron Microscope

As with all inventions the microscope was developed in steps over a long period of time starting with the use of lenses by the Romans. Microscope improvement and application accelerated when Anton van Leouwenhoek perfected his microscopes during the second half of the 1600's. These microscopes could magnify about 300x, and he was famous for all of his many observations of very small objects including cells. The modern day electron microscope magnifies up to 200,000x. It is equivalent to aiming the Hubble Telescope at the atomic level. The following discussion will introduce you to microscope basics. Also, you may hear a short lecture or watch a film on the microscope before answering the review questions.

Materials

✓ compound microscope. *Be sure to carry it with both hands!*

Image Size

The light rays from a big object take up more of the surface at the back of the eye than the light rays from a small object. This is one way that our brains distinguish between small and large objects. Likewise, the light rays from a near object take up more of the surface at the back of the eye than the light rays from a distant object. This is why distant objects appear small compared to close objects.

It is the ***angle*** of light entering the eye that determines the amount of light spreading on the back of the eye. Big objects have a greater angle of spread than small objects. And near objects have a greater angle of spread than far objects. See Figure 3.1.

Magnified Objects

A magnifying lens bends the light rays traveling through it and ***increases the angle*** of light entering the eye, resulting in a bigger spread of the image at the back of the eye. Because the image is spread over more of the back surface of the eye, more nerve cells are activated and much more clarity of image is sent to the brain. We see the object as bigger and we have excellent resolution at the same time. See Figure 3.2.

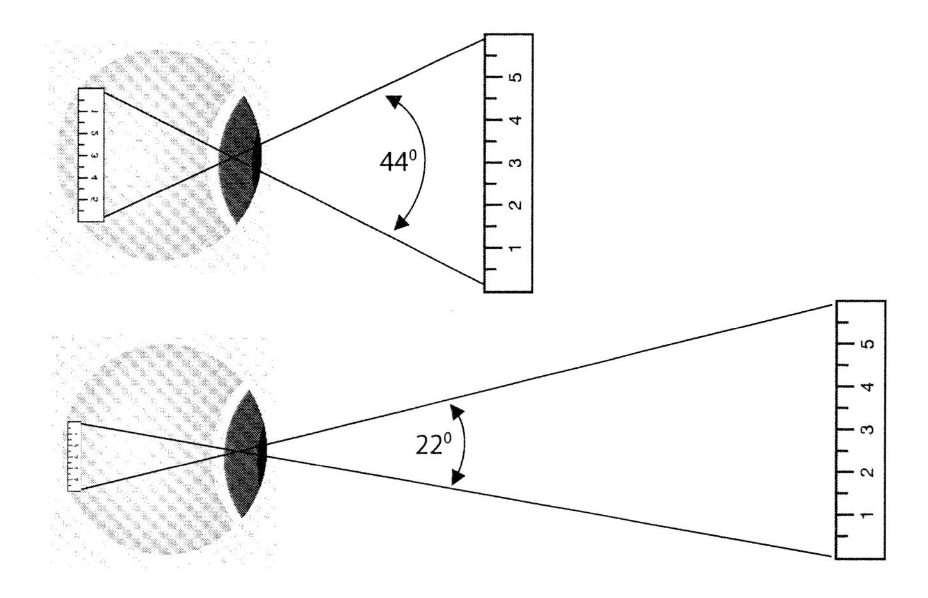

Figure 3.1. The spread of light rays coming into the eye from Close vs Far Away objects. The angle of light rays coming from the object depends on how far away the object is from the eye. The object appears small because its image is spread over a small part of the retina in the back of the eye.

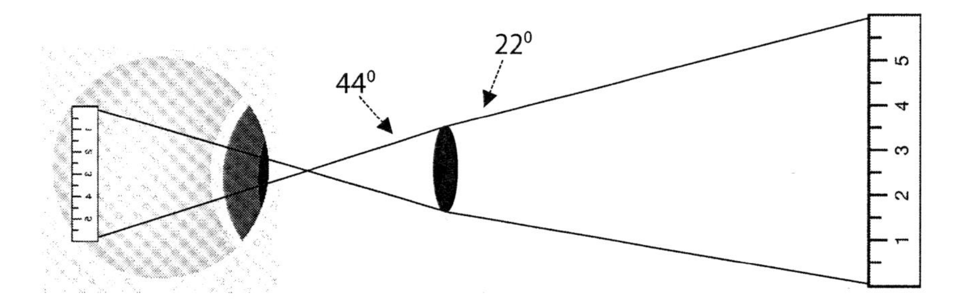

Figure 3.2. Increased spread of light by a magnifying lens. The lens is bending the light rays and increases the angle of light entering the eye. A magnified object appears larger because its image is spread over more of the retina.

Resolution

Resolution:
"If you can read this, you have far too much time on your hands."

Resolution is the clarity of image produced by the microscope lenses. It can be improved in several ways. The shorter the wavelength of the energy that is projected through the object, the better the resolution. Electrons have a much shorter wavelength than visible light, which is why an electron microscope produces the best resolution. Applying a thin layer of oil between the object and the lens produces less random scattering (blurring) of light rays. And finally, excellent quality lenses are designed to refocus "stray" light rays from each point of the viewed object, which greatly improves the resolution.

Compounded Images

When there are *two* lenses in a sequence, the image of the object is magnified *twice* before entering your eye. For example, if the first lens magnifies the object 43x, and the second lens magnifies that 43x image by 10x more, then the total magnification of the object is 430x the original size.

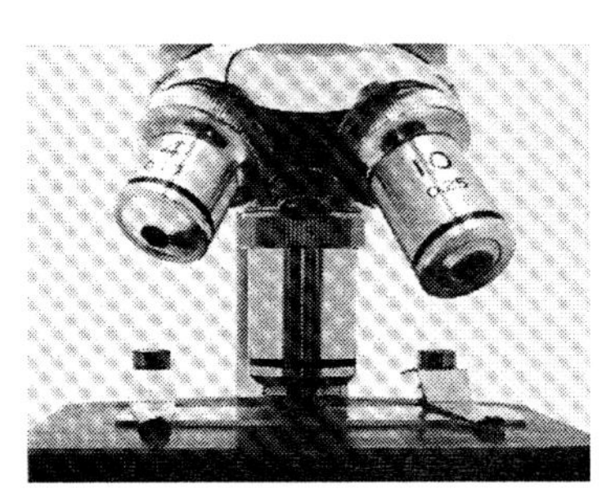

43x
Objective
Lens

10x
Ocular
Lens

43x times 10x = 430x (Total Magnification of the Image)

Figure 3.3. Compounded magnification of the image by a microscope. The image is magnified twice – first by the objective lens (closest to the object), and then that image is magnified again by the ocular lens (next to your eye.

Field of View

When looking through the microscope lens, you see only part of the object that is on the slide. It appears in a "circle of view" that is called the ***field of view***. Field of view is inversely proportional to magnification. Less magnification results in a larger field of view, and more magnification results in a smaller field of view. In other words, you will see less of the object when viewing under higher magnification, but that part you do see will appear much larger.

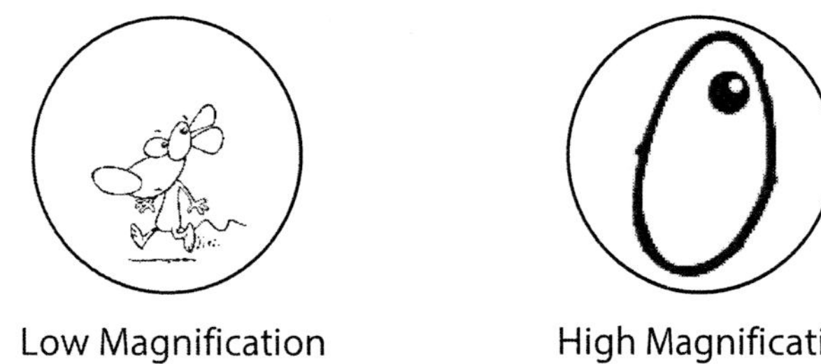

Figure 3.4. Field of View. The field of view is inversely proportional to magnification.

Depth of Focus

Seeing through a microscope is similar to viewing through a camera. Objects that are very close or very far away can be out of focus compared to the clarity of objects in between. In a microscope, that zone of clear focus is called the ***depth of focus***. Depth of focus is inversely proportional to magnification. The more the magnification, the smaller the depth of focus. For example, a magnification of 430x has a shallow depth of focus compared to a magnification of 40x.

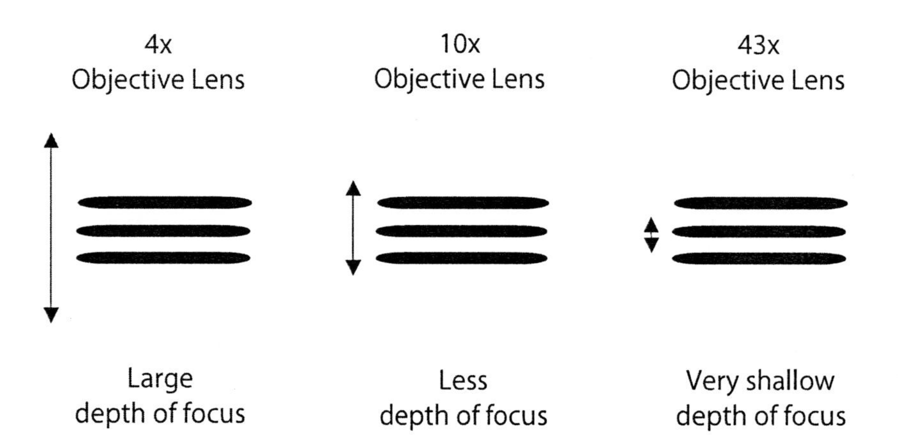

Figure 3.5. Side view of three overlapping threads showing Magnification vs Depth of Focus. The greater the magnification results in less depth of focus.

? Question

1. What happens to a ray of light when it passes through a lens?

2. If the light rays from an object enter the eye at a small angle, then the object will cover a __________ (smaller or larger) portion of the back of the eye.

3. The closer you move an object to your eye, the __________ (smaller or larger) the angle of light from that object entering your eye. Therefore, the object appears to be __________ (smaller or larger).

4. The magnifying lens __________ (increases or decreases) the angle of light coming from the object viewed. This results in __________ (increased or decreased) spread of the image at the back of the eye.

5. Define resolution. Be specific.

6. Is resolution the same as magnification?

7. What is the total magnification if the objective lens is 4x and the ocular lens is 10x?

8. What is the field of view? Be specific.

9. What happens to the size of the field of view as you rotate the nosepiece and change magnification?

10. Which one of the objective lenses gives the greatest magnification?

11. Which one of the objective lenses gives the greatest field of view?

12. When searching for a specimen, which lens should you use first?

13. What is the depth of focus? Be specific.

14. How does depth of focus change with magnification? Be specific.

Microscope Parts and Functions

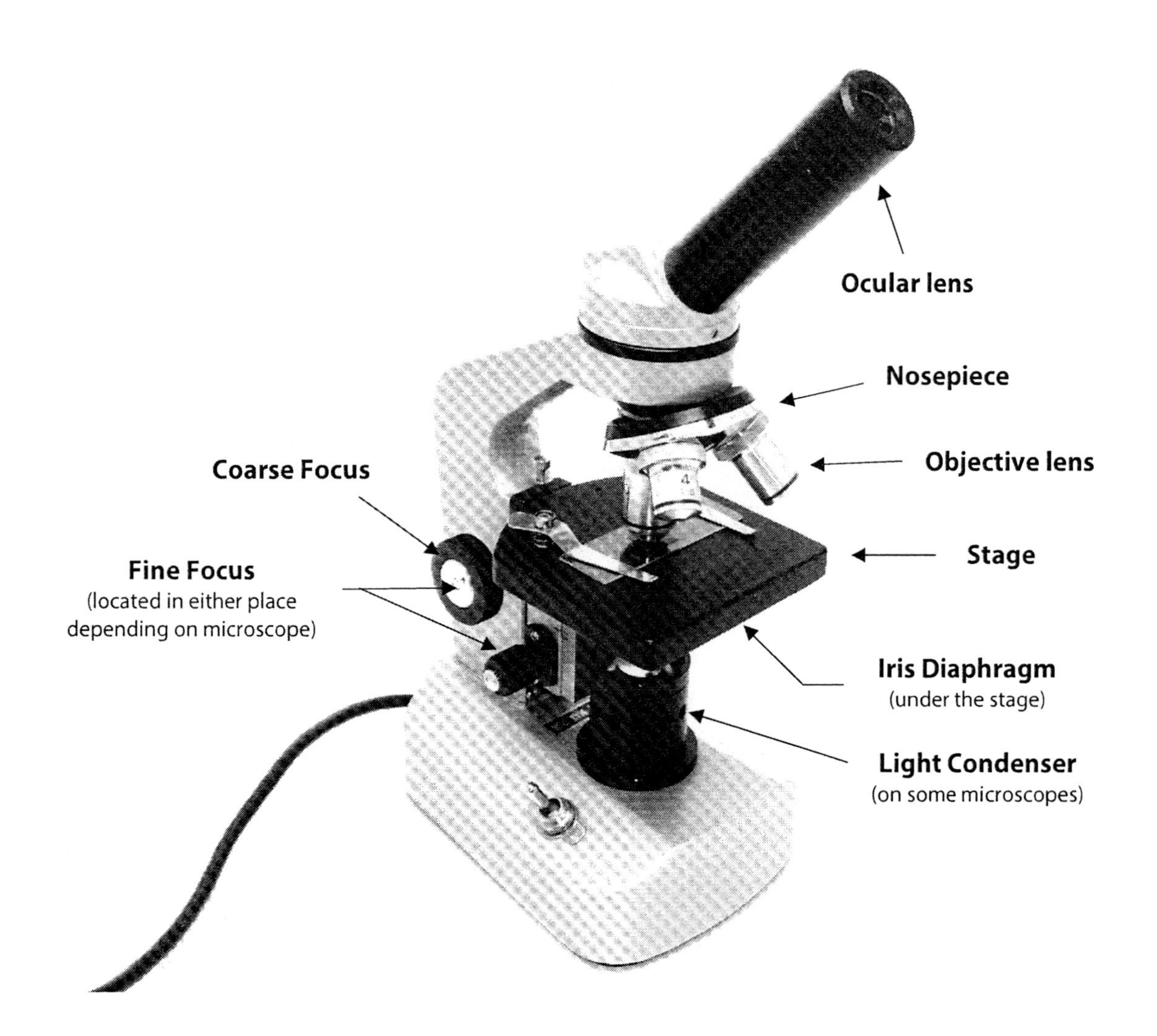

Figure 3.6. Parts of the Microscope. Find these on your microscope.

1. **Nosepiece:** A circular plate with 3 objective lenses that can be rotated into viewing position for different magnifications.

2. **Objective Lens:** This is the first lens that magnifies the image, and it is positioned just above the object being viewed on the microscope slide.

3. **Ocular Lens:** This is the second lens that magnifies the image, and it is the one closest to your eye when you are looking through the microscope tube.

4. **Coarse Focus Knob:** Turning this knob decreases the distance between the objective lens and the stage which allows you to focus the image entering your eye.

5. **Fine Focus:** This knob is for fine focusing the image. Use it whenever you are viewing through the higher magnifications. Usually it is the inside knob of the coarse knob. Sometimes it is a separate knob.

6. **Iris Diaphragm:** This lever (or rotating disk) adjusts the amount of light shining through the object. Use just enough light to illuminate the object and give good contrast. Some microscopes have a light condenser with an iris diaphragm lever. The condenser focuses light on the microscope slide. It should be positioned the same distance below the slide as the objective lens is above the slide. To achieve the clearest image, adjust the condenser each time you change the objective lens.

7. **Stage:** This platform holds the microscope slide. Some microscopes have an immovable stage, and you must carefully move the slide with your fingers. Other microscopes have two additional knobs to move the slide from side to side and forward and backward.

Exercise #2
A Pinhole Microscope

The pinhole microscope will demonstrate how the control of light rays determines your ability to see very small and close-up objects.

Materials

- ✓ 8" x 8" piece of black paper with a pinhole in the center
- ✓ piece of paper with typed words on it
- ✓ metric ruler

Procedure

- In the first trial, don't use the pinhole paper. Hold the typewritten page in your left hand and slowly move it towards your eye. At some point the words on the page will become blurry. Have your lab partner measure the distance from the typed page to your eye. _____ cm

  As you moved the printed page closer to your eye, did the size of the letters appear to change?

  The words became blurry at some point near your eye. What quality of your vision is being lost up close? ______________ (We used this term to describe the clarity of the image.)

- Repeat the experiment using the pinhole paper. Hold the pinhole paper with your right hand, and position the hole as close to your eye as possible. Move the typed page closer to your eye, and notice that the words which were blurry at _____ cm in the first trial are now readable at _____ cm this time.

  What quality of your vision has been improved by the pinhole paper?

  Can you figure out how the pinhole microscope works?

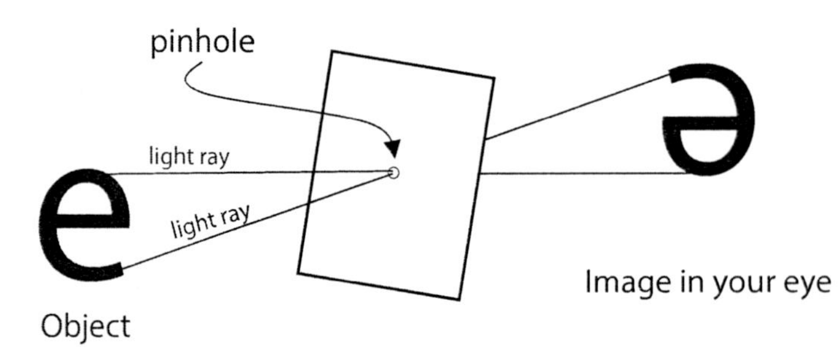

Figure 3.7. The Pinhole Microscope.

Exercise #3
Finding and Focusing

Materials

- ✓ compound microscope, one microscope slide, and one coverslip
- ✓ cutout of the letter "e" from a piece of newspaper

Procedure

- Turn on the microscope light.
- Use the coarse focus knob to move the nosepiece all the way "up," and "click" the 4x objective lens into position. (Some microscopes raise the nosepiece, and others lower the stage. Check to see which microscope type you have.)
- Get a prepared slide of the letter "e" or make a slide with the letter "e". Cut out a magazine letter "e" and place the letter on the glass slide. Slowly lower a coverslip over the letter to hold it in place. Use no water.
- Place the slide on the stage, position the letter "e" over the stage hole, and secure the slide.
- Look through the ocular lens and slowly turn the coarse focusing knob so that the distance between the lens and the slide decreases. The letter "e" will come into view.
- Center the letter "e" in your field of view, and then use the fine focus.
- Adjust the iris diaphragm until you have the proper amount of light with good contrast. It is very important that you reduce the light so that the details of the object can be seen. *Using too much light is the*

most common mistake that students make when working with the microscope. Note: Some microscopes have a dimmer switch to change light intensity. When using those microscopes, adjust light with both the dimmer switch and the iris diaphragm to get the clearest image. Your instructor will help you.

? Question

1. Draw the letter "e" as it is positioned on the stage. Then draw the orientation of the letter "e" as it appears when viewed through the microscope.

Normal View

2. What has happened to the orientation of the image as it passes through the lenses of the microscope?

3. When you move the slide forward on the stage, in what direction does the "e" appear to move when viewed through the microscope?

4. When you move the slide to the right, in what direction does the "e" appear to move?

Microscope View

5. What are your conclusions about the appearance and movement of an object when viewed through a microscope?

Exercise #4
Depth of Focus: Crossed-Threads

Think back to the photographs that you have taken and looked at. There always seems to be a zone of sharp focus in the picture with objects out of focus in front of and behind the sharp focus zone. That sharp focus zone is called the ***depth of focus***. This can become a serious problem when using a microscope as you increase the magnification. In fact, when looking through a microscope, the zone of sharp focus can be a major limitation in viewing the whole object. But you can make use of that restriction as a tool to inspect different layers of the object being viewed.

Depth of focus is inversely proportional to magnification. In other words, as the magnification of the object increases, the depth of focus decreases. (Don't confuse this with resolution, which is a function of lens quality.)

Materials

✓ prepared microscope slide of three crossed and colored threads

Procedure

Crossed- Threads

- Review the directions in Exercise #3 for finding and focusing with the 4x objective lens.
- Once you have the crossed threads in focus—with proper contrast, be sure the point where the threads cross is centered in your field of view.
- Turn the nosepiece until the 10x objective lens "clicks" into place. If you did a good job of centering the cross of the threads at low power, then the cross should be in your field of view at higher power. If not, go back to step #1, and try again.
- Focus using the fine focusing knob, and adjust the contrast if necessary. Again, center the crossing threads in the field of view.
- Turn the nosepiece until the high power (40x or 43x, depending on microscope brand) objective lens "clicks" into place, and focus using the fine focusing knob. Adjust the contrast if necessary. Properly position the condenser if your microscope has one.
- Now, use the fine focusing knob to move the zone of sharp focus below the threads so that all the threads are blurred and slightly out of focus.
- Then, reverse the direction you were turning the fine focusing knob,

watch the crossed threads until the first thread comes into focus. That will be the bottom thread. **Hint:** If you start out of focus above the threads and move the lens closer to the slide, then the first thread to come into focus would be the top thread. However, if you start out of focus below the threads and move the lens away from the slide, then the first thread to come into focus will be the bottom thread. Focus back and forth until you can determine the overlap of the crossed threads.

- Normally, focusing the highest power objective lens should be started slightly below the object (out of focus) and moved away from the stage so that you don't accidentally crunch the lens into the glass slide, cracking it and scratching the lens. After you become more practiced with a microscope, you won't make this mistake, but for now please be very careful.

? Question

1. What color is the thread that is on the bottom, the middle, and the top?

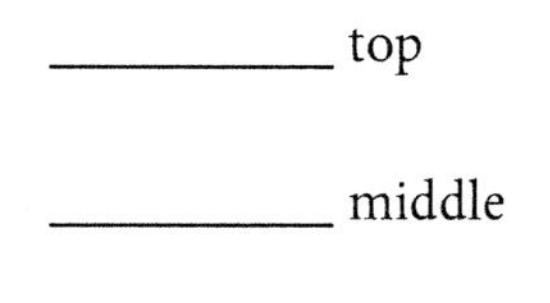

__________ top

__________ middle

__________ bottom

2. What microscope part is used to change the light intensity? (Answer may differ depending on your microscope model.)

3. What happens to the light intensity as you increase or decrease magnification of the objective lens?

4. When you switch to higher magnification, what should you do to the light intensity?

5. Describe the proper focusing technique in terms of moving the lens (especially the high-power lens) up and down when you first look at an object.

6. Why is this focusing technique so important?

Exercise #5
Estimating the Size of an Object

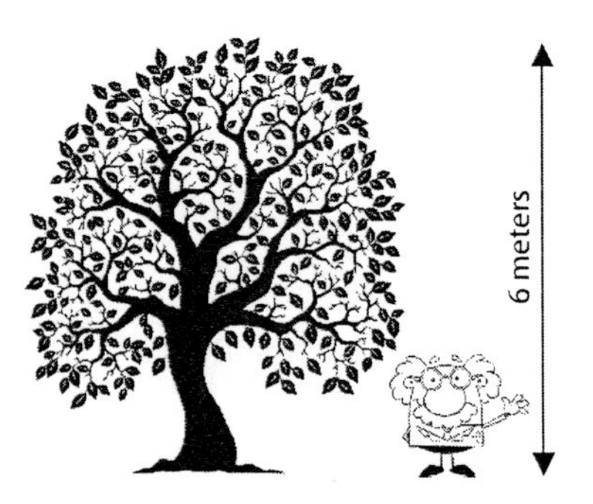

In order to estimate the size of an object, it is absolutely necessary to have an idea of the size of your field of view. This applies to estimating the size of any object. For instance, if a person was standing next to a tree and you knew the size of the tree, then you would have a standard against which you could estimate the size of the person. The person is about 1/_____ (what fraction) the size of the tree. Therefore, the person is about _____ meters tall.

The standard we use to estimate the size of an object when viewed under the microscope is the diameter of the field of view. The ***diameter of the field of view*** is the length of a straight line through the center of the field.

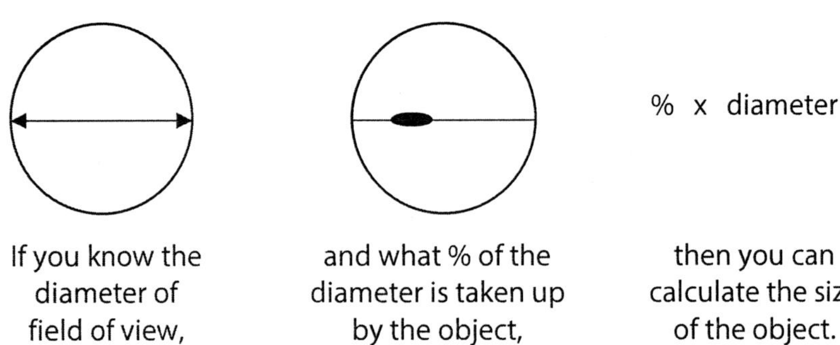

Figure 3.8. Using the diameter of the field of view to estimate the size of the object under view.

Materials

✓ transparent 15-centimeter ruler

Procedure

- Each dash on the metric side of this ruler is a millimeter unit. Under the microscope these dashes appear wide. To properly measure the diameter of the field of view under a microscope, you must measure a millimeter as the distance from one side of a mark to the same side of the next mark.

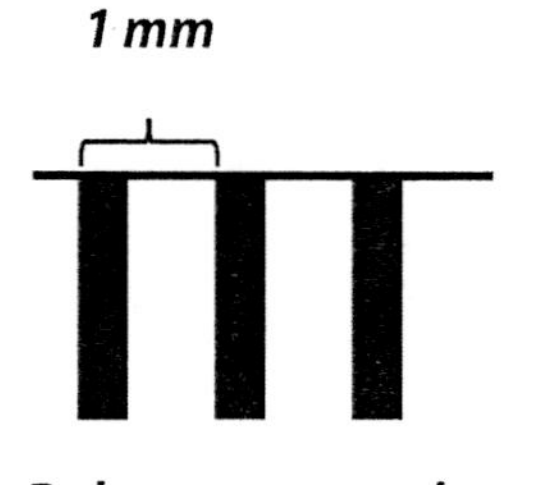

Ruler as seen under the microscope.

- Using the ruler, measure the diameter of the field of view with the 4x objective lens in position. Estimate this diameter to the nearest 0.1 of a millimeter.

  Diameter of 4x field of view = _____ mm.

- Since the diameter of the field of view is inversely proportional to the magnification, we can use this knowledge to calculate the diameter for the other two higher power lenses. **Remember:** As magnification increases, the field of view decreases.

? Question

1. If the diameter of the field of view through the 4x lens is 4mm, then under 10x, the field will be . . .

   Larger than 4mm or Smaller than 4mm

2. What is the diameter of the field of view you measured through the 4x lens?

   _____ mm

3. Instead of physically measuring the diameter of the 10x view, you can use a calculation based on your measurement of the low power diameter to give you the answer for the diameter of the next magnification. Because the 10x objective lens provides 2½ times more magnification than the 4x objective lens, its field of view is 2½ times smaller. 10X (next maginification) ÷ 4X (lower magnification) = 2½ times more magnification at the 10x objective lens.

   Use the above calculation to figure out how big the diameter of the field of view would be through the 10x lens. **Remember:** The field will be 2½ times smaller than at 4x.

   _____ mm

Whatever the amount of magnification increase by the microscope, the field of view is that much smaller. You see less and less of the object as you increase magnification.

4. The calculation method used for the 10x lens in the previous question above also applies to the 43x objective lens. 43x is _____ times more magnification than the 4x objective. So, the diameter of the field of view using the 43x objective lens will be _____ times smaller than at 4x.

Note: If your microscope has a 40x objective or some other number, substitute the magnification of the lens that you do have for all of the calculations above where the number "43x" occurs.

5. What is the diameter of the field of view for your microscope through the 43x lens? _____ mm

Materials

- ✓ a glass slide and a coverslip
- ✓ The small letter "i" cut from a newspaper. Be sure to include the dot.

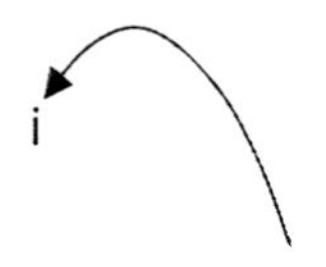

When you look through high magnification (430x), the field of view is about the size of the "dot" over the letter i.

Procedure

- Make a slide of the "i" like you did in Exercise #3. Look at it under low power and center the dot in your field of view. Switch to the 10x objective lens, center the dot, and then switch to the 43x lens. Estimate how much of the field of view is occupied by the dot through the 43x lens. _____ %
- If the dot is about _____ % of the 43x diameter of the field of view, then the diameter of the dot is approximately _____ mm. (You calculated this in question #5 above.)

Exercise #6
Searching the Fly Wing

This Exercise will test your ability to look for something specific under the microscope.

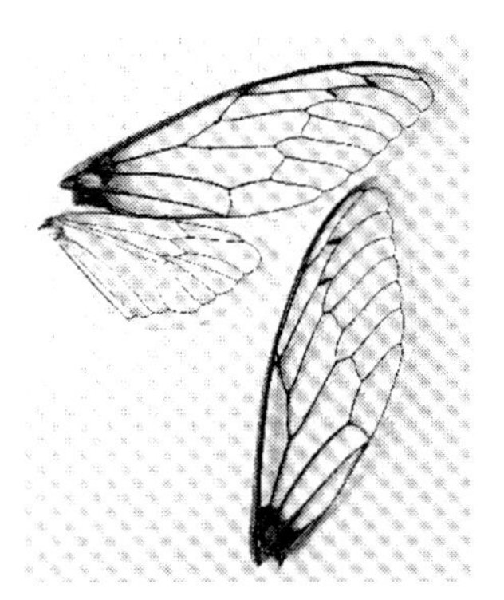

Fly Wings

Materials

- A prepared microscope slide of a fly wing.

Procedure

- Look at the picture of the fly wing at the front of the lab. Study it carefully. A particular vein intersection is marked on the picture.
- If it helps you to remember, draw a simple sketch from the picture. Be sure to indicate the field of view as it appears in the picture.

- Find the exact vein intersection in your microscope slide under the same magnification as you see in the picture. Show it to your instructor.

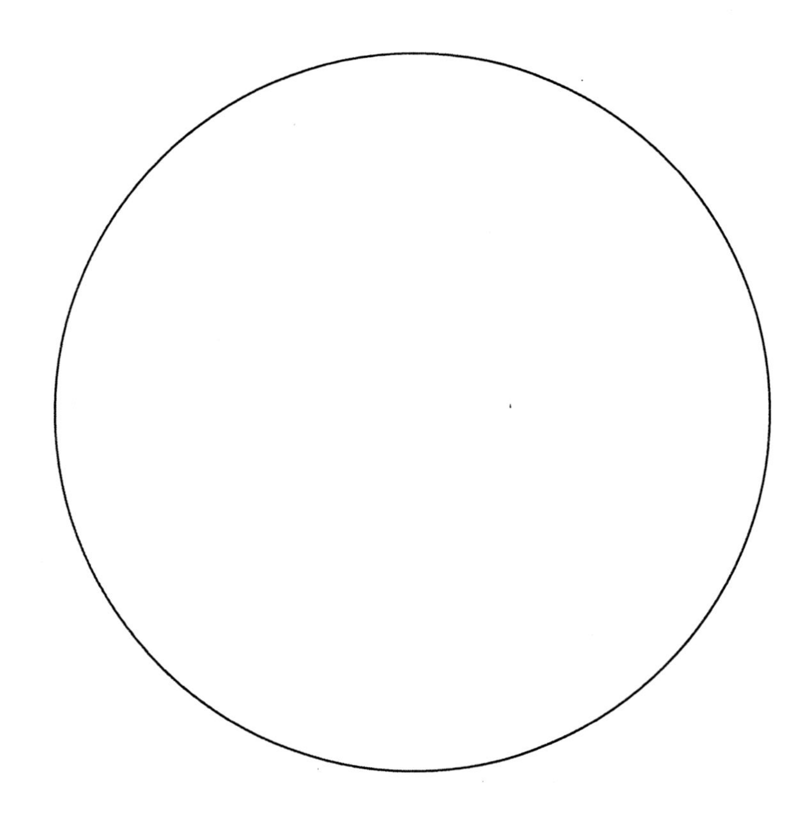

Fly Wing Veins

LAB

4

Cells

The Cell Law states:

1. ***All of life consists of cells.***
2. ***All cells come from previous cells.***
3. ***All life processes derive from cellular activities.***

Many educated people of the early 1800's had the belief that an invisible force was responsible for life in all its growing and changing forms. In 1839, Matthias Schleiden and Theodor Schwann presented the idea that ***cells*** were the creative unit responsible for life. Scientists and laymen soon came to agree about cells. Many puzzling questions were now answered. Life forms are related to each other at the cellular level, and the differences among life forms is determined by the chemical processes of those cells.

Cells may have first evolved from simple chemical structures known as "fat bubbles" which are formed when fatty substances such as phospholipids are added to water. Researchers are able to assemble a prototype cell by adding pieces of DNA to an empty bacterial cell until that cell can produce daughter cells that also self-replicate. Discoveries like this suggest that the synthesis model of cell evolution is one possibility. This week you will look at cells and learn some of their general features.

Exercise #1 **How to Make a Wet Mount Slide**
Exercise #2 **Human Cheek Cells**
Exercise #3 **Onion Cells**
Exercise #4 ***Paramecium***
Exercise #5 ***Elodea* Leaves**
Exercise #6 **Location of Color in Plant Cells**
Exercise #7 ***Zebrina* Leaf Epidermis with Stomata**

Exercise #1
How to Make a Wet Mount

In order to observe cells, you will have to become good at the technique of making a slide. This requires patience and careful handling of equipment. Take your time. Whenever you make a slide of something during this semester, you should use the wet mount method. It is the very best way to get a clear view of the object, and it prevents the specimen from drying out.

Materials

- ✓ microscope slide
- ✓ coverslip
- ✓ water dropper bottle

Procedure

- Put a drop of water on the slide.
- Put the object into the drop of water. The object must be very thin. You will see the importance of this when you make a wet mount of onion cells.
- Place the coverslip over the object by first placing one edge down, and then slowly lowering the other side so that you don't trap air bubbles. Air bubbles will look like discarded tires and are actually quite interesting in appearance, but they will interfere with your view of the object you really want to see.

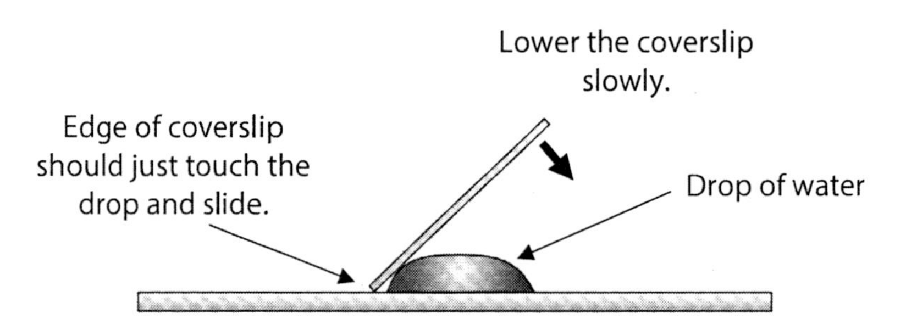

Figure 4.1. How to make a successful wet mount slide. Practice this method a couple of times with water only (no specimen) until you have some confidence.

Exercise #2
Human Cheek Cells

There are two types of cells: prokaryotic and eukaryotic. All of the cells you look at in the lab today are of the type called ***eukaryotic*** (true nucleus). The eukaryotic type of cell is the basic component of all multi-celled life forms. Some eukaryotic cells, such as the *Paramecium*, are single-celled organisms. However, the largest number of single-celled organisms are the bacteria and their relatives, and they are of another cell type called ***prokaryotic*** (before nucleus). You will have another opportunity to investigate them later in the semester during the lab "Surrounded by Microbes". Now you will begin your journey into the world of cells by looking at human cheek cells.

Materials

- ✓ compound microscope
- ✓ microscope slide and coverslip
- ✓ toothpick
- ✓ Methylene Blue stain

Procedure

- Put a small drop of water on the microscope slide.
- Gently scrape the inside of your cheek with the blunt end of the toothpick. This technique will collect hundreds of eukaryotic cells on the toothpick.
- Pay attention to exactly where you put the cells on the slide. They are hard to find. *Throw away the toothpick into the special waste container or disinfectant solution.*
- Cover the drop with a coverslip.
- Look at the cells under high power with your compound microscope. **Remember:** They will be very hard to see.
- Now, put a drop of methylene blue stain at the edge of the coverslip.
- Use a small piece of tissue paper at the other edge of the coverslip to absorb the excess fluid and pull the stain across the slide. This method allows you to apply a stain without removing the coverslip. Applying stain to the object will darken some cell structures, allowing you to see them better. Experiment with the iris diaphragm, light intensity, and condenser (if your microscope has one) to get more contrast and clarity.

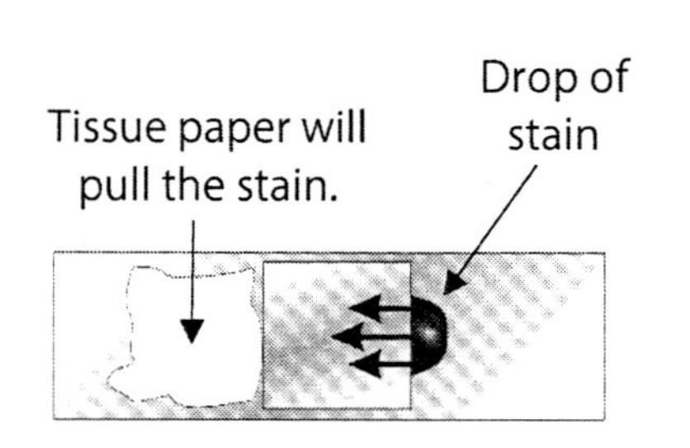

Method for staining.

- Look at the cells again under high power. You should be able to see the ***nucleus*** and the ***cell membrane***. The nucleus controls the cell functions, and the cell membrane controls what molecules go into and out of the cell. The advantage of stains is that we can see structures better. The disadvantage is that stains kill the cells. *Never use a stain if you want to see living cells!*
- Draw a simple sketch of your cheek cells. Label the cell membrane and the nucleus.

Human Cheek Cells

? Question

1. How do you know that your cheek cells are eukaryotic cells?

2. You may be asked on a Lab Test to make a wet mount of cheek cells and find them under the microscope. Can you do it?

3. Point out the cell membrane and cell nucleus to your lab partner. Be able to do the same for your instructor on a test. Put the slide and coverslip into the disinfectant solution. You might not reuse these slides. Follow the directions from your instructor.

Important: *Use alcohol or disinfectant solution to clean your slide before reuse.*

Exercise #3
Onion Cells

In this Exercise, you will be examining a plant cell. As you work through the procedures, notice differences between plant cells (onion) and animal cells (human cheek cells).

Materials

- ✓ cutting board and knife
- ✓ Cut an onion into "onion rings."

Procedure

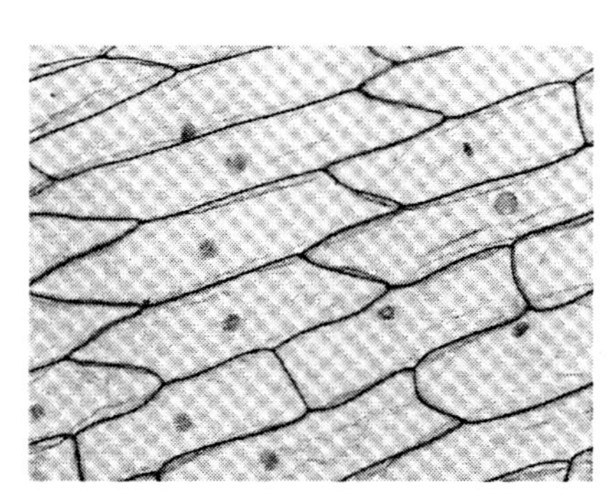

Onion Peel

- The cut onion will come apart into 1/8"- thick rings. Make your peel from the inside (not the outside) of one of the onion rings. You should be able to peel off a one-cell-thick layer of tissue. It will look like a piece of plastic wrap. You can use a razor blade or forceps to start the peel. But don't slice off a piece! That would be many cell layers thick. You need a one-cell layer.
- Place the onion peel into a drop of water on the slide, trying not to fold it over on itself.
- Finish the wet mount, and look at the cells under the compound microscope (low power first, then high power).
- Now, put a drop of iodine stain at the edge of the coverslip. Use the tissue paper to draw the stain across.
- Look at the cells again. You should be able to see the ***nucleus*** and the ***cell wall***. The nucleus controls cell functions, and the cell wall is a little box made of cellulose (wood) produced by the cell for support. Draw a simple sketch of the onion cells at high power. Label the cell wall and the nucleus.

Onion Cells

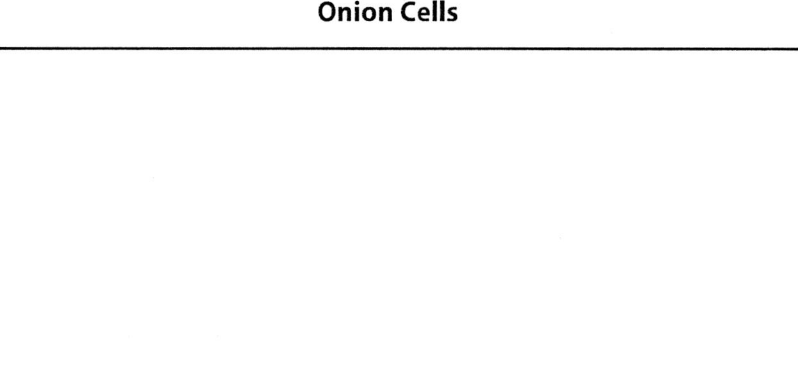

- *Wash off the slide and coverslip so that you can use them for the next Exercise. Don't throw them away. They can be used over and over again.*

? Question

1. What differences and similarities did you observe between plant cells (onion) and animal cells (human cheek)?

Exercise #4
Paramecium

A *Paramecium* is a one-celled organism. Because it must do everything in its life as only one cell, it is far more complex than any single human cell.

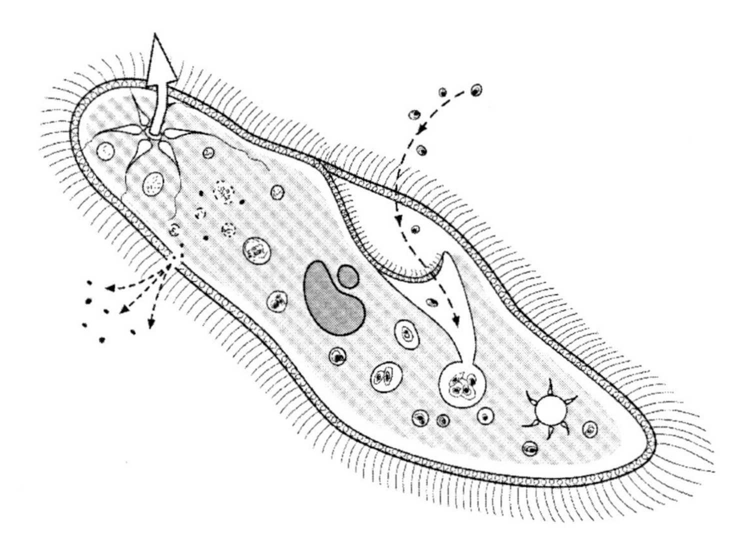

Figure 4.2. Structure of *Paramecium*. The process of feeding involves the engulfing of particulate food which is then contained and processed in digestive vacuoles. After digestion the waste products are excreted to the outside of the cell.

Materials

- ✓ A drop of water from the bottom of the *Paramecium* culture (these organisms usually settle near the bottom). Use an eyedropper to collect a sample, and put the drop on your slide.

Procedure

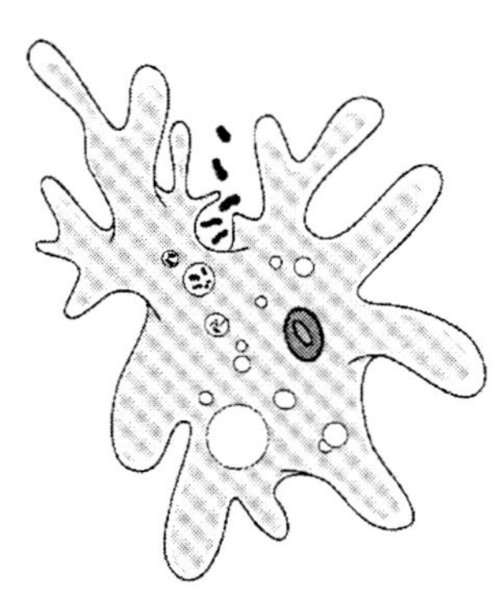

You may also see Amoeba in the pond water sample.

- Make a wet mount, and find the *Paramecium* under the 10x objective lens (that will be 100x magnification).
- Your job is to train your hands to be able to follow this organism under the microscope. **Hint:** Don't think! Let your hands work by themselves.
- Practice your quick-vision skills by making a simple sketch of the *Paramecium*. Label the ***nucleus*** and the ***cell membrane***.

Paramecium

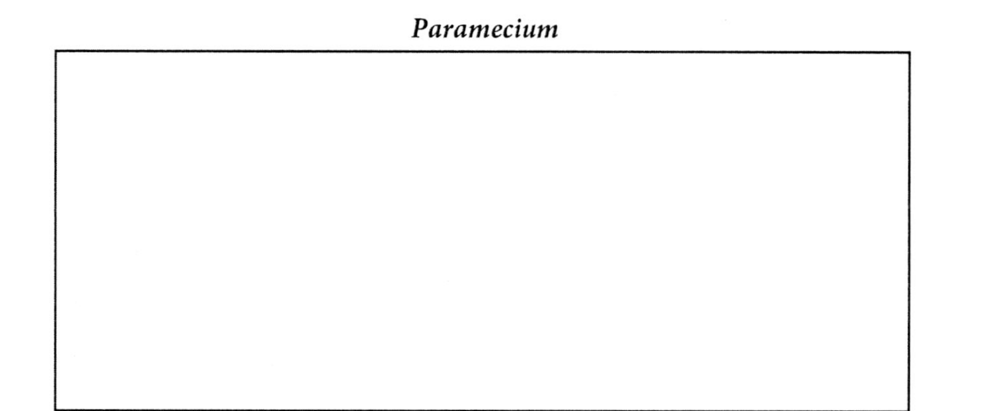

Optional: We have a product called Protoslo® that can be added to the water drop sample. It will dramatically slow the movements of the Paramecium because it thickens the water. However, Protoslo® will also push the one-celled organisms to the outside perimeter of the water drop. You have to first mix the Protoslo® with the drop of sample.

- *Wash off the slide and coverslip in preparation for the next Exercise.*

? Question

1. What obvious differences did you observe in the one-celled organism as compared to one cell of a multi-celled organism (cheek or onion cells)?

2. If you were asked on a test to find a *Paramecium* in "pond water," could you do it? **Remember:** Pond water will contain many different organisms.

3. How does a *Paramecium* eat?

Exercise #5
Elodea Leaf

Elodea is found in freshwater ponds and is commercially grown and sold as an aquarium plant. Pay attention to the differences between the *Elodea* cells and the onion cells that you observed in Exercise #3.

Materials

✓ An *Elodea* leaf from near the tip of a healthy plant. Use forceps to pluck the leaf. Keep track of which side is the upper side of the leaf.

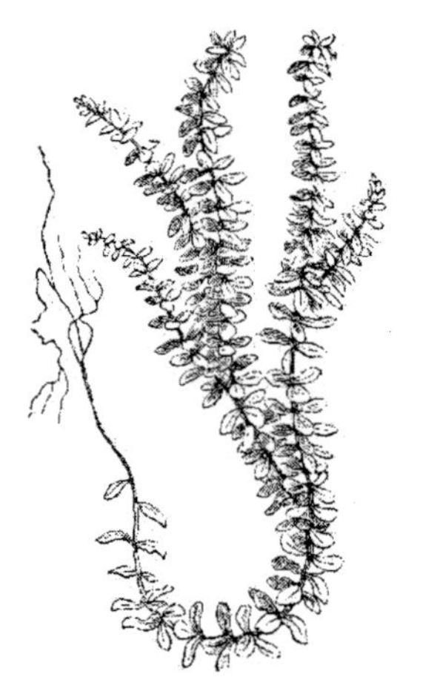

***Elodea* Plant**

Procedure

- Make a wet mount with the upper side of the leaf facing up, and find the *Elodea* cells under the 10x objective. Look near the tip of the leaf. These are often very active cells. Then switch to high power magnification.
- You will know that the cells are alive and active if you can see ***chloroplasts*** moving around the cell. It may take a few minutes of warming under the microscope light before the chloroplasts will begin to move. Chloroplasts are the cell structures (organelles) that do photosynthesis (food production) in plants.
- There is a large sac of fluid inside the *Elodea* cell called the ***central vacuole***. Imagine a swimming pool that has a huge clear sac of water floating in it. You can't actually see the sac of water, but the movements of everyone in the pool will be influenced as they bump into that large clear sac. Now watch the movement of chloroplasts. See if you can observe the indirect evidence that the central vacuole is in the cell and is influencing the movement of those chloroplasts.
- Draw a sketch of the *Elodea* leaf cell, and label the central vacuole and the chloroplasts.

Elodea Leaf Cell

- During the Microscope Lab you used the fine focus at high power to determine which color of thread was on top of the others. An *Elodea* leaf is two cells thick. You should be able to decide whether the top layer is made of bigger or smaller size cells than the bottom layer. Also, one of the layers has thicker cell walls. Compare the layers now.

- *Save this Elodea slide for the next Exercise. You will want to look at the cells one more time.*

? Question

1. What color are the chloroplasts?

2. When you see the color of plants, what structures are you actually seeing?

3. What does the movement of the chloroplasts tell you about the cell?

4. If you can't actually see the central vacuole inside the *Elodea* cell, then how do you know that it is there?

5. *Elodea* leaves are two cells thick. One of the layers has thick cell walls. Those cells are in the . . .

   Top layer or Bottom layer

6. What is the most obvious difference you observed between the *Elodea* cell and the onion cell?

7. What does that observation tell you about the activity of food production in the onion cell?

8. Where is food produced in the onion plant?

Exercise #6
Location of Color in Plant Cells

The color of plant parts is determined by various pigments in the cells, and it is either inside organelles called ***plastids*** or dissolved in the water of the central vacuole. Chloroplasts are one kind of plastid, and there are other plastids that contain different colored pigments. Your task in this Exercise is to determine and compare where the color is located in an *Elodea* leaf, red-onion skin, and yellow flower petal. In order to determine whether a pigment is in the central vacuole or inside the plastids, you must look at a one cell layer and note the distribution of the color within the individual cells.

Materials

- ✓ a small piece of red onion
- ✓ a yellow flower petal (Your lab may substitute red bell pepper.)
- ✓ Your *Elodea* leaf slide from Exercise #5
- ✓ two more slides and coverslips

Procedure

A color distribution like this indicates that the pigment is inside the plastids.

- Make a wet mount of a *one-cell-thick layer* of the red-onion skin. Peel a layer from the inside red part of the onion (not from the dry outside skin).
- Determine whether the red color is inside the plastids or distributed throughout the water of the central vacuole.
- Make a tearing peel of the yellow flower petal. The ragged edge will be one cell thick.
- Make a wet mount of the yellow flower petal, and look at the one-cell-thick area on the ragged edge of the tearing peel.
- Determine whether the yellow color is inside of the plastids or is distributed throughout the water of the central vacuole.

  Note: If your lab is using red bell pepper, then cut a very small piece (without the skin) and crush it on your slide with a razor blade. Apply a drop of water and coverslip.

A color distribution like this indicates that the pigment is in the water of the central vacuole.

? Question

1. Look again at your slide of the *Elodea* leaf. Where is the green color in *Elodea* cells?
2. Where is the red color in the red-onion skin cells?
3. Where is the yellow color in the yellow flower petal cells (or red bell pepper)?
4. If you were asked on a Lab Test to determine where the color is in red rose petal cells, or orange flower petals, could you do it and show your evidence (including your skill at making a one-cell-thick wet mount)?

Wash off your slides and coverslips. Save one for the next Exercise.

Exercise #7
Zebrina Leaf Epidermis with Stomata

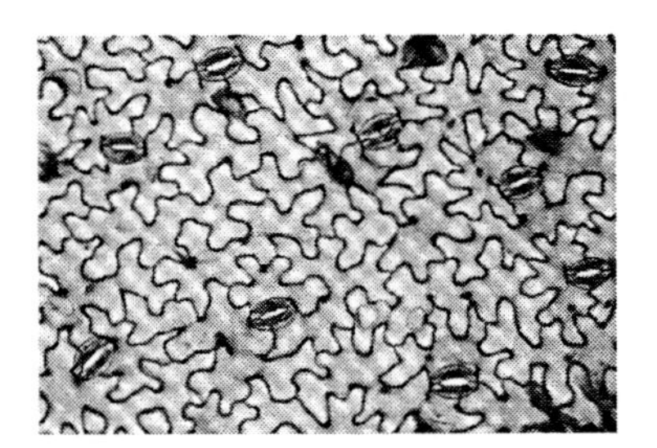

Scattered Stomata

Most animals have some method of breathing. Do plants have any such equivalent process? Scattered throughout the underside skin of the *Zebrina* leaf are small openings called ***stomata***. Stoma is the Greek word for "mouth." These openings look like green lips. The stomata regulate the flow of air into and out of the leaf. Your job is to find these stomata.

Materials

- a *Zebrina* leaf

Procedure

- Make a leaf peel with the leaf upside down so that you can get a one-cell-thick peel of the underside of the *Zebrina* leaf. (The top side of the leaf does not have stomata.) A thin layer will peel off the bottom of the leaf as you tear if you are doing the procedure correctly. Cut off the thin layer piece with a razor blade. Don't try to make a thin slice with your razor blade. You won't get a single-layer slide. Always use the tear/peel technique.
- Make a wet mount and look for the stomata.
- Look closely at the structure of the stomata. Notice whether you can identify the organelles inside of the two cells that make up the stomata. These two stomata cells are called ***guard cells*** because they "guard" the opening. (Your textbook discusses the details of the chemical processes by which the stomata are opened and closed.)
- Look at the skin cells around the stomata. Notice what cell organelles they don't have.
- Draw a simple sketch of the *Zebrina* leaf stomata and the cells that surround it.

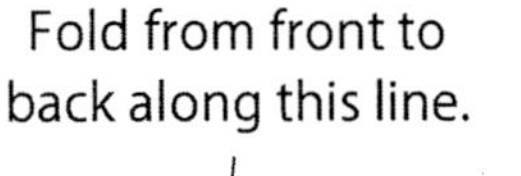

Fold the leaf back on itself, and peel a single layer off the bottom side of the leaf.

Stomata

? Question

1. *Zebrina* leaf stomata perform a specific function. What is it?

2. What organelles do the guard cells contain that are absent in the skin cells of the leaf?

3. Why would the guard cells have chloroplasts when the other skin cells of the leaf don't have chloroplasts?

4. If you were asked on a Lab Test to make a one-cell-thick wet mount of leaf stomata, could you do it?

Wash off your slide and coverslip, and return them to the supply table. Return your compound microscope to the cabinet.

LAB

5

Chemistry of Water

There are so many unexpected properties of water. If you think that there are only three phases of water (vapor, fluid, and solid), you would be only partially correct. There are five known different phases of "liquid" water and perhaps fourteen different phases of ice. For example, at -120^{0}C water becomes ultra-thick like molasses. Water is even stranger at the molecular level. Quantum experiments suggest that water has 25% fewer Hydrogen protons than expected. Its molecular formula may actually be $H_{1.5}O$ instead of what we learned in a science class (H_2O). Weird indeed!

Life first evolved in water, and most life on this planet lives in water. Furthermore, plant and animal forms are 70% to 90% water. These facts clearly establish the importance of this substance to all living beings. Water is rare in the universe, and because of this, life is also rare. In this lab we will investigate some of the unique chemical properties of water: the most essential substance for living organisms.

Exercise #1 **Properties of Water**
Exercise #2 **What the Heck is pH?**
Exercise #3 **Molecular Motion**
Exercise #4 **Diffusion of Water Into and Out of Cells**

Exercise #1
Properties of Water

In this Exercise you will investigate some of the physical properties of water. A more thorough discussion of water's characteristics can be found in your textbook.

Hydrogen Bonds

When two hydrogen atoms react with an oxygen atom to form water, there is an unequal sharing of electrons among the atoms. This creates a slightly negative charge on the side of the water molecule that has more of the "electron cloud."

Water molecules act like a bunch of magnets holding on to each other by the attractions between the + and – ends. This attraction between water molecules is a called a ***hydrogen bond***. In physics, the term for the force that holds molecules of substance together is called ***cohesion***. Hydrogen bonds are the cohesive force holding water molecules together as a liquid, and that is the secret to water's special properties.

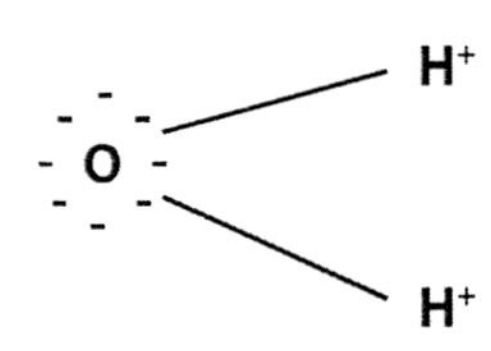

The negative electron cloud is closer to the oxygen atom.

? Question

1. As water falls from the clouds, what force keeps the water in drops?

2. In order for the liquid water to evaporate and become steam, heat must be added. In a pan of boiling water, what bond is being broken by the heat of the stove?

3. So, if heat is required to evaporate water, then what is released when water condenses as hydrogen bonds form?

4. On a calm but rainy day, the temperature rises slightly when it starts to rain. Explain.

Water Adhesion

Adhesion occurs when two or more different substances are stuck together as if by glue. Water has some interesting adhesion properties.

Materials

- ✓ a small container of water
- ✓ an eyedropper
- ✓ four microscope slides

Procedure

- Put several drops of water on one slide, then place the other slide directly on top of the wet slide.
- Try to pull apart the glass slides without sliding them past each other.
- Repeat this experiment with the two dry slides.

? Question

1. How strong is the force of attraction between the two dry slides?
2. How strong is the force of attraction between the two wet slides?
3. What is the name of the force that holds the two slides together?
4. A freeze-dried anchovy is fairly easy to break between your fingers. Yet, when the fish is allowed to sit in water for a while, it only bends with the same effort. Explain.
5. Based on your answer to #4, what is one important role of water in living organisms?

Capillarity

Materials

- ✓ a glass capillary tube
- ✓ a small container of water

Procedure

- Hold the capillary tube vertically between your fingers, and put the bottom end just below the surface of the water. Water holds onto itself (***cohesion***) and holds onto the glass (***adhesion***). The process you have just observed is called ***capillarity***.

? Question

1. What happened when you did this experiment?

Capillarity

2. Draw a simple sketch of the results.
3. Explain your results. What is it about water that makes it do this? (Be specific about the forces of attraction.)
4. How is this property of water important to plants?

Heating Properties of Water

The difficulty of heating water reveals the strength of the hydrogen bonds and the temperature stabilizing role of water within living organisms. When a substance is heated, the rate of temperature increase depends on how easy it is for heat energy to increase the speed of molecular motion in that substance. If it doesn't require much heat to increase the motion, then we say that the substance is "easy to heat up."

You can apply this idea to the heating of water and answer the question: *"Are the hydrogen bonds between water molecules strong enough to make water a substance that is hard to heat up?"*

Materials

- ✓ a hot plate
- ✓ a chunk of metal
- ✓ a beaker (the 100-ml size is best)
- ✓ a container of water
- ✓ tongs to remove weight

Cooling Rule: *The cooling rate of an object is directly related to the amount of heat energy absorbed by that object. If an object cools quickly, then it didn't absorb much heat energy to start with. If water is slow to cool, then it absorbed a lot of energy to heat it up.*

Procedure

- Weigh the piece of metal.
- Put an amount of water equal to the weight of the metal object into the beaker.
- Put the piece of metal into the water, and heat the container until it just begins to steam.
- Immediately remove the piece of metal from the beaker, and put both the beaker of water and the piece of metal onto the table. (Your results are more accurate if you pour the hot water into a room-temperature beaker at the same time you put the metal on the table.)
- Repeatedly touch both the water and the piece of metal until both are approximately the same temperature. Keep track of how long it takes for each to cool.

Time for the water to cool = __________

Time for the metal to cool = __________

? Question

1. Which substance cooled the slowest?

   Metal or Water

2. Which substance would require more heat energy to heat it up? Remember that the amount of heat given off by a substance equals the amount of heat absorbed by that substance when it was heated.

   Metal or Water

3. We experience sitting in 70°F water as being more chilling to the body than sitting in room air at 70°F. Explain why.

4. It is estimated that you lose heat to cold water four times faster than to cold air. Your body is mostly water. What if your body had much less water in it, and the rest of it consisted of substances that heated and cooled as easily as metal. Would a cold environment be more chilling or not? Explain.

5. What bond gives water its unique property of being slow to heat up and slow to cool down?

6. Is water a temperature-stabilizing substance for living organisms? Explain.

Evaporation of Water

Heating water from 0° to 100°C just "speeds" the molecules. As you continue to heat water at 100°C, hydrogen bonds begin to break. The ***heat of vaporization*** is the amount of energy needed to break hydrogen bonds. It can be measured if you know how much heat is used to boil away 10 ml of water already at 100°C.

Materials

- ✓ a hot plate.
- ✓ three equal-size beakers (the 250-ml size is best).
- ✓ a thermometer. *Be careful, please!* This equipment is fragile.

Procedure

Heat Meter

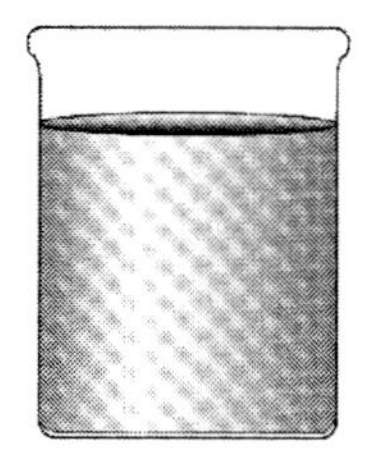

Beaker A
100 ml of ice cold water
(0° C)

- Fill beakers **A** and **B** with the proper amounts of water. Put them in the freezer or into special "ice tubs" in order to stay cold until you start the experiment.

Beaker A is the heat meter. It will be heating during the experiment. Each 1°C increase in **Beaker A** means that 100 calories of energy has left the hot plate and entered each of the three beakers.

Comparison of Heating vs Boiling

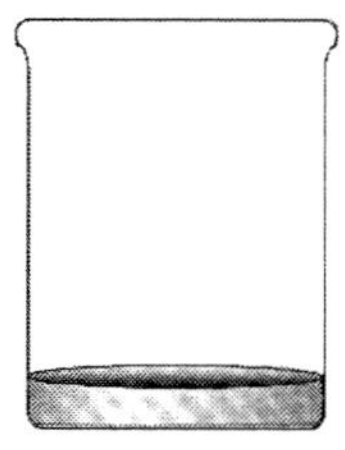

Beaker B
10 ml of ice cold water
(0°C)

Beaker B is a comparison to see which takes more energy— heating water or boiling water. **Note:** Heating water increases the movement of water molecules. This is difficult because the hydrogen bond is so strong. Boiling liquid water into steam requires actually breaking hydrogen bonds.

Beaker C is the experiment. How much energy is required to break (boil away) all of the hydrogen bonds in 10 ml of water?

Experiment

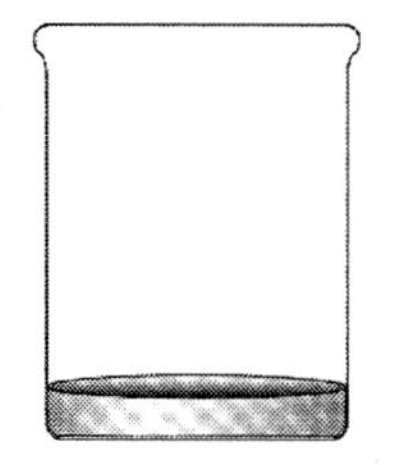

Beaker C
10 m of boiling hot water
(100°C)

- Get boiling water from the Instructor's Table (or you may be asked to boil your own). Use the special measuring pipettes for safely removing 10 ml of boiling water. You must prepare **Beaker C** at the last minute just before you start the experiment. Read on.
- Preheat your hot plate at a setting that you know will boil water moderately. (Probably not the highest setting!)
- It is important that you start all three beakers at exactly the same time, without time for the beakers to change temperature before heating on the hot plate. Record the starting time as soon as all beakers are on the hot plate.
- Continue reading the next page before starting the actual experiment.

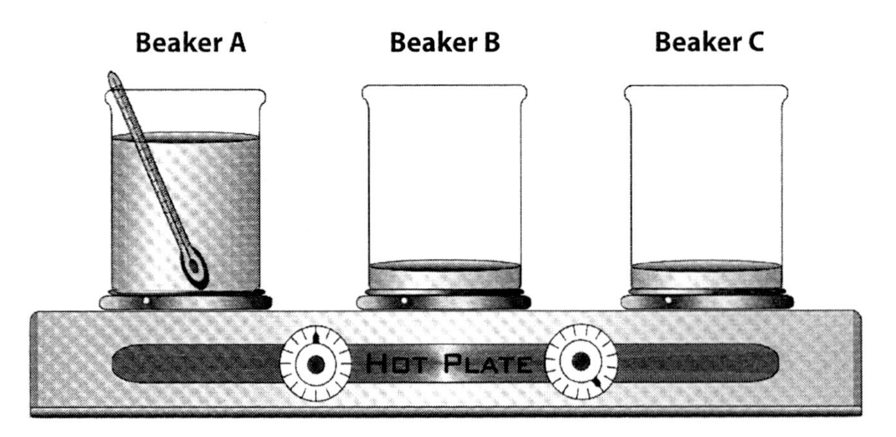

Figure 5.1. There are three events that you must record during this experiment.

There are three events that you must record during this experiment.

1. How many minutes does it take for **Beaker B** to go from 0°C to 100°C (little bubbles form at the bottom)?

2. How many minutes does it take for all of the water in **Beaker C** to boil away?

3. Record the temperature in **Beaker A** at the exact time when all of the water finally boiled out of **Beaker C**.

_____ °C in **Beaker A**

? Question

1. What takes more energy? (circle your choice)

To heat 10 ml of water from 0°C to 100°C.

or

To boil away (evaporate) 10 ml of water that is already at boiling temperature (100°C).

2. When water evaporates (boils away), explain what happens to the hydrogen bonds between molecules.

3. **Beaker A** (Heat Meter) tells us how much heat energy left the hot plate and entered into Experiment **Beaker C**. One calorie of energy is the amount of heat required to increase the temperature of 1 ml of water 1°C. How much did the temperature change in **Beaker A**?

   _____ °C. (Subtract the starting temperature if it was above 0°C.)

4. Now, calculate the total calories of heat recorded by the Heat Meter (**Beaker A**) considering both the amount of water in it and the temperature change of the water.

   _____ calories

5. The answer to question #4 is the amount of energy to vaporize 10 ml of water in **Beaker C**. How many calories would be required to evaporate only 1 ml of water?

   _____ calories

6. Based on your results, discuss how effective is the evaporation of water at removing excess heat from your body (sweating).

Exercise #2
What the Heck is pH?

Hydrogen Ions

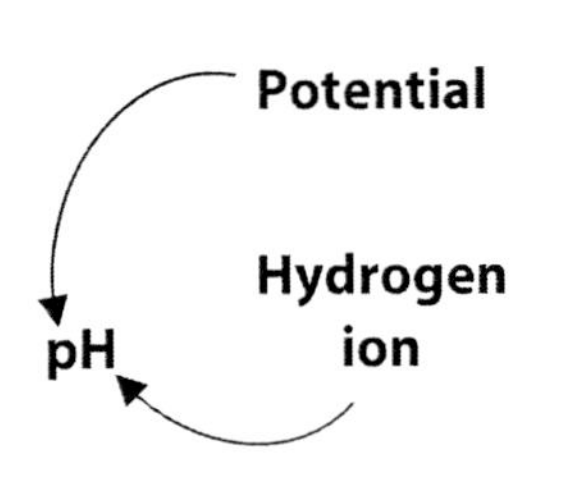

An ***ion*** is an atom that has lost or gained electrons and thereby has become electrically charged (either + or –). Ions act differently than uncharged atoms. (It's like comparing magnets with non-magnets.) These ions are very important in our life processes, and we would die without them. Table Salt is an example of two essential ions—sodium and chloride.

Of all the ions in your body, none is more important than the ***hydrogen ion***, H+. The term "pH" refers to the concentration of H+ ions in water. Biologists are interested in H+ concentration because it affects chemical reactions so greatly. A small change in the hydrogen ion can dramatically affect life. Acid rain and acid stomach are two expressions of the concentration of H+ ions. Also, the blood of a human is so sensitive to H+ concentration that a small pH change from your normal of 7.4 can result in your death.

We monitor the pH of our fish aquariums and our swimming pools in order to avoid potential problems. In the case of the aquarium, we are trying to maintain a good environment for micro-organisms, whereas in the swimming pool we are trying to prevent micro-organisms from growing.

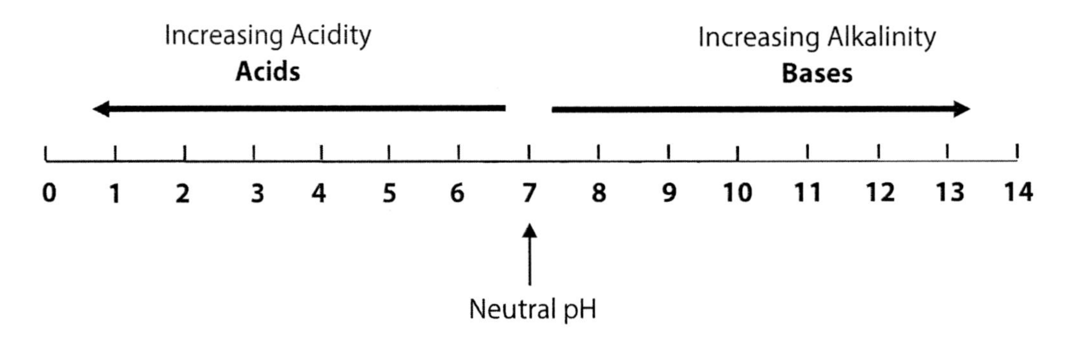

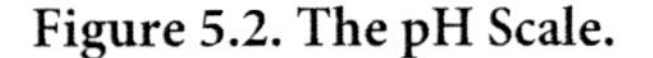
Figure 5.2. The pH Scale.

- Each H_2O molecule can form two opposite ions: H+ and OH–. When acids or bases are added to water there will be a change in the balance of H+ and OH–.
- The pH scale ranges from 0 to 14.
- Each step up the pH scale means that there is 10 times more OH– (base) and 10 times less H+ (acid) than the step below.

- Each step down the pH scale means that there is 10 times more H+ (acid) and 10 times less OH− (base) than the step above.
- A water solution with a pH of 7 is ***neutral***. The concentration of the H+ ions (acid) is equal to the concentration of the OH− ions (base).
- If the pH is less than 7, then there are more H+ (acid) ions than OH− (base) ions, and the solution is called an ***acid***.
- If the pH is more than 7, then there are more OH− (base) ions than H+ (acid) ions, and the solution is called a ***base***.

? Question

1. Knowing how important pH can be to living organisms, what effect does acid rain have on forests?

2. What does the "p" stand for in the term pH?

3. pH is a measure of __________ ion concentration.

4. A pH of 3 would be . . . Acid or Base

5. A pH of 11 would be . . . Acid or Base

6. What is the relationship between H+ and OH− at a pH of 7?

7. How much more H+ is in water at a pH of 3 when compared to a pH of 6?

8. How much more OH− is in water at a pH of 11 when compared to a pH of 7?

How to Measure pH

The pH can be measured with a machine or with special color indicators. A pH machine directly reads the H+ concentration and displays the pH on a screen. A color indicator is a special molecule that changes color at a particular pH level. If you are using a pH machine, follow the directions given by your instructor. Otherwise, follow these instructions for use of paper test strips.

Procedure

- Use the pH paper test kit to determine the pH of the three unknown solutions on the demonstration table.
- Tear off a 1" strip of pH test paper, and squirt a drop on it from the test solution. Compare the color of the pH test paper to the pH color chart. That's the pH!

? Question

1. Is Solution **A** acid or base or neutral?

2. Is Solution **B** acid or base or neutral?

3. Is Solution **C** acid or base or neutral?

Buffers

A ***buffer*** is a chemical substance that can be added to water, and will make that solution resist a pH change.

Materials

- ✓ 25 ml of Sample X in a small beaker
- ✓ 25 ml of Sample Y in a small beaker
- ✓ a dropper-bottle of phenol red
- ✓ a dropper-bottle of acid

Procedure

- Set up two beakers: one with 25 ml of Sample X and one with 25 ml of Sample Y. Label each.
- Put 5 drops of phenol red into each beaker. Phenol red is a pH color indicator. It turns yellow in acid, and it turns red in base.
- Counting the drops, add acid one drop at a time until each beaker turns yellow. Gently shake the beakers after each drop in order to mix the acid into the test solution. If either solution hasn't changed to yellow after 25 drops of acid have been added, then stop adding acid and assume that it will take many more drops to change the pH.

? Question

1. Which solution contains a buffer?

   Sample X or Sample Y

2. How many more drops of acid did it take to change the buffered solution compared to the non-buffered solution?

   _____ more drops

3. Why do you think that one of the brands of aspirin is called "Bufferin"?

Exercise #3
Molecular Motion

All atoms and all molecules move! They are bouncing off of each other at an incredible speed. (It's a good thing that O_2 and N_2 molecules are so small, because they would "sandblast" your skin if they were bigger.) Chemists discovered that the speed of molecular motion is influenced by several factors, and we will investigate two of them. Also, we will look at a couple of special effects created by molecular motion.

Can You See Molecular Motion?

Actually, molecules are too small to see. But a clever physicist calculated the energy in moving water molecules and has determined that those molecules have enough energy to bump into and move some very small particles, like carmine dye. When viewed under a microscope, this movement can be seen.

Materials

- ✓ a slide and coverslip
- ✓ a compound microscope
- ✓ a drop of carmine dye particles

Use a couple of the smallest pieces of carmine dye, and squash them on the slide.

Procedure

- Make a wet mount of the carmine particles.
- Look at the very smallest particles that you can see. Show your instructor.
- The vibrating motion of these particles suspended in water is called ***Brownian Motion***. (Named after guess who?) The tiny particles move whenever a water molecule (which you can't see) bumps into the carmine particle. Observe.

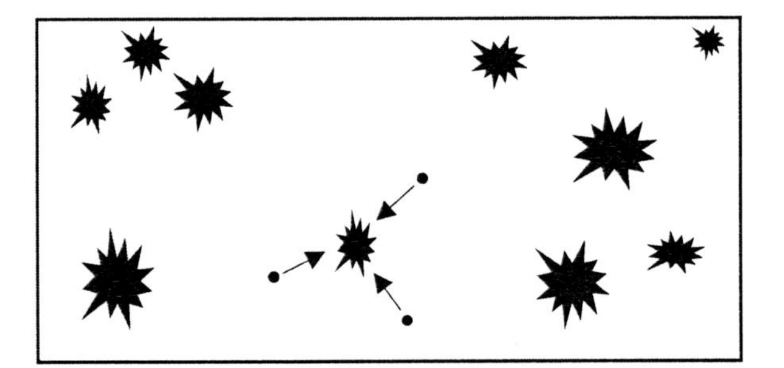

Figure 5.3. Carmine dye particles being hit by invisible water molecules.

? Question

1. When you watch Brownian Motion are you actually seeing molecules move? Explain.

2. What do you think would happen if you held a flame under the slide? Explain.

3. What would happen if you held an ice cube under the slide? Explain.

Light Molecules vs. Heavy Molecules

$C_{16}H_{18}ClN_3S$

Methylene Blue

The weight of all the atoms in one molecule of a substance is called the ***molecular weight***. Methylene Blue dye has a molecular weight of 374. The purple dye Potassium Permanganate has a molecular weight of 158. If molecular weight makes any difference in the speed of motion, then we should be able to measure that difference.

You will need to work as a class for this experiment. Assign one member of your group to work with members from other groups.

Materials

$KMnO_4$

Potassium Permanganate

- ✓ Two agar plates.
- ✓ Put one crystal of Methylene Blue in the middle of the agar plate. (Your instructor may have you put a drop of Methylene Blue solution into one of the small depressions in the agar.)
- ✓ Put one crystal of the Potassium Permanganate in the middle of the other agar plate. (If you are using solutions, put a drop of Potassium Permanganate into the other agar depression.)

Procedure

- Record the starting time for this experiment.
- At 30 minutes and again at one hour, come back to the agar plates and measure the diameter of the spreading colors. Record your measurements below.

Table 5.1. Results of Rate of Spread of Different Size Molecules.

Rate of Spread of Color (Diameter of Color)		
Test Molecule	**In 30 Minutes**	**In One Hour**
Potassium Permanganate Molecular Weight = 158		
Methylene Blue Molecular Weight = 374		

? Question

1. Which molecules move faster?

   Potassium Permanganate or Methylene Blue

2. What does this experiment tell you about the speed of movement of different-sized molecules?

3. Draw your one-hour spread in the agar plate circles below. Label each plate.

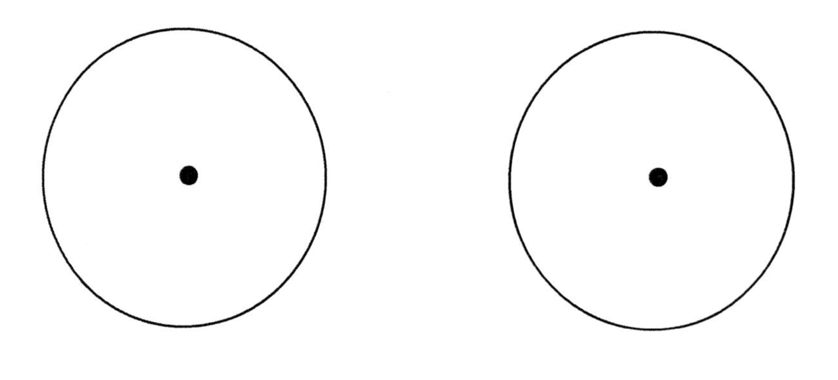

Exercise #4
Diffusion of Water Into and Out of Cells

Everything diffuses from high to low.

The movement of molecules from where they are in high concentration to an area where they are in low concentration is called ***diffusion***. Because all molecules move, they also diffuse!

When a cube of sugar is put into a cup of coffee we know that the sugar will diffuse from where it is in high concentration (the cube) to where it is in low concentration (the hot coffee). However, we don't normally think about what the water molecules are doing in that same cup of coffee. The rules of diffusion apply to water concentration just like they do to the sugar. Water will diffuse from where it is in high concentration (the hot coffee) to where it is in low concentration (the sugar cube).

We can observe the effects of water diffusion in living cells. The diffusion of water through a cell membrane is called ***osmosis***. This is a very important process involving the movement of water throughout the cells of all living organisms.

Materials

- ✓ a slide and coverslip
- ✓ a leaf from the *Elodea* plant
- ✓ an eyedropper of salt water

Procedure

- Make a wet mount of the *Elodea* leaf, and look at the leaf cells under high power. Draw a picture of the distribution of chloroplasts within a typical cell.
- Work with another lab group. One group is to leave their normal *Elodea* slide under the microscope. The other group is to add salt water to their *Elodea* slide.
- Put one drop of salt water at the edge of the coverslip. Use a piece of tissue paper on the opposite side of the coverslip to absorb and pull the salt water across the slide. This salt water will soon surround all of the leaf cells.
- Wait 10 minutes. Look back and forth between the two microscopes. Draw a picture of the distribution of chloroplasts in the normal *Elodea* cell. Then, draw the distribution of chloroplasts in the salt water cell.

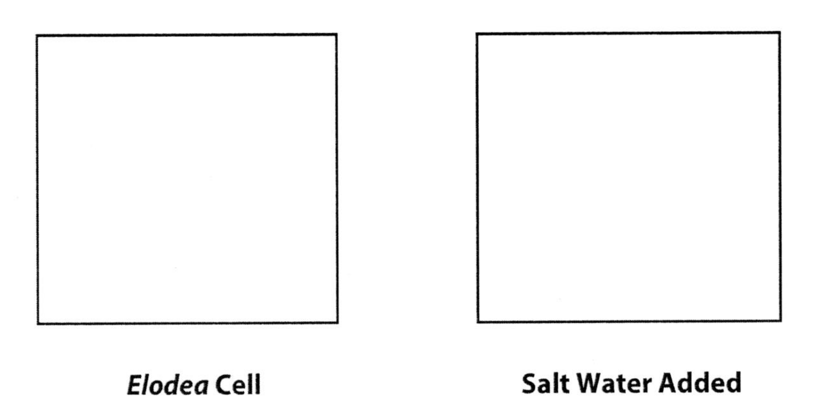

Be sure to clean the microscope stage if you got any salt water on it.

Let this picture represent one *Elodea* cell surrounded by salt water. The small circles are the water molecules and the larger ovals are the salt molecules.

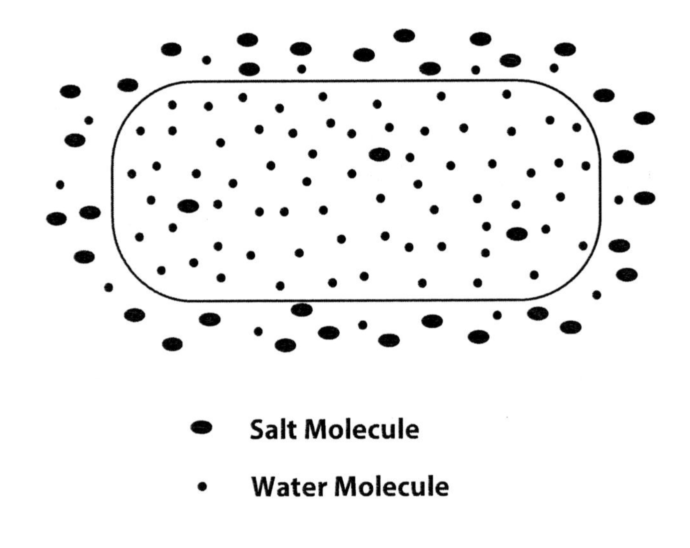

? Question

1. Where is the water in high concentration?

   Inside of the cell   or   Outside of the cell

2. Where is the water in low concentration?

   Inside of the cell   or   Outside of the cell

3. In what direction will osmosis (water diffusion) occur?

   Into the cell   or   Out of the cell

4. Explain what you saw in your second picture of the Elodea leaf cells.

   What is the membrane surrounding the grouped chloroplasts?

5. When you eat a lot of highly salted food (a bag of potato chips, for example), what happens? Why?

6. People with high blood pressure or heart problems are told to be careful about their intake of salt. Why?

Water Movement into the Plant Root

Under normal conditions root hair cells are relatively low in water molecules and high in other kinds of molecules (cell salts and nutrients). Draw arrows to show water movement between the soil water, the root hair cell, and the water transporting tube.

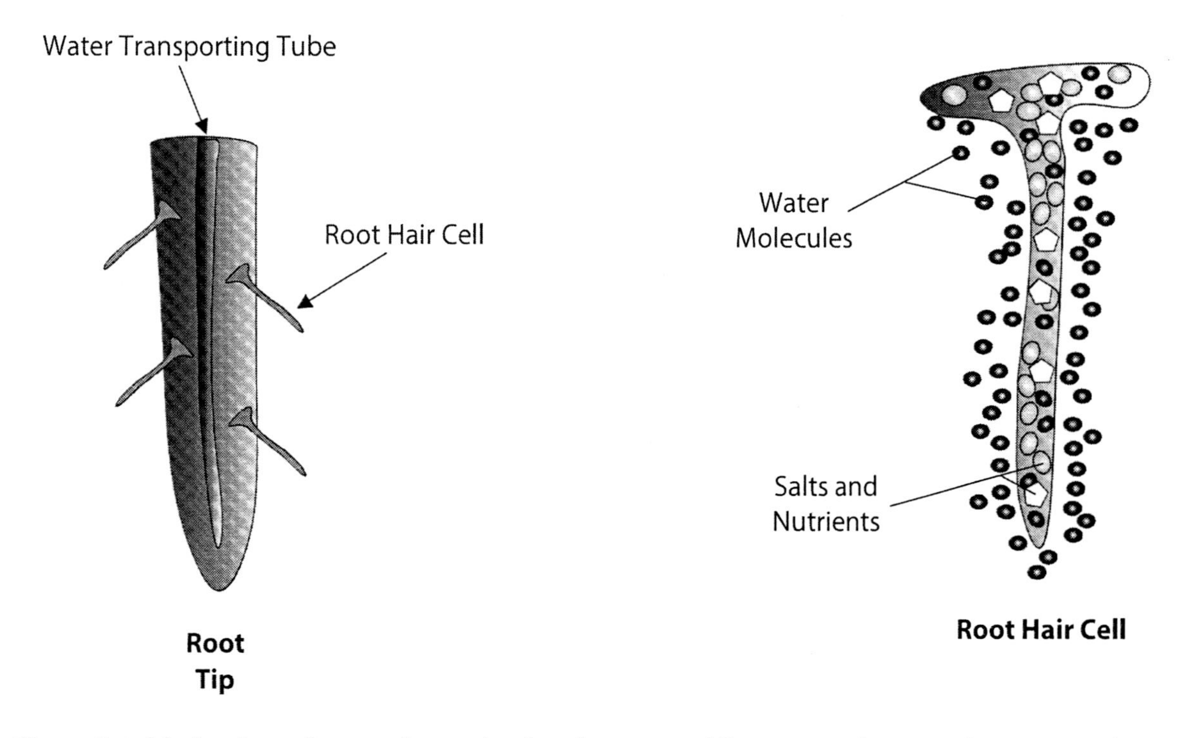

Figure 5.4. Mechanism of water absorption by plant roots. Water moves by osmosis.

? Question

1. What would happen to the water movement if you put salt or a lot of fertilizer in the soil?

2. Explain how the properties of water investigated in Exercises #1 and Exercise #4 assist a plant in the movement of water from the soil, into the root hair, and throughout the structure of the plant.

Optional Activity: Water Density at Different Temperatures

Water is a most unusual substance. At some temperatures it acts like you expect, and at other temperatures it is unusual. Try this surprising experiment.

Procedure

- Very accurately mark the water level in a beaker of cool water.
- Heat the beaker of water to just before steaming (when bubbles start to appear) and record the water level.
- Refill the beaker with ice cubes and enough water to match your first water level mark. (Push all the ice below the water surface.)
- Allow the ice to melt, and record the water level.
- Compare your results. Surprised?

LAB

6

Enzymes

"Enzymes are things invented by biologists that explain things which otherwise require harder thinking."

- Jerome Ysroael Lettvin
Cognitive Scientist; 1920-2011

Living cells need to build molecules and break molecules. ***Enzymes*** are special proteins made by the cell that greatly speed up the chemical processes of molecular making and breaking. If cells did not have enzymes, then high amounts of heat energy would be required to perform cell chemistry. But that quantity of heat destroys structures in the cell and is inconsistent with life. It turns out that enzymes are the key to all chemical activity in the cell. Without them life would not be possible. Because they are so important to life, our genetic systems have evolved many genes to build them and control their activity. Biochemists tell us that the differences between one species and another species are the kinds of enzymes produced. Metabolic diseases result from missing enzymes caused by genetic defects. Understanding the particulars of metabolic defects is key to the treatment of these disorders. The ***human genome project*** is currently developing methods for discovering all human enzymes and the metabolic pathways they control. Improved health methods in the future will follow this research. For example, most pharmaceutical drugs are involved directly or indirectly with enzyme activity.

We have a health food industry that markets enzyme supplements. What are the known values of these supplements? The study of human enzymes is new and complex, and marketing claims are based mostly on guesswork. They may be right about some claims, but we know that they are wrong about most. What is known, what is likely to become known, and the health implications of enzyme research will be our focus this week. How do enzymes operate, and what factors influence their operation?

Exercise #1 **Enzyme Action**
Exercise #2 **Experimental Design of an Enzyme Controlled Reaction**
Exercise #3 **Experimental Factors Affecting Enzyme Action**
Exercise #4 **Enzymes in Nutrition, Sport, and Medicine**

Exercise #1
Enzyme Action

The Role of Heat in Chemical Reactions

Heat speeds up molecular motion and increases the frequency of molecular collisions. Collisions of small molecules can result in the spontaneous formation of larger molecules. Heat energy also breaks the chemical bonds holding big molecules together which produces smaller molecules.

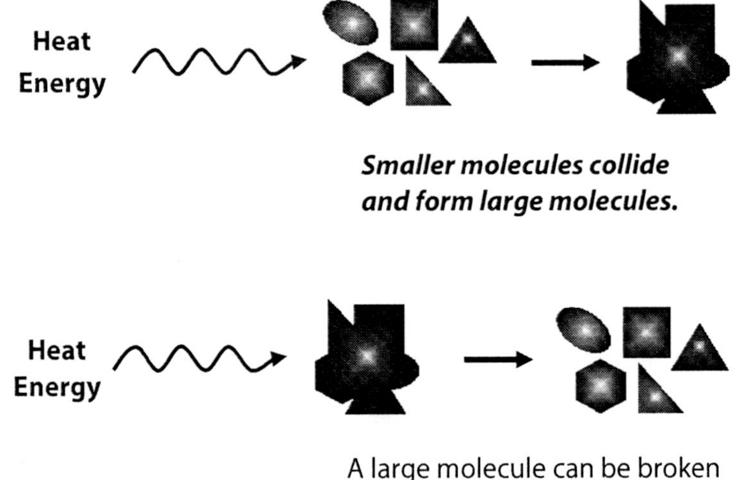

Figure 6.1. The role of heat energy in chemical reactions.

But, heat is not enough.

Because of these two effects of heat energy, molecules can be created and restructured into new substances. However, the amount of heat required to break most molecules apart is too high for a living cell. And the amount of time required for random molecular movements to collide small pieces together is much too long. Cells needs a solution for both of these problems. They must do reactions much faster and with much less energy than would be required from heat.

Enzymes and the Energy of Activation

All reactions need a "push" to get started. This push is called the ***energy of activation***. Sometime the push involves quite a lot of energy. If a cell was dependent only on heat to supply the energy of activation, then that cell would soon overheat and molecules would be destroyed. Enzymes radically decrease the "push" required to start a reaction. They decrease the energy of activation.

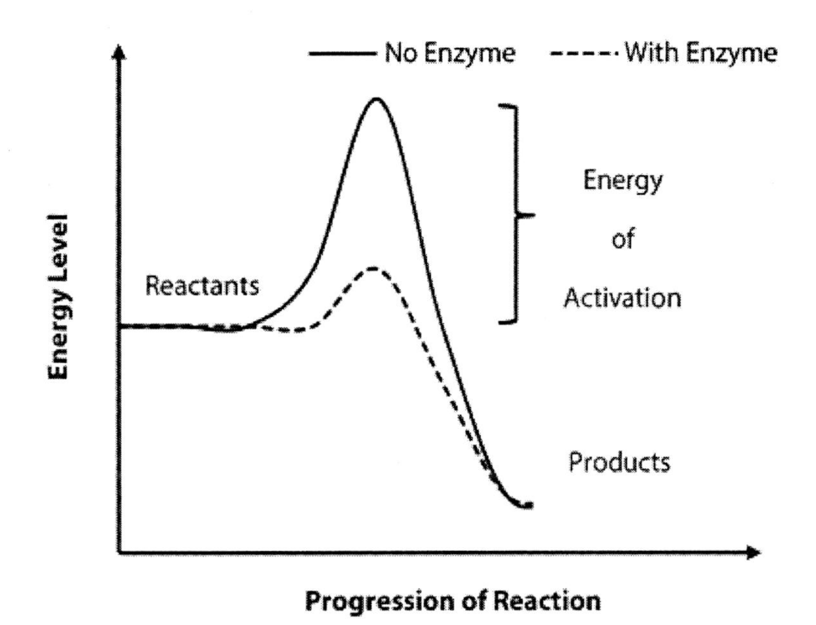

Figure 6.2. Enzymes lower the energy of activation.

Energy of Activation

- This is a "hill of energy" that the reactants must be pushed over before they will react and form products.
- The size of the "hill of energy" is reduced by enzymes.
- When the energy of activation is lowered, the speed of reactions is greatly increased.

Enzymes can be used over and over because they aren't actually changed by the chemical reaction. Any substance that speeds reactions and is not used up during the process is called a ***catalyst***. Enzymes are catalysts. The molecule that is to be changed during a reaction is called the ***reactant*** or ***substrate***, and the molecule produced at the end of the reaction is called the ***product***. There are complex interactions between substrates and enzymes that greatly lower the energy of activation for the reaction. As a result the reaction speeds much faster with enzymes.

An enzyme reduces the normal amount of work that it takes to get the job done.

? Question

1. What are enzymes?

2. When you add heat energy to a chemical system, what happens to the speed of molecular reactions?

3. But what happens if too much heat energy is added to that chemical system?

4. What are two problems in heat driven chemical reactions that the cell needs to solve?

5. Define "energy of activation".

6. What do enzymes change about the required activation energy?

7. Are enzymes destroyed or used up during the chemical reaction?

8. What is a term (other than enzyme) for a substance that speeds a reaction but is not part of the end product?

How do Enzymes Reduce the Energy of Activation?

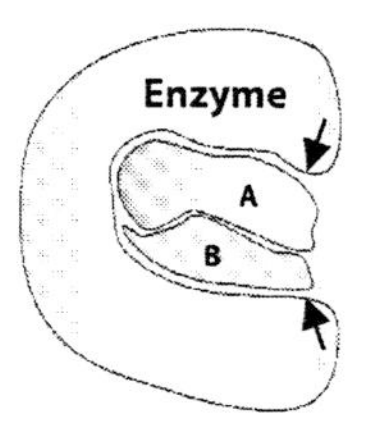

There are enzymes that put molecules together.
A + B → AB

Enzymes are proteins that have ***unique shapes*** and ***reactive sites*** allowing them to break molecules apart or put molecules together without using much energy. The shape is dependent on the kinds and sequence of amino acids in that enzyme. Amino acid sequence is controlled by a gene, and each enzyme is built under the direction of several genes. The special shape and reactive sites of some enzymes attract and hold two molecules close to each other until they chemically bond. Other enzymes break large molecules into smaller molecules. See Figures 6.3 and 6.4.

Sometimes the functional shape of an enzyme requires that another molecule or ion attach to it. Many enzymes bind with metal ions like Mg++ to create the active form of the enzyme. Other enzymes attach to special "coenzymes" that carry chemical groups (like phosphate, acetyl, methyl, or others) to the reaction. Vitamins are coenzymes. We can't make vitamins with our own genes, so they must be in our diet. An example is the vitamin Niacin. It is a coenzyme to at least 200 different enzymes.

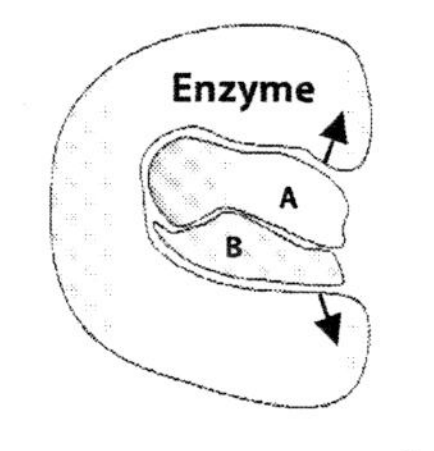

There are enzymes that break molecules apart.
AB → A + B

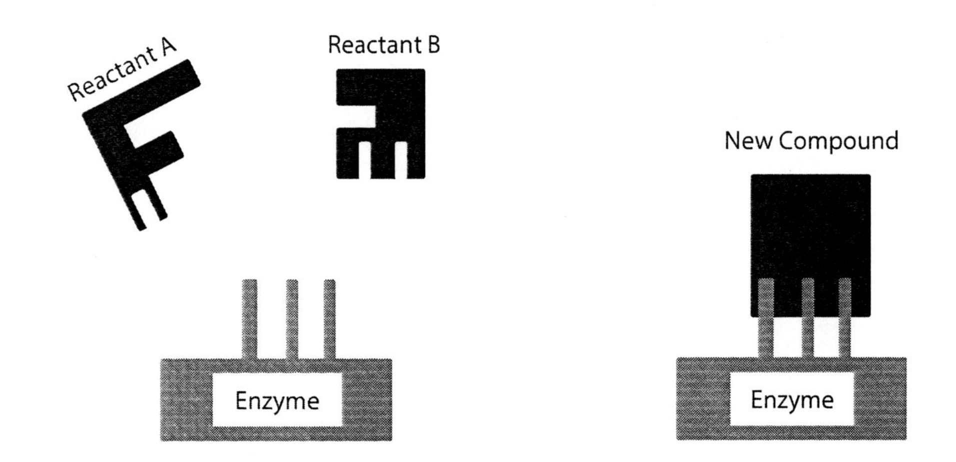

Figure 6.3. The unique shape of an enzyme correctly positions the reactants in a way that allows them to form a new compound.

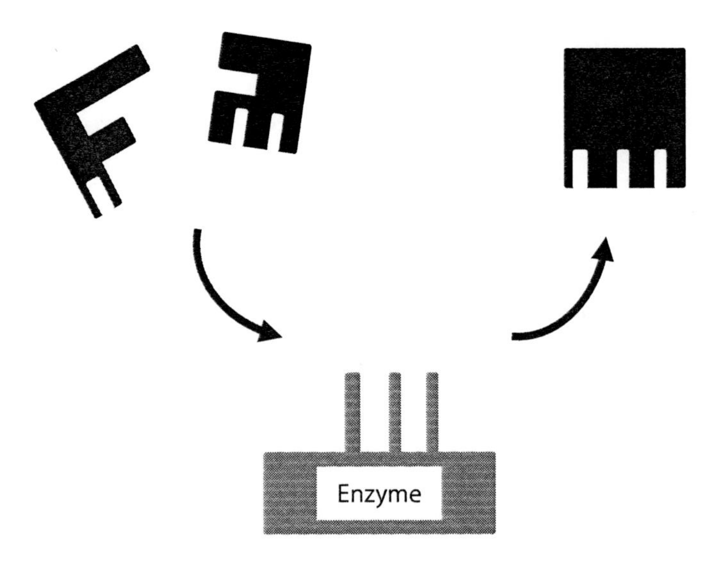

Figure 6.4. The process continues until the reactants are used up or until some other control mechanism shuts it down.

? Question

1. What creates the unique shape of an enzyme?

2. Why does it take so much time for two reactants to react by heat alone?

3. What would happen to a normal chemical process in your cells if the shape of the enzyme controlling that process was changed by an environmental toxin?

4. What does gene mutation do to the shape of the enzyme?

Metabolic Pathways, Enzymes, and Genes

One way to better understand human physiology is to discover all of our metabolic pathways.

Metabolic pathways are a series of cellular reactions necessary to convert a starting substrate into a finished product. There can be many steps in a pathway, but the next example has four steps. Each step is catalyzed by its own special enzyme produced by a particular gene (or group of genes). Refer to Figures 6.5 and 6.6.

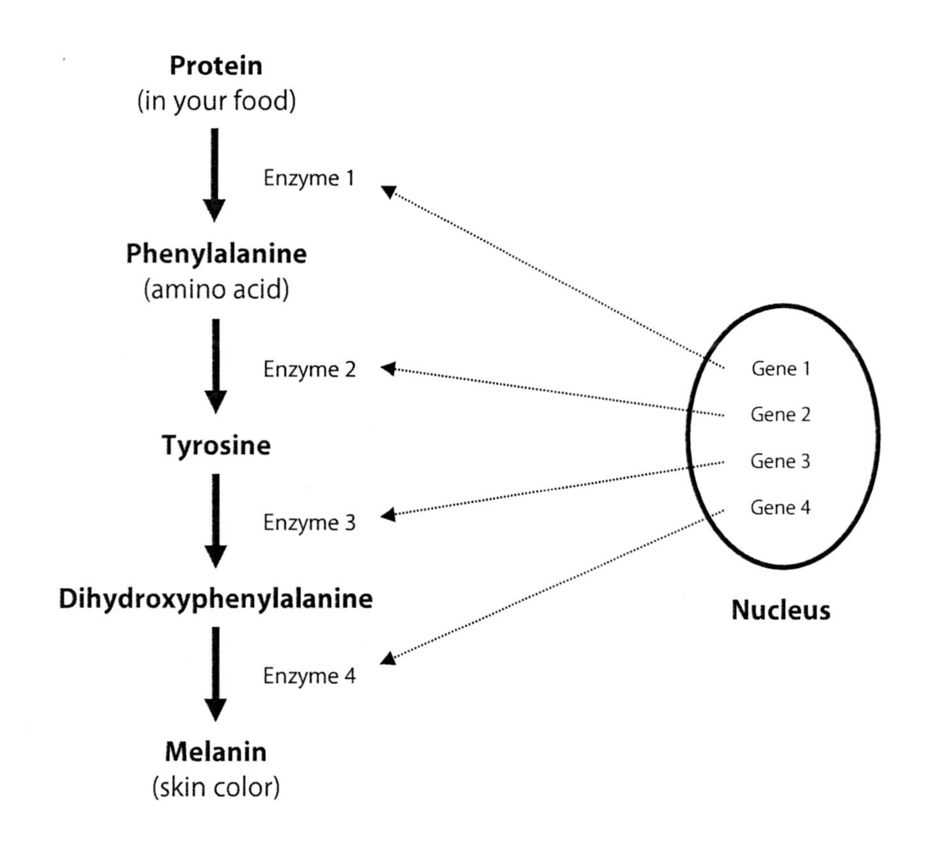

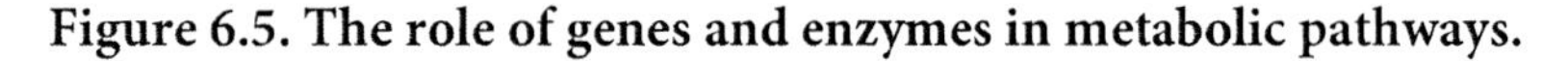
Figure 6.5. The role of genes and enzymes in metabolic pathways.

DNA and RNA control the construction of enzymes.

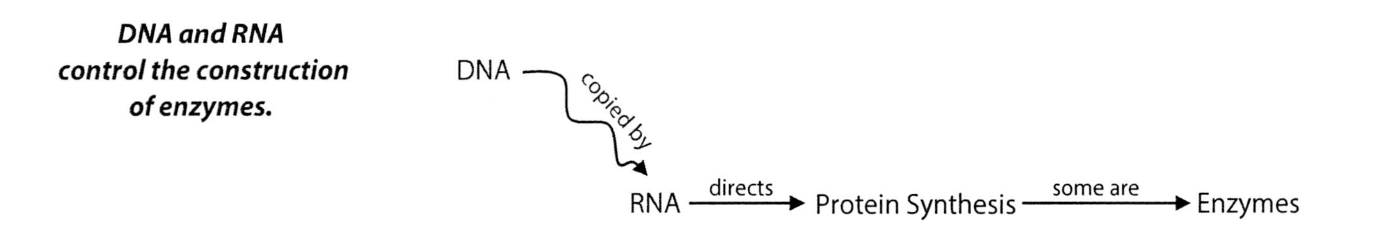

Figure 6.6. Mechanism of enzyme production.

The ***Human Genome Project*** currently estimates that there are over 100 metabolic pathways and nearly 1000 individual biochemical reactions operating in us. And there may be 3000 enzymes involved in all of this. Obviously, every organism must possess the ability to manufacture these enzymes. Each enzyme is created under the direction of a cluster of genes, and you have many genes (30,000 or so in a human). You need lots of genes because enzymes are necessary to perform all of the biochemical processes that make up your life as a complex organism.

? Question

1. Why does your body need so many different enzymes?

2. What component in the nucleus is responsible for synthesizing a particular enzyme?

3. What kind of molecule is an enzyme?

4. If a gene was changed by "mutation," then what would happen to the shape of the enzyme built by that gene?

5. Based on your answer to question #4, what would be the consequence to the organism affected by those "mutations"?

6. 98.3% of all human genes also occur in the chimpanzee. What does this suggest about the similarity of chemistry in these two species?

7. There is only a 0.4% genetic difference among all humans. What does this suggest about the similarity of chemistry among humans as compared to that between humans and chimpanzees?

Exercise #2
Experimental Design of an Enzyme Controlled Reaction

In science we try to make sure that the work done during the day gets us a little closer to understanding what is really going on in the world. ***Experimental design*** is the planning of an experiment so that we get a yes or a no answer to the question we are asking.

The Experimental Control

The scientific method begins with a possible explanation of something we are observing. That explanation is a called the ***hypothesis***. It generates the questions we are trying to answer in an experiment. For example, maybe adding more water to a lawn will make it grow faster. We could design an experiment to test that hypothesis. In the experimental design we might decide to water the lawn twice as much as normal rainfall. Let's suppose that we do this experiment for a month, and we observe that the lawn actually does grow quite a bit faster. Can we safely conclude that we have tested the hypothesis?

What if someone were to point out that the weather was warmer than normal during our experiment, and that is why our experimental lawn grew faster than normal? Or suppose that someone else said that our lawn really didn't grow any more than her lawn which was watered only by rainfall during the same month?

If you fail to include the experimental control, your critics won't fail to notice.

Because we can't answer these questions,
our "experimental design" has failed. Why?

We did not control our experiment! An ***experimental control*** is a duplicate procedure that is set up exactly like the experiment except that the factor being tested (more water) is left out. So, in our experiment we should have monitored a nearby lawn that received only normal rainfall during the same warm month. Then we could compare the growth rate of that "control lawn" to our "experimental lawn" which was receiving extra watering. The results from observing the control lawn would have answered both questions from our critics.

All experiments need a control!

A Simple Enzyme Reaction

Time for another experiment!

Simply stated, an enzyme speeds up the conversion of a ***substrate*** into a ***product***. Potato juice has an enzyme that will change a colorless substrate (called catechol) into a yellow-brown product (called quinone). Forget the fancy names; we will call them substrate (colorless) and product (yellow-brown).

Materials

- ✓ two test tubes. Rinse them out with tap water. There may be some contamination from the last class.
- ✓ Colorless substrate
- ✓ Potato extract enzyme

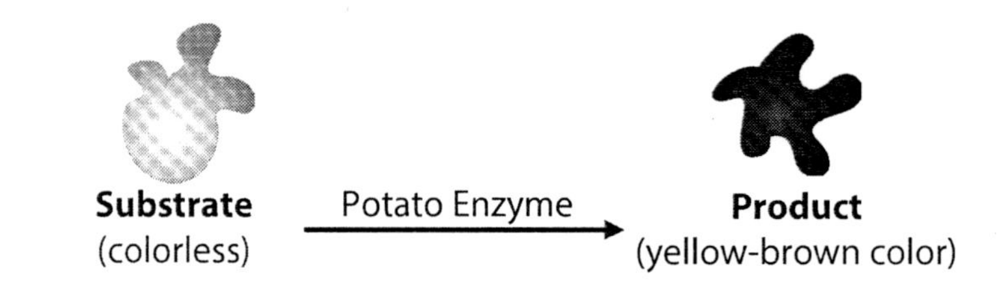

Procedure

- Work in groups of 3 or 4 people.
- Fill each tube half full with distilled water
- Add 10 drops of the colorless substrate to one of the test tubes, and shake the tube. This is the "control" for the experiment. It does not have the enzyme.
- Add 10 drops of the colorless substrate plus 10 drops of the potato extract enzyme to the other test tube, and shake the tube. This is the "experimental" tube. It has the enzyme.
- Record how long it takes (in seconds) for experimental tube to turn yellow-brown.

? Question

1. Why are we adding enzyme to the second test tube?

2. Did you observe any product forming? (Remember: A yellow-brown color means that the product has been made.)

3. If so, how long did it take to form the product?

4. What observations can you make about the first test tube?

5. Why is the first test tube necessary for the validity of your experiment?

6. What are your conclusions?

Now, it's your turn.

Exercise #3
Experimental Factors Affecting Enzyme Action

This part of the lab will help you to practice with the design of experiments and observe some of the factors that affect enzyme action.

Procedure

- Divide into groups of 3 or 4 people. Your instructor will assign two of the six questions to your lab group and to another lab group in the class. Each question will be tested by two lab groups in order to help validate the experimental results.
- Your group is to design an experiment with a control to answer two of the six questions. (Your instructor may require that you test more than two of the questions. Be prepared to do so.)
- Once your group has worked out an experimental design for the two questions, check with your lab instructor for possible suggestions. (For example, if you didn't time the reaction in Exercise #2, you must do it in these experiments.)
- Perform the experiments that you have designed.
- Your group will report its results and conclusion to the rest of the class later in the lab period. Fill out your lab report. Be sure it is complete.

The Six Enzyme Questions

After your lab group has designed experiments to test your two questions, check with your instructor to get additional hints.

#1 ***Is the speed of the reaction changed by variation in temperature (freezing to 120 °F)?***
Note: Run the experiment in an ice bath, and at room temperature, and in a warm water bath below 120 °F (48 °C).

#2 ***Is the speed of the reaction influenced by the amount of enzyme?***
Note: Try different amounts of enzyme: 10 drops, 5 drops, and 1 drop.

#3 ***Is the speed of the reaction influenced by the amount of substrate?***
Note: Try different amounts of substrate: 10 drops, 5 drops, and 1 drop. However, less substrate means that the color change will be lighter because there is less substrate to turn color. How will you handle that problem?

#4 ***Is the speed of the reaction influenced by pH?***
In Exercise #2 you ran the experiment at pH = 7. Use the prepared acid and base solutions instead of water to test the effect of pH on the enzyme. In both cases, leave the enzyme for at least 10 minutes in the test pH solution. This will allow enough time for the pH of the environment to act on the enzymes, if it does. Then add the substrate to see if the enzyme is still active.

Be very careful when handling acids or bases. Wash your hands or eyes immediately if any solution touches them.

#5 ***Are there natural substances, like phenylthiourea, that can inhibit enzyme action?***
Note: Use the same approach as in #4 leaving the enzyme in the tube of phenylthiourea for at least 10 minutes before adding the substrate. Be very careful. Phenylthiourea is a poison.

#6 ***Are enzymes destroyed by high heat?***
Note: Heat the enzyme to steaming (not boiling) for 15 minutes. Let it cool to room temperature. Then do your testing.

? Question

1. Sometimes raw vegetables and fruits are put into the refrigerator to slow the "browning" effect. Is this action supported by the results of experiments during this lab?

2. Sometimes lemon juice is put on raw vegetables or fruits to slow the "browning" effect. How can you explain this effect based on results of experiments during this lab?

3. When fruits or vegetables are left exposed to the air after they have been boiled, they don't turn brown. How can you explain this based on results of experiments during this lab?

LAB REPORT

Question ___

Hypothesis:

Experimental Design:

Results:

Conclusions:

Question ___

Hypothesis:

Experimental Design:

Results:

Conclusions:

Exercise #4
Enzymes in Nutrition, Sport, and Medicine

Change the enzymes - change your chemistry.

The biochemical model of physiology has transformed therapies and understanding of health. When the psychological model was dominant fifty years ago, the hot topics were confidence, attitude, focus, etc. Today, even when these are mentioned, the description of them is usually biochemically based. People want to know whether a training or nutritional program can change their biochemistry. Or even more directly, they want to know if drugs, diets, or supplements can "push" biochemistry to their specific goal. Enzymes are becoming part of this new focus because they are the catalysts of biochemistry. This interest in enzymes follows from the impressive medical success of using enzyme inhibitors in biochemical pathways. Most all drugs used in medicine are inhibitors of specific enzymes or enzyme receptors, and that effect is why they are used.

It is important to remember that Medicine has been focusing on specific biochemical pathways for very particular therapeutic purposes. Consider that we have 30,000 or more genes, maybe 3000 or more enzymes and bio-reactions, more than a hundred biochemical pathways, plus there are many more that are not yet described. Understanding all of human biochemistry is a huge undertaking for the future. In the meantime, there will be gradual successes on very particular health problems. To better appreciate the future direction of the enzyme/biochemical model we will start with the underlying goals and assumptions of the particular health areas that will use new breakthrough discoveries.

Medicine

Medicine's basic approach is to identify serious medical problems in patients (diagnosis) and to intervene with treatments that cure the cause or relieve the symptoms. The medical model is to use drugs that create the desired effects. This means that the focus is on a part of the body physiology that directly produces the disorder, and then find a drug that will interfere with it. When the problem was high cholesterol, the first areas researched were the biochemical pathways that directly produce cholesterol. The pharmaceutical industry adapts to this clinical need by researching those pathways and developing drugs that can inhibit some part of the process. The government (FDA) plays its role by funding research to investigate biochemical pathways and monitoring the effectiveness of drugs developed by the pharmaceutical industry.

Usually the first biochemical therapies target the production of the defined "problem" (cholesterol is an example). It took a few years before clinical investigations progressed to study the absorption of cholesterol from food in the intestine and to develop inhibition strategies for that method of reducing cholesterol in the patient. In Medicine it is accepted that most biochemical therapies will have limited effectiveness and perhaps some serious side effects. Most of these side effects will be short-term, and others are long-term considerations. The responsibility of the clinical community is to monitor and counter-treat those side effects and complications. To the outsider this general approach may appear to be "chasing the mouse around the room", and there is a general frustration of, "Why can't they cure me without any problems?"

Do you get tired of waiting for Medicine to help you with a health problem?

In addition to identifying medical problems and treatments, the medical community has an ongoing commitment to the research of potential health risks and the description of healthy behavior. Medicine is highly regulated, and its focus is on scientific experimentation. Because of this and other factors, Medicine is slow and expensive. It is reasonable to assume that this approach will continue into the foreseeable future. Now we can compare Medicine's approach with that used by the Nutrition and Sport Industries.

Enzyme Supplements in Sport and Nutrition

The health fields of Sport and Nutrition depend on research being done by the academic and medical communities. When information is lacking, Sport and Nutrition do pursue their own investigation. Usually that investigation is focused on correlation studies and limited lab experimentation. Typical questions are, "Does this diet give me more energy or help me to lose weight?" or "Does this training method improve strength, endurance, or coordination?" Sport and Nutrition are not regulated like the Medical fields. This greatly affects how money is spent and the reliability of information generated by these industries. Money is spent primarily on promotion and advertising of products and training methods rather than on research and testing as in Medicine. Reliability of Sport and Nutrition recommendations is more dependent on the judgment of the consumer. There are Sports and Nutrition associations and publications that present some debate on study findings, but government and academic oversight are small influences.

Digestive Enzyme Supplements

What can be said about the claims of enzyme supplements? These claims can be divided into two categories – those that improve digestion and those that change metabolic processes in the body. First, let's consider the general claims made about digestive enzyme supplements by the Supplement Industry.

- Enzyme deficiency is widespread.
- Cooking destroys natural enzymes in the food.
- Not taking enzyme supplements puts unnecessary strain on the body.
- Taking enzyme supplements helps with many medical complaints.

Are these claims valid? There are several rare genetic mutations that result in non-production of digestive enzymes. An example is lactose intolerance which renders the person unable to digest milk sugar. That person could take the enzyme supplement lactase, but most people choose to avoid milk products instead. So the claim that enzyme deficiency is widespread can be called false.

The second claim is true, but it assumes that you need enzymes that are in the food in order to be able to digest that food. This has not been validated by scientific experimentation. Medical studies have shown that digestion is not easier when additional enzymes are added except if the patient has a genetic disorder. The digestive process is more complex than a few additional supplemental enzymes can influence.

The last claim presents an opportunity for the placebo effect. If you advertise that there are general improvements produced by a supplement, there is a dependable 20% improvement attributable to the placebo effect alone. The scientific conclusion at this time is that digestive enzyme supplements do not seem to matter in the health of the general public.

Metabolic Enzyme Supplements

A second category of enzyme supplements is claimed to target the actual metabolic processes in the body cells of the customer. Let's consider what we know about the digestive system and what must happen in order for the supplement claims to be valid. At least six challenges exist:

Who do you trust?

- The supplement must be what it says it is.
- It must not be destroyed by normal digestion.
- Is it absorbed into the blood?
- Once in the blood, does the liver immediately remove it? (Blood from the intestine goes first to the liver and then to the rest of the body.)
- Can it be transported from the blood into the body cells?
- Does the enzyme actually change the cell biochemistry in the desired way with no serious side effects?

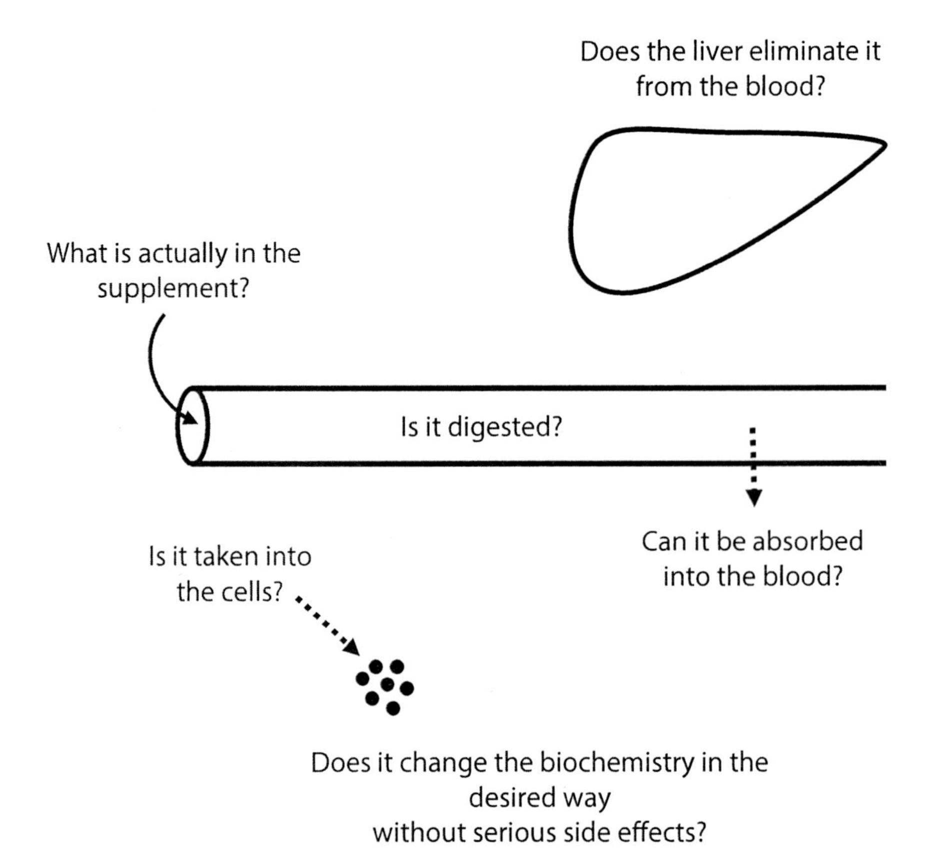

Figure 6.7. Questions about Metabolic Enzyme Supplements.

Metabolic enzymes are protein and are easily digested which makes them unavailable from our food. Their source for us is inside the body cells under the direction of genes. Current research suggests that metabolic enzymes could not make it from oral intake to the body cells. The only long term research of the body's metabolic processes is being done by the Academic and Medical communities and not by the supplement industry. The supplement industry is not required to have confirmatory research. It is only required to not fraudulently misrepresent claims. This is a legal definition – not physiological. They can use loosely controlled correlation studies and conjecture based on known processes as their evidence. That does not mean that they are always wrong – only that there is no substantial evidence at this time that they are right.

In conclusion, we see that the Supplement Industry has very different methods and oversight compared to Medicine. If they did promote enzyme inhibitors as used in Medicine, the industries would be required to operate like Medicine. This will not happen in the foreseeable future. Sport,

Nutrition, and the Supplement industries will continue to depend on advertising, promotion, effective argumentation, and some correlation studies to convince consumers that theirs is a reasonable alternative to Medicine or to no treatment at all. What do you think?

? Question

1. Most drugs in medicine are _______________ of specific enzymes or ____________________.

2. What is the role of the pharmaceutical industry in that process?

3. What is the role of government in that process?

4. In Medicine it is expected that most biochemical therapies will have limited _______________ and some serious _______________ .

5. Sport and Nutrition primarily depend on what type of research?

6. How is most of the money spent by the Supplement Industry?

7. List four general claims made about digestive enzyme supplements.

8. List six general challenges to the claims made about the effectiveness of metabolic supplements.

Go on the Internet and search for "enzyme supplements". What are three dominant themes?

Discussion Group

- Take 10 minutes to discuss the following questions about digestive and metabolic supplements.

1. What were the three dominant themes about enzyme supplements that you found on the Internet?

2. Does your lab group agree that Medicine, Nutrition, and Sport are very different industries?

3. What is the key difference?

4. Do you think that the difference is necessarily good or bad for public health? Why?

See you next week.

LAB

7

Photosynthesis

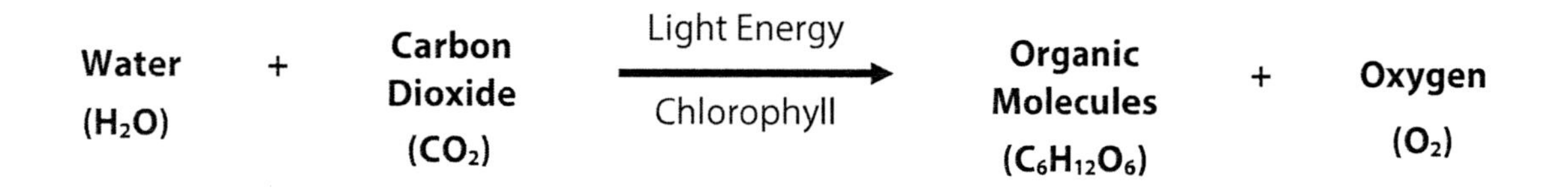

Photosynthesis is the process by which plants use sunlight energy to make new plant tissue, and in doing so they create the by-product of oxygen which animals need. All chemical reactions involve changes in the electrons of atoms, and the key to photosynthesis is a special molecule called ***chlorophyll***. This large and complex molecule has electrons that become "activated" when light shines on it. This is unusual because most substances only heat up when exposed to sunlight; that is, their electrons aren't "activated" by light energy. But the electrons in chlorophyll are activated by light, and this electron energy is used to make organic molecules (new plant tissue).

Plants convert the energy of light into food energy and provide us with our oxygen. In the early history of our planet there was no oxygen in the atmosphere. Oxygen was released into the air only after photosynthesis evolved. This week's lab focuses on the processes of photosynthesis and dependence of animals on plants.

Special Note: Exercise #1 is a tutorial intended to help you in lecture class. It may not be covered in lab this week. Ask your instructor. Exercises #3 and #4 will require more than an hour of your time. You should set up these two experiments at the beginning of lab so that you won't run out of time to finish them.

Exercise #1	**Tutorial on Energy and Photosynthesis** (optional for this week)
Exercise #2	**Leaf Pigments**
Exercise #3	**CO_2 Uptake by Plants**
Exercise #4	**O_2 Production by Plants**
Exercise #5	**Oxygen Demand for Humans**
Exercise #6	**How Big of a Plant Does it Take to Keep You Alive?**

Exercise #1
Energy and Photosynthesis (tutorial for lecture)

Energy

Energy can move and change matter, and can exist in different forms. Some of these forms include heat, light, electrical, chemical, and mechanical energy. In addition to describing the form of energy, we can measure the amount of it. The amount of energy can be measured using various experimental methods. For example, we can estimate the amount of heat energy in a flame by observing how fast that flame can "move" water molecules (heat them up).

Several basic principles of energy change have been discovered. Two of these, the First and Second Laws of Thermodynamics, are of particular value in understanding life processes.

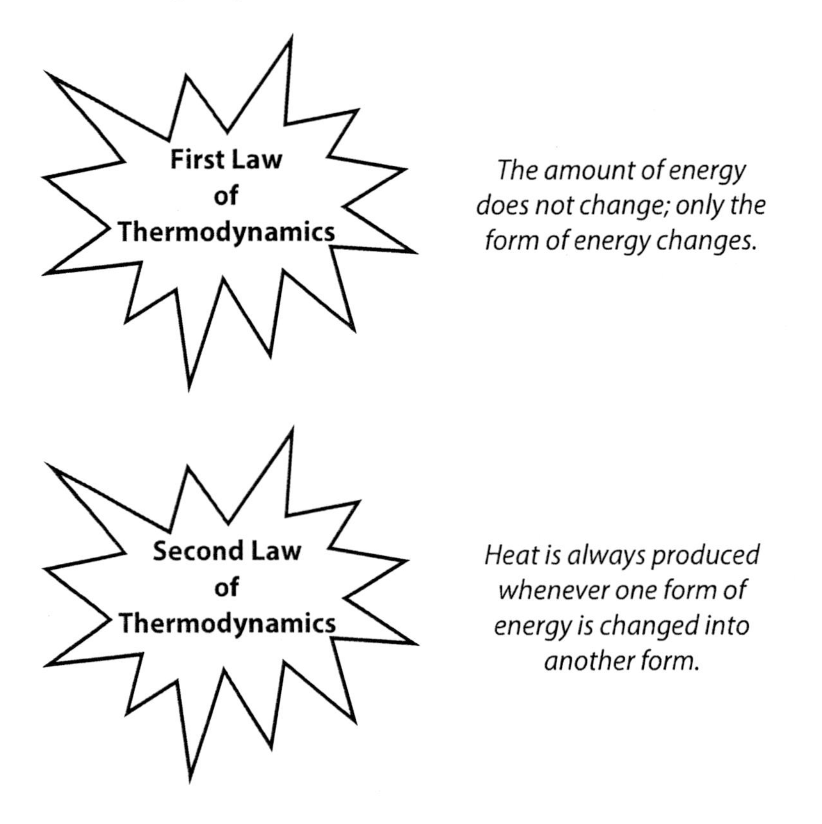

Imagine that the Engineering Department decided to build a campus wind generator for capturing and storing energy in batteries that could be used later to run electric motors. The size of the energy boxes in the diagram below represents the amount of energy available at each step along the way. The numbers in the boxes represent energy units.

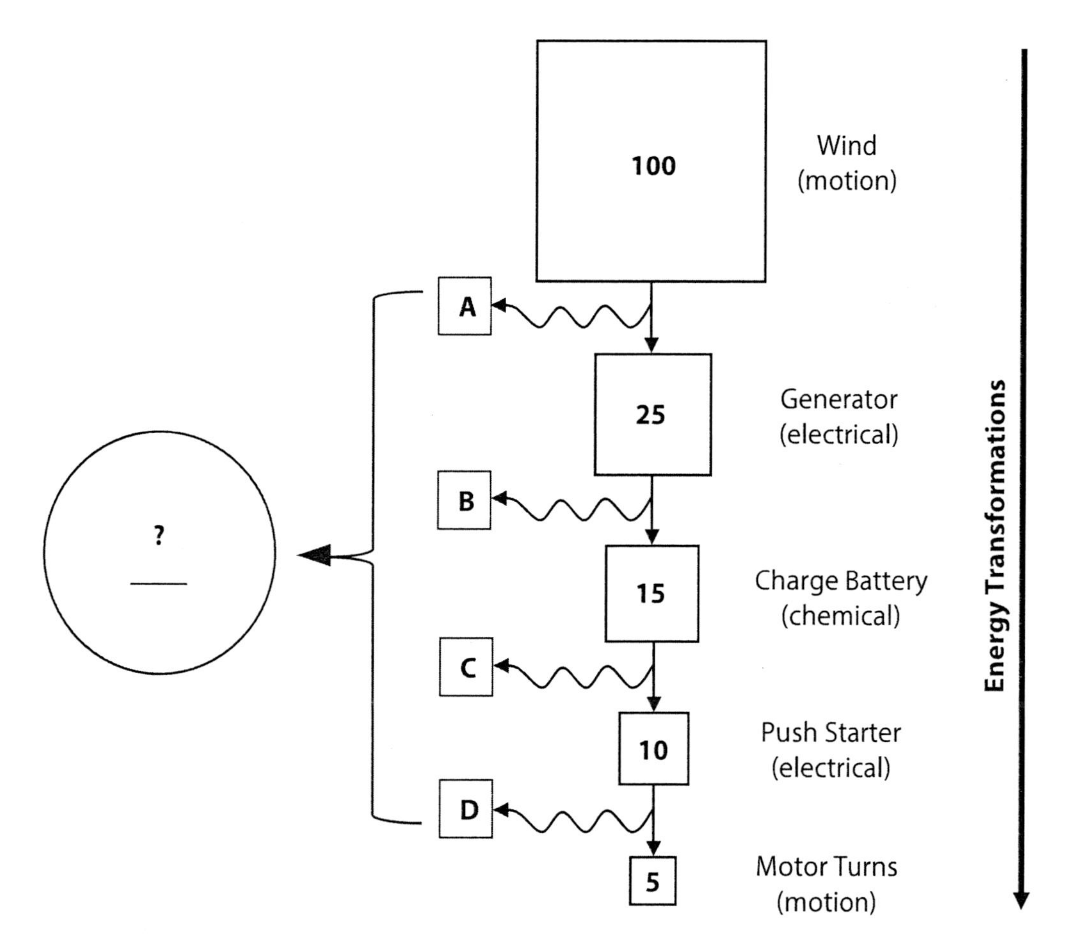

Figure 7.1. Example of an Energy Transformation System. Useable energy decreases and heat energy is released.

? Question

1. What form of energy is represented by A, B, C, and D?

2. Write the name of this form of energy inside the circle on the left side of the diagram.

3. Which Law of Thermodynamics is used to answer question #1?

4. Using the First Law of Thermodynamics, calculate the amounts of energy released at: A_____ B _____ C _____ D _____

5. What is the total amount of heat in the big circle? _____

6. What would happen to efficiency if the wind directly turned the campus motors?

7. Imagine there is an "unknown thing" giving off heat. What can we conclude is going on inside the "unknown thing"?

Energy in Plants

Photosynthesis is the transformation of light energy into chemical energy.

The process of transforming SUN ENERGY ⟶ PLANT is called ***photosynthesis***. We will describe the substances that are changed during photosynthesis in the next exercises. We can measure both the amount of light energy absorbed by a plant during a day and the amount of heat released by the plant. Assume that the size of the boxes represents amounts of energy.

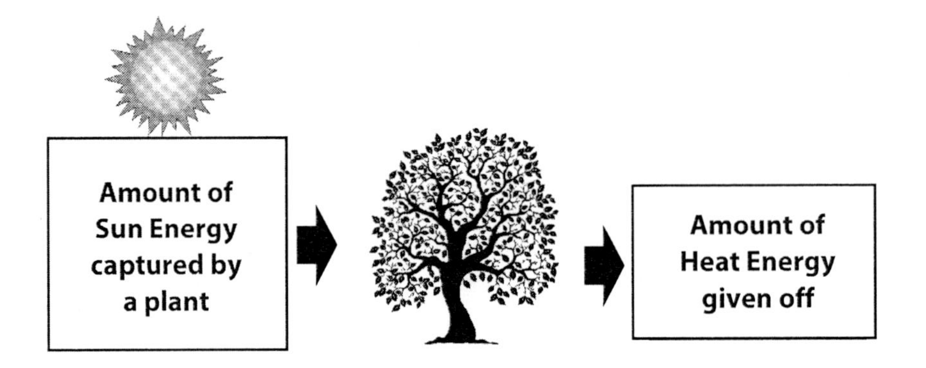

? Question

Considering the First Law of Thermodynamics and the difference in the sizes of the diagram boxes, where is the rest of the energy? (Hint: What does a healthy plant do during the day?)

Energy in Ecosystems

The Laws of Thermodynamics help us to understand conversions of energy within ecosystems. Energy of sunlight is first converted into the chemical energy of plants. This conversion occurs with about 1% efficiency. This means that photosynthesis requires about 100 units of sun energy to produce one unit of sugar energy. The animal that eats plants is called a ***herbivore***. The animal that eats herbivores is called the ***first carnivore***. The ***second carnivore*** eats the first carnivore, and so on. At each step from the plant outward, the energy conversion efficiency is about 10%. (For example, 100 kg of plant are required to produce 10 kg of rabbit.) The next diagram shows the amounts of energy at each food level in the ecosystem.

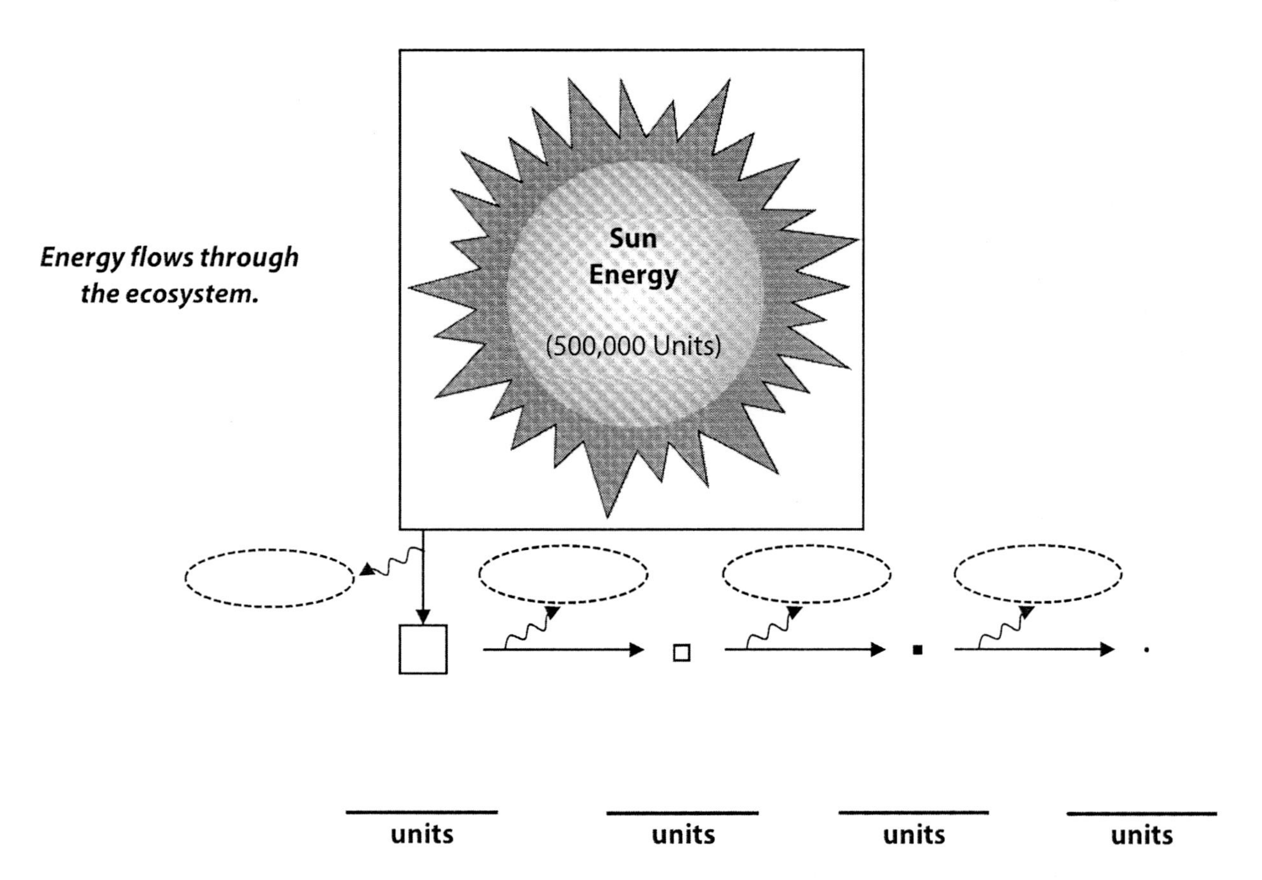

Figure 7.2. Energy flow through an Ecosystem. The amount of energy decreases at each energy conversion from one food level to the next.

? Question

1. Each of the four boxes represents a food level. Write the name of each food level (from plants to second carnivores) below the boxes.

2. Starting with 500,000 units of sunlight energy, calculate and record the number of energy units (on the lines below each box) for each of the energy levels in the diagram.

3. What form of energy is given off during the conversion from one ecosystem level to the next? (Hint: Remember the Second Law of Thermodynamics.)

4. Record the amount of energy (question #3) given off at each step in the ecosystem diagram. (Put your answers in the dotted ovals at the ends of the wavy arrows.)

5. Assume that humans are the second-level carnivores and that each energy unit in their level represents 25 people. How many people can be supported in the ecosystem?

6. Assume that we eat herbivores only. As first-level carnivores, how many humans can now be supported in the ecosystem?

7. If we ate plants only, how many humans could be supported by the ecosystem?

8. If we ate plants only, what would happen to the amounts of energy available for the other herbivores and carnivores in the ecosystem?

9. Let's look at energy conversion another way. It takes about 0.5 kg of meat to support a human for one day. Assuming that beef is the only available nutrient, how much meat energy (kilograms of meat) will it take to get you to 20 years old?

10. How much plant energy (kilograms of plant) will it take to feed the beef that gets you to 20 years old? (Refer to question #9.)

11. Your digestive system is very efficient at digesting meat, but is not as efficient at getting all of the available nutrition from plants. Assume that you were raised on plants instead of meat. Plant material has about 40% of the per-kg food value as meat. How much plant energy would it take to get you to 20 years old?

Photosynthesis

The discussion so far has presented photosynthesis from the energy perspective. Now we consider the changes in substances (matter) during photosynthesis. Chemical changes that occur during photosynthesis have been investigated for the past 300 years. The general highlights of these discoveries help us to understand the basic process of photosynthesis.

A large plant was monitored to see where its weight came from during the year.

Three centuries ago people wondered where plants came from. They knew that plants grew out of the ground, but how that happened was a complete mystery to them. The first step in answering the question was to plant a small tree in a large pot supported off the ground. They did this so that the soil of the container was separated from the soil of the earth. Only the dirt in the pot was available to the plant. The people watered the plant during one year. At the end of the year the small tree had gained 100 kg. (The actual data from this experiment have been changed to simplify the discussion.)

? Question

1. What do you think the experimenters considered as two possible sources for the substances (matter) that became incorporated into new tree growth? (This experiment was performed nearly 300 years ago. People didn't know very much about chemistry, and air was thought to be a non-substance.)

   ____________________ and ____________________

2. When the experimenters measured the weight of the two substances (question #1) they thought went into the plant, they found 35 kg of one and 2 kg of the other. Where do you think they thought the rest of the plant's weight came from?

People guessed that there must be some invisible substance in light or somewhere else that was added to the plant. It took another century to discover the chemical makeup of air. Once accurate scales were invented, small changes in the weight of air could be measured. Experimenters put a plant into a sealed jar with air that had been weighed. After the plant was exposed to light for a few hours, the air lost weight. They had discovered another source of matter for the new plant growth—the air! The basic equation for photosynthesis was complete.

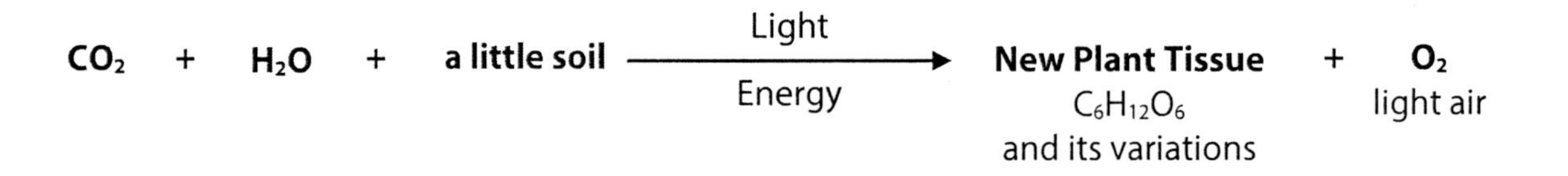

? Question

1. Look at the molecular formula for plant tissue. Where does the carbon come from?
2. Where does the hydrogen in $C_6H_{12}O_6$ come from?
3. When radioactive isotopes of oxygen atoms are put into CO_2 molecules and the plant is allowed to photosynthesize, only new plant tissue ($C_6H_{12}O_6$) is radioactive and the oxygen gas given off is not radioactive. Draw a dotted line in the photosynthesis equation to show where the oxygen atoms in CO_2 go.
4. Draw a dotted line in the equation to show where the oxygen atoms in H_2O go.

Chloroplasts

All of the unicellular algae and multicellular plants are eukaryotic and have specialized organelles called ***chloroplasts***. Chloroplasts contain chlorophyll and enzymes for the photosynthesis process. The exceptions to this rule are the photosynthetic bacteria (cyanobacteria). They are prokaryotic; that is, they don't have cell organelles like chloroplasts. However, cyanobacteria do photosynthesis and they have chlorophyll distributed through the cytoplasm.

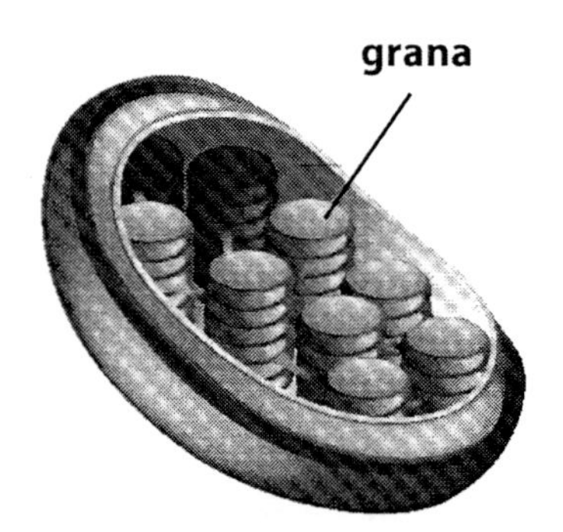

Structure of the chloroplast

The chloroplast has many specialized structures. Stacks of discs, called ***grana***, contain the chlorophyll and act like photoelectric cells. When light shines on these grana, the electrons of the chlorophyll are "activated." This is the first stage of photosynthesis. The second stage of photosynthesis involves the building of sugar molecules from carbon dioxide and water. These two ingredients will not combine unless chemical energy is provided. That energy comes from the "activated" electrons in the chlorophyll reaction. Sugar molecules are made in the fluids of the chloroplasts that surround the grana. Simple sugars made during photosynthesis are modified into all of the other organic molecules needed by the plant.

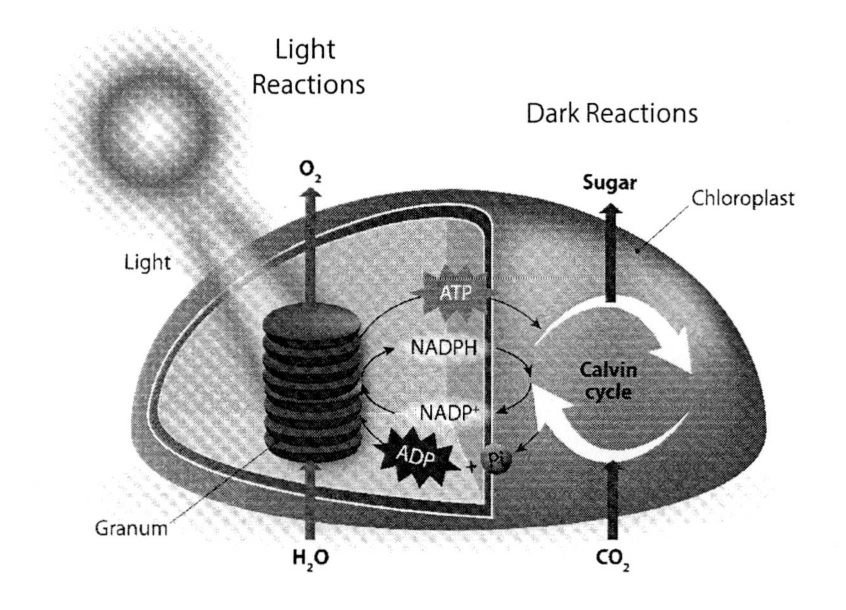

Figure 7.3. Processes of photosynthesis. The light reactions directly require light energy from the sun. The "dark reactions" are driven by the high energy electrons produced by the light reactions.

? Question

1. When light shines on chlorophyll, the light energy is transformed into...

2. Some biologists refer to the two parts of photosynthesis as the light reactions and the synthesis reactions. Which part do you think that they call the light reactions?

3. Which part of photosynthesis do you think they call the synthesis reactions?

4. The synthesis reactions happen in the daytime, but are sometimes called "dark" reactions because light energy is not directly required. What kind of energy is required to run the synthesis reactions?

   Where does that energy come from?

Characteristics of Light

Our description of energy includes anything that is non-material, travels in waves, and having the capacity to move and change matter. There are many different kinds of energy, and each has its characteristic wavelength. When all of these energy waves are arranged on a scale from short waves to long waves, the scale is called the ***electromagnetic spectrum***.

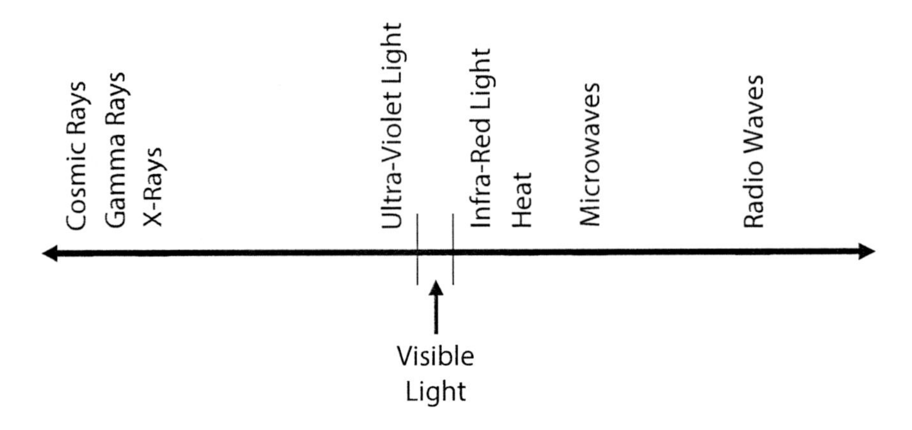

Figure 7.4. The Electromagnetic Spectrum. All energy exists as "waves" described by this scale from very short to very long waves.

Energy waves can be too strong or too weak for life as we know it.

You can see that visible light is only a very small slice of the electromagnetic spectrum. An important feature of electromagnetic energy is that shorter waves have more energy than longer waves. This is the key to understanding why only a narrow segment of waves in the electromagnetic spectrum is ideal for biological reactions like photosynthesis and vision. Some people humorously refer to this idea as the "Goldilocks" principle—the wavelength has to be "just right." The energy waves just shorter than violet light are called ***ultra-violet***, and the waves just longer than red light are called ***infra-red***. Ultra-violet waves are damaging to life. These waves have too much ________________. The molecules exposed to ultra-violet waves become over-activated and are chemically changed or destroyed. When molecules are exposed to infra-red waves, they are "warmed up," but these waves do not have enough ________________ to activate electrons (essential for biochemical reactions).

White light (visible light) consists of several different wavelengths of energy (different colors). You can remember where the various colors occur in the visible light spectrum by recalling the image of a rainbow in your mind.

A rainbow is the visible light spectrum.

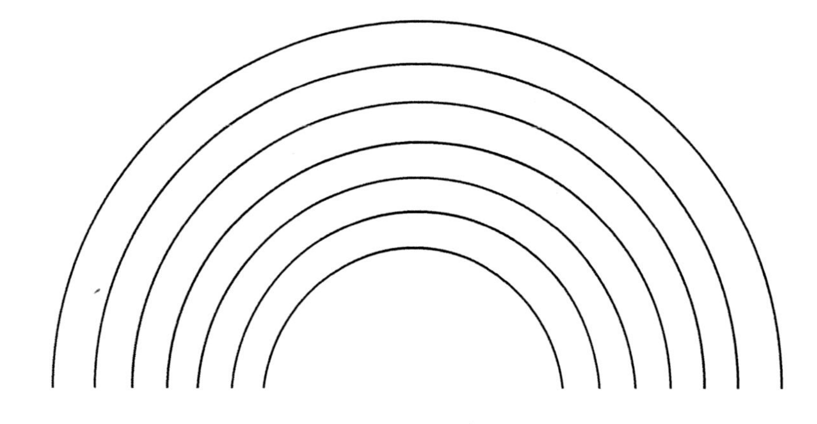

? Question

1. Which color is on the short side of the rainbow?

2. Which color is on the long side?

3. Which color is about in the middle?

4. Put these colors in the rainbow diagram above. (The long side of the rainbow is the longer wavelength.)

5. Which color do you see when you look at plants? (The color that you see is the color that plants do not use in photosynthesis.)

6. Which two primary colors do you not see when looking at a plant? (These are the two wavelengths of energy that plants mostly use for photosynthesis, so they are absorbed by the plant rather than reflected back to your eye.

_______________ and _______________

There is much more to photosynthesis than is covered in this Exercise. You will have to get those details from lecture or your textbook.

Exercise #2
Light Activation of Chlorophyll

Botanists tell us that the electrons of the chlorophyll molecule are "charged up" by light energy, and those electrons release that energy immediately to make organic molecules (food) during photosynthesis. In this Exercise, you will see for yourself whether chlorophyll can be "charged up" by light.

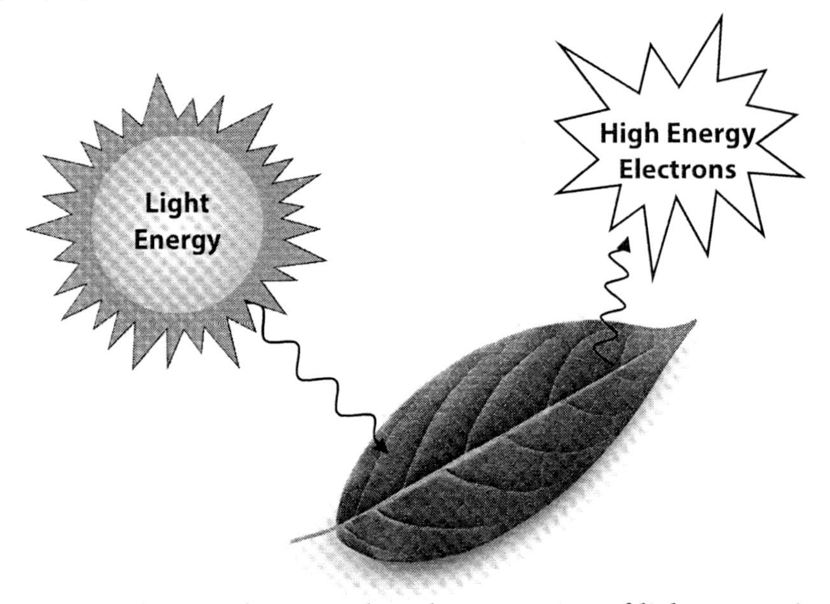

Figure 7.5. Photosynthesis involves the conversion of light energy into electron energy. That electron energy is then used to run the processes of plant life.

Procedure

- Your instructor will take you into a dark room and shine a blue light onto a pure solution of chlorophyll. Blue light contains no other light colors in it. *Note:* Your instructor may use a long-wave UV light that also produces a lot of blue light.
- Your group is to observe. Then, go out of the room and discuss what you saw. (Your instructor may shine the blue light on green food coloring as a control for this experiment.)

  Hint: Pure chlorophyll cannot pass any energy to the rest of the photosynthesis process (to make food) unless the chlorophyll is contained in a chloroplast. That does not mean that the chlorophyll can't react to light. It only means that it can't make food.

? Question

1. When light activates the electrons of chlorophyll, then those electrons have

   Less energy or More energy

2. Physics tells us that if a substance absorbs energy, then eventually it will lose that energy in one form or another. What did you observe about the chlorophyll solution when the light was shined on it?

3. Based on the results of this experiment, fill in the empty box.

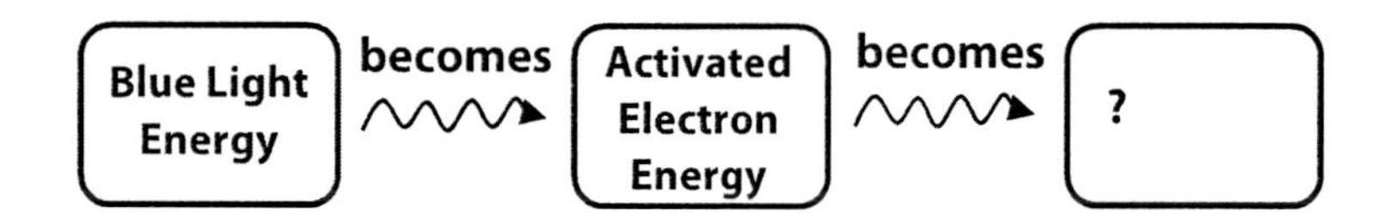

4. Under normal conditions, plants convert activated electron energy into what?

Leaf Pigments

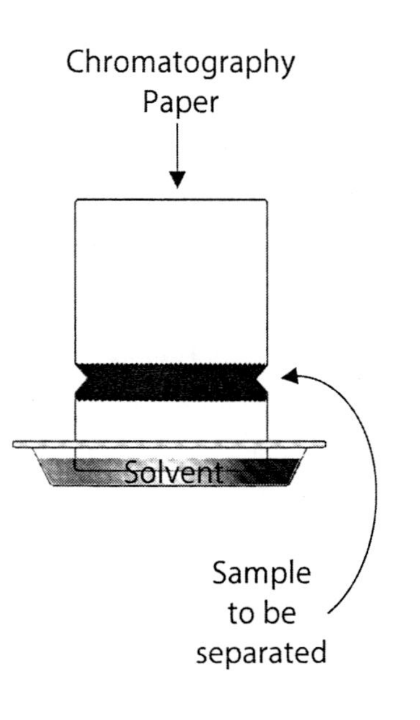

There are many different kinds of organic molecules, and sometimes they can be all mixed together in something we want to analyze. A sample may appear to be only one substance, but it is often a mixture of many different substances. ***Chromatography*** is a very basic chemical process used to separate organic molecules from each other. During this process a solvent passes through a sample that has been impregnated on a piece of paper.

As the solvent travels up the chromatography paper, heavier or more chemically charged molecules will be left near the bottom. The other molecules (lighter or less chemically charged) will be carried up the paper. The secret to chromatography is that each different kind of organic molecule in the sample will be picked up by the solvent at a different rate. It depends on the individual characteristics of each substance. Therefore, the organic molecules will be spread out along the paper.

Materials

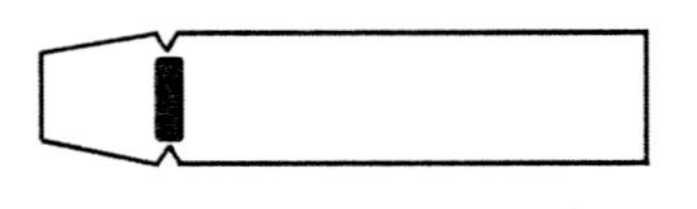

Use a quarter to roll a thick green line of spinach juice onto the chromatography paper.

- ✓ a chromatography jar and cork
- ✓ a chromatography paper and scissors
- ✓ a spinach leaf and a quarter

Procedure

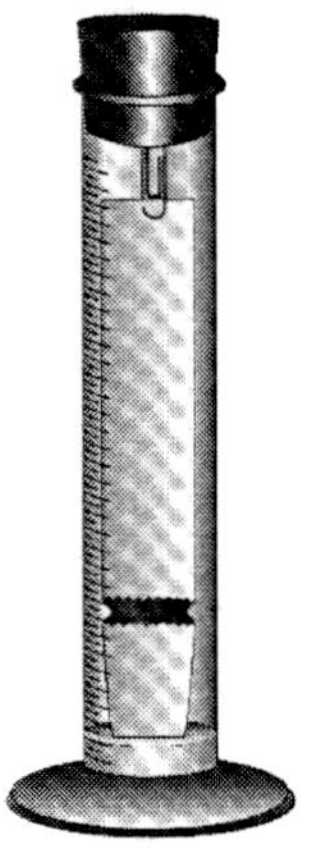

Chromatography setup for leaf pigment separation.

- Wash your hands with soap and water so that the substances normally on your hands (food oils, lunch residue, etc.) don't become part of the chromatography separation.
- Cut a point on the end of the chromatography paper. Cut two small notches in the sides about 1.5 cm up from the point. These notches force the solvent to go through the spinach juice.
- Roll a quarter across a spinach leaf to squash a line of juice between the two notches. Make sure this line is dark green. Go over it several times.
- Set up the chromatography jar. Place the notched paper so that when it hangs from the cork, the point just touches the bottom of the jar.
- You must do the rest of the experiment under a fume hood or outside in the open air. Be careful! The solvent you are using is highly flammable!
- Pour the solvent into the chromatography jar to a depth of about 0.5 cm. Plug the cork with the hanging paper into the jar. Leave the jar under the fume hood and don't move it.
- During the next 10–30 minutes, the spinach juice will be separated into its individual pigments. Determine how many pigments are present. Each may be a slightly different shade, or might be the same color but at a different location on the chromatography paper. Present your answer, and show the evidence to your instructor.

When you have finished the chromatography separation, pour the solvent into the waste jar in the fume chamber. Return the chromatography setup to the lab classroom. Do not wash out the setup! Solvents collect in the air spaces of city drain systems and can be dangerous to sewer workers.

Exercise #3
CO_2 Uptake by Plants

The photosynthesis equation says that $H_2O + CO_2$ become $C_6H_{12}O_6 + O_2$. Carbon dioxide is used to make part of the organic product (food). If this is true, then we should be able to observe CO_2 being used up during photosynthesis. There is a very simple way to show changes in CO_2 level. Phenol red is a red solution that turns yellow when CO_2 is added to it, and then it turns back to red when CO_2 is removed.

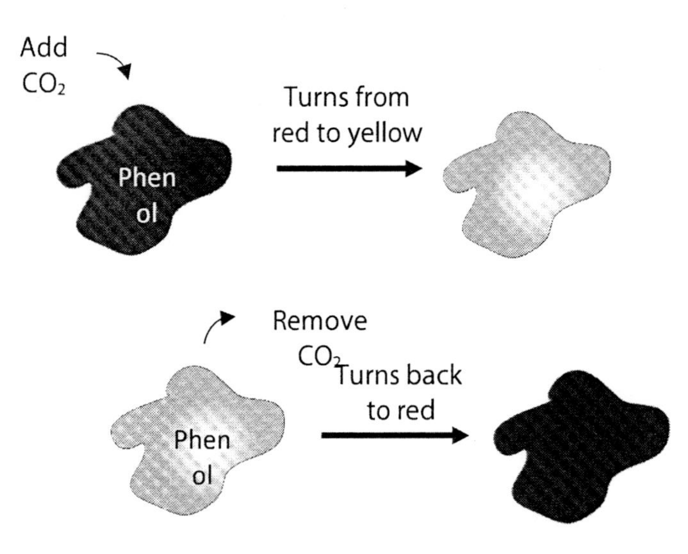

Figure 7.6. Phenol red can be used as an experimental tool to answer the question: "Do plants use CO_2 during photosynthesis?"

Materials

- ✓ test tube
- ✓ phenol red solution
- ✓ small piece of *Elodea* plant
- ✓ straw
- ✓ small beaker

Procedure

- First, "charge up" the phenol red with CO_2 using your own breath. The easiest way is to pour one test tube full of phenol red solution into a small beaker. Blow very carefully through a straw into the solution until it turns yellow. (Don't blow so forcefully that you make a mess.)

- Pour the yellow solution back into your test tube. Add a small piece of *Elodea* plant (about 10 cm).
- Carefully put the bent glass tube cork into the test tube, leaving no air bubbles.
- Put the experimental setup in front of a light source for 30 minutes.
- What happens?

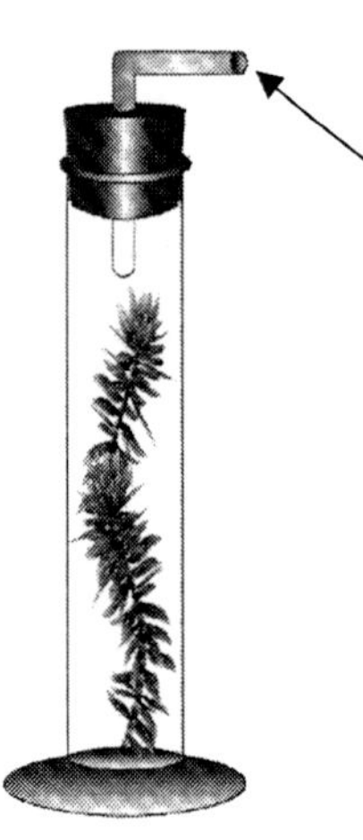

Figure 7.7. Experimental setup to test whether plants use CO_2 during photosynthesis.

- Now, your group is to design a simple experiment that will test whether light is required by the plant during photosynthesis. Be sure to include a control.
- Check with your instructor when you think you have a good design for the experiment.
- Now, do the experiment.

Please put the used *Elodea* plants into the special container! They have some phenol red on them that will contaminate the rest of the *Elodea*. These plants will be rinsed and used again for the next lab class.

? Question

1. If CO_2 is removed from the phenol red solution, then what process is going on in the *Elodea*?

2. Is light required by the plant during photosynthesis?

3. Describe your controls. (There are two.)

4. What is the purpose of having a control?

Exercise #4
O_2 Production by Plants

The photosynthesis equations says that oxygen is produced. If this is true, then we should be able to observe that happening.

How much oxygen is produced by the Elodea plants in one hour?

Procedure

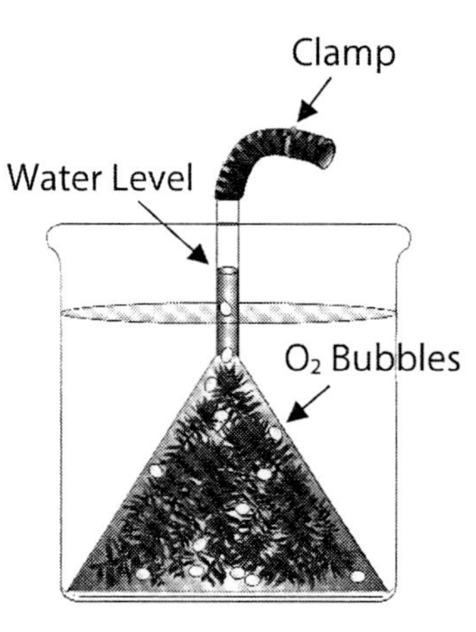

Experimental Setup

- You need ten 2" pieces of healthy *Elodea* plants. Trim 1/8" off each stem. A fresh cut will allow oxygen to bubble out of the stem.
- You have to suck water up the funnel and up the tube, and then clamp the hose at the top to keep the water level from falling.
- If the experiment is working, the oxygen bubbles will collect at the top of the tube and push the water level down. This drop in water level is what you will measure during the experiment.
- Set up a light source shining from the side, but make sure to put a beaker of clear water between the light and the *Elodea* container. (This beaker prevents the *Elodea* from overheating. The clear water container will absorb the heat from the light bulb, yet still allows light to pass through for photosynthesis.)
  Note: If you can take the apparatus outside into the sunlight, your plants will photosynthesize much faster.
- We will do only one of these experimental setups for the class to observe. Select one person in your group to work with the instructor to set up the apparatus.
- Have your group's representative record the O_2 production every 15 minutes for one hour.

- Record the total milliliters (ml) of oxygen produced during one hour.
- You will use this production value during Exercise #6.

? Question

ml of O_2 produced by the plant in one hour = _____

Exercise #5
Oxygen Demand for Humans

Next week we will actually measure the O_2 consumption of a mouse under different temperature conditions and compare different animals and plants. However, this week we can borrow an estimate of human oxygen demand from experimental research. How much oxygen does a human need to survive one hour of biology lab class?

The O_2 used by a human in one hour can range from ¼ of a liter of O_2 per kg of body weight to as high as 8 liters per kg. (Although that high rate of metabolism could be maintained for only about 2 minutes without total exhaustion.) A person in biology lab class uses about 0.4 liters of O_2 per kg of body weight in one hour as long as they aren't walking around.

Procedure

- Assume that the O_2 used by a person during one hour of biology lab is about 0.4 liters (400 ml) per kg of body weight.
- Assume that the average human weighs 60 kg.
- What is the oxygen demand (in ml) for an average person during one hour of biology lab?

? Question

O_2 used by a human in one hour = __________ ml

You will use this calculation again in Exercise #6.

Exercise #6
How Big of a Plant Does it Take to Keep You Alive?

You have an estimate of the amount of O_2 produced during one hour by the *Elodea* Plant (see Exercise #4), and you have an estimate of the amount of O_2 used by a human in one hour (see Exercise #5).

How big of a plant does it take?

Procedure

- Record the amount of O_2 that the *Elodea* plant produced in one hour.

  O_2 produced by plant per hour = __________ ml

- Measure the cross-sectional area of the *Elodea* plant container. If light is shining from the side, then the cross-sectional area of light on the plants is a triangle shape. (The funnel looks like a triangle when viewed from the side.) Area = ½ Height x Width. The cross-sectional area represents how much light can be captured by the plant. Perform the calculations in centimeters and use whatever formula is appropriate for the shape of your experimental setup. What is this value for your experiment?

  Light-catching surface of the plant = __________ cm^2

- How much oxygen (in ml) is produced by 1 cm^2 of the plant cross-sectional area?

  O_2 produced per cm^2 of plant in one hour = __________ ml

  Hint: You would get the correct answer by dividing your answer from procedure #1 above by the answer for procedure #2.

- What was your calculation of the oxygen demand for a human being during one hour of lab class? (Refer to Exercise #5.)

  O_2 consumed by student per hour = __________ ml

Use this formula to answer the next questions.

$$\text{Size of Plant Needed to Keep You Alive (in cm}^2\text{)} = \frac{\text{Human Oxygen Demand (in ml)}}{\text{Plant Oxygen Production (in ml) per cm}^2\text{ of Plant}}$$

? Question

1. What size of plant is required to keep you alive?
   __________ cm^2

2. Change this plant size to m^2 by dividing your answer above by 10,000. (There are 10,000 cm^2 in one m^2).
   Size of plant required = __________ m^2

3. If you determine the square root of the plant area above, then you will have calculated the side measurement of a square shape representing the plant area.
   Side = _____ m

4. In the above calculations, you have determined how big a plant is required to keep you alive during daylight hours, but what will keep you alive at night? Remember, plants don't photosynthesize at night. Does this change your estimate of how big a plant it takes to keep you alive both day and night?

   By how much?

5. Go outside and mark off on the ground how big of a plant is required to keep you alive during both day and night.

LAB

8

Cell Respiration and Metabolic Rate

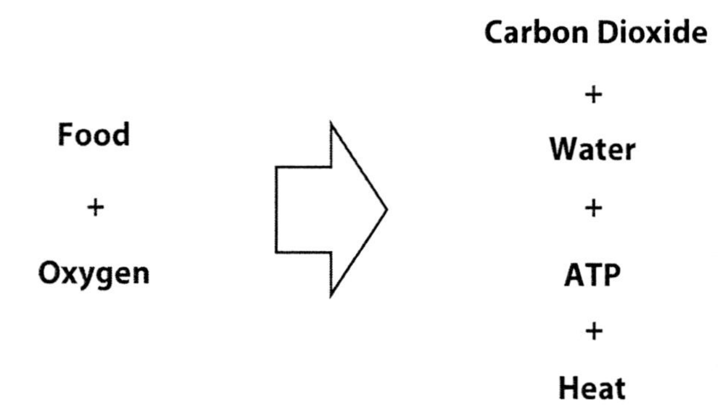

Metabolism is derived from a Greek word meaning "change" – food is changed into life. It includes all of the biochemical reactions in the body that are responsible for growth, activity, reproduction, and everything else discussed in this course. The specific extraction of energy from food is usually called ***cell respiration***, and the organic building reactions (making proteins, fats, nucleotides, and others) are called ***synthesis***. Living organisms extract electron energy from the chemical bonds in organic molecules (food), and they convert that energy into a more useful form of energy (called ATP) to run cell activities. This cell process uses oxygen and produces carbon dioxide. This week's lab will investigate factors affecting metabolism and metabolic rate. Exercise #1 is a tutorial on Cell Respiration intended to help you in lecture class. It may be skipped as part of this lab.

Exercise #1	**Tutorial on Cell Respiration** (optional for this week)
Exercise #2	**What is Metabolism?**
Exercise #3	**Metabolic Rate and Environmental Temperature**
Exercise #4	**Comparison of Endotherm and Ectotherm**
Exercise #5	**Food Demand for Humans**
Exercise #6	**Dietary Supplements for Increasing Metabolism**

Exercise #1
Tutorial on Cell Respiration (Optional)

Chemical Breakdown of Food

The part of metabolism that is involved with the chemical breakdown of food is called ***cell respiration***. You can understand much about cell respiration by using basic rules of chemical reactions.

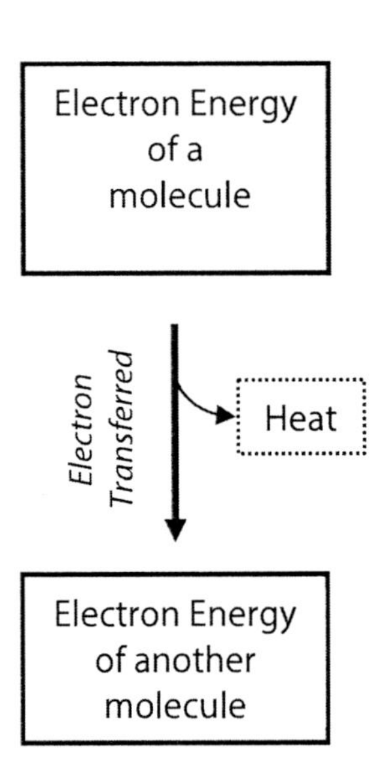

Rule: You can best understand what is happening during the metabolic process by following the electrons.

Rule: Electrons have energy.

Rule: Sometimes electrons can be moved from one molecule to another. When that happens the molecule gaining the electron is gaining energy. The other molecule is losing that energy.

Rule: Whenever hydrogen atoms are shown to be added to or removed from a molecule in a chemical reaction, assume that the same number of electrons are being moved (one hydrogen added = one electron added).

Rule: The Second Law of Thermodynamics applies to all situations where the electron energy of one molecule is transformed into the electron energy of another molecule. Some heat is always released. This is the same heat released during your metabolism.

Example of a "break down" reaction during metabolism.

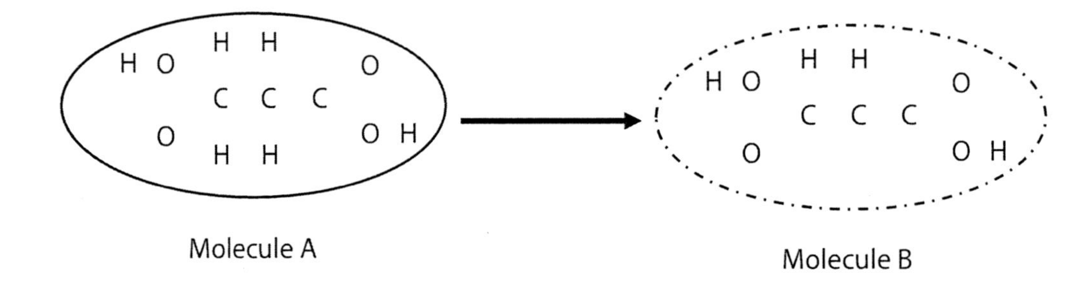

? Question

1. In the reaction shown above, is Molecule **A** gaining or losing electrons?

2. Is Molecule **B** gaining or losing energy?

3. Which has more electron energy?

   $C_6H_{12}O_6$ or $6CO_2$

4. How many electrons are removed from sugar ($C_6H_{12}O_6$) during cell respiration when it is broken into $6CO_2$ and water?

The energy of one electron from food can be transformed into the electron energy of 3 ATP molecules.

5. Cells need a special molecule called ATP to do the work of life. Assume that the energy of one electron from food ($C_6H_{12}O_6$) can be transformed into the energy of 3 ATP molecules. How many ATP molecules are generated by the breakdown of one sugar molecule during cell respiration? Refer to your answer for question #4.

ATP

ATP (adenosine triphosphate) is a special high-energy molecule in the cell. This molecule can also exist in a low-energy form called ***ADP*** (adenosine diphosphate). ATP has more high-energy electrons than ADP. That extra electron energy comes from food molecules. ATP is like a "cell battery" that is charged up by food.

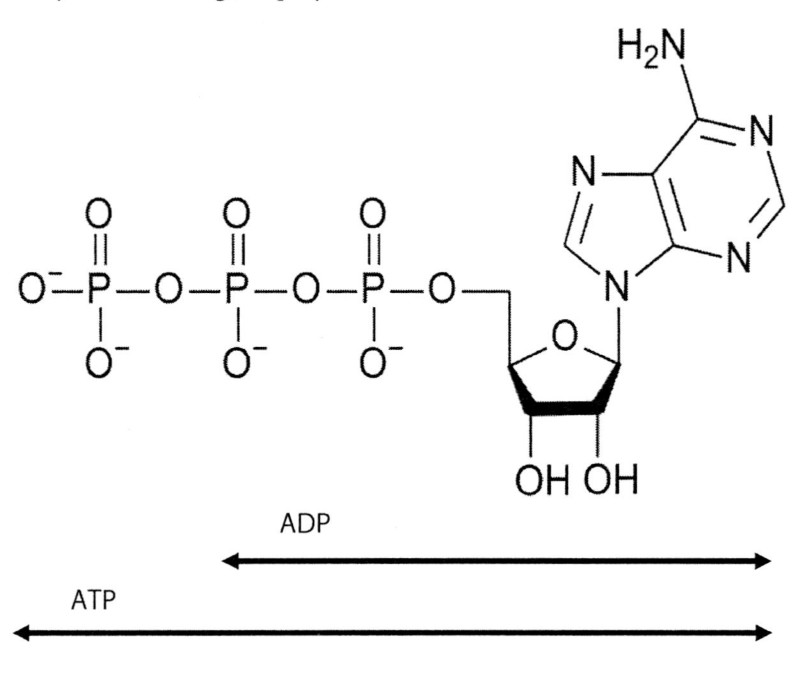

Energy Exchange System

Figure 8.1. The chemical structure of ATP has one more phosphate group than ADP. That phosphate group has high energy electrons.

ATP delivers high-energy electrons to other energy-requiring processes in the cell. The two processes (ADP "charged" to ATP) and (ATP "discharged" to ADP) create an energy exchange system in the cell.

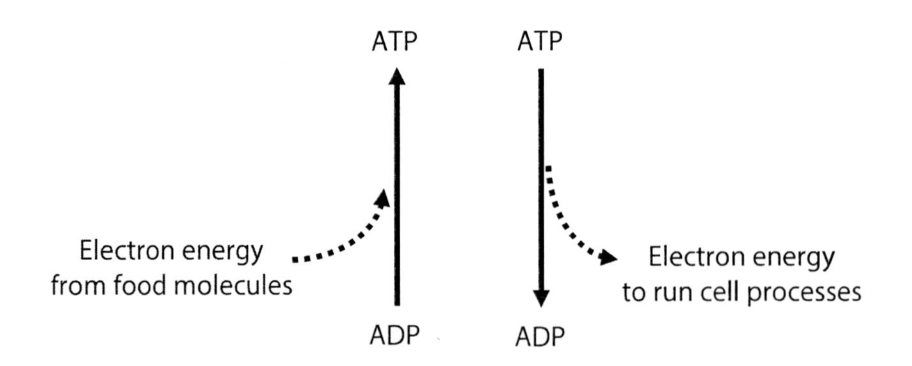

Figure 8.2. The energy exchange system of metabolism.

Cell Respiration – Anaerobic and Aerobic

Chemists tell us that some chemical reactions can happen without oxygen. Can any part of the breakdown of food proceed without oxygen? The answer is yes. Respiration can occur without oxygen, and there were primitive cells living only by this process before photosynthesis or aerobic respiration evolved on Earth.

ATP is like a "cell battery".

Respiration without oxygen present is called ***anaerobic*** (without air). Respiration with oxygen is called ***aerobic***. Aerobic respiration occurs inside a specialized organelle called the ***mitochondrion***, whereas anaerobic processes (also called *fermentation*) are associated with other membranes in the cytoplasm. The sugar molecule is only partially broken down during anaerobic respiration. Have all high-energy electrons been removed from the sugar molecule below?

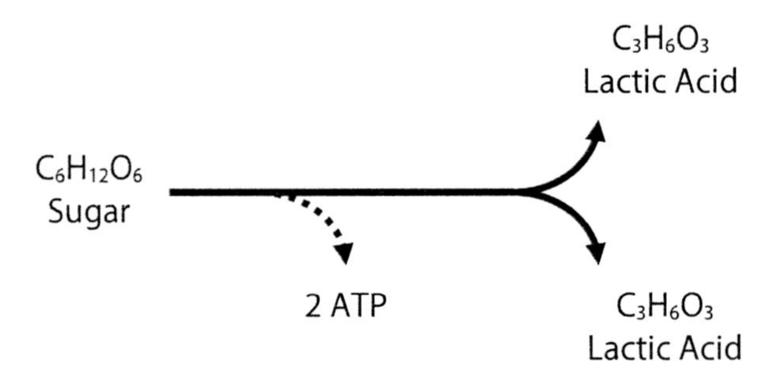

Figure 8.3. Anaerobic Respiration. A sugar molecule is split into two lactic acid molecules, and only 2 ATP of energy is captured.

The energy of two ATP molecules is captured by the cell when sugar is "split" into the two lactic acid molecules. These two ATP molecules are the only energy captured from the food molecule during anaerobic respiration. High-energy electrons remain in the lactic acid, and they have not been captured by the cell. Previously you calculated that 36 ATP are generated during aerobic respiration. Anaerobic respiration is very inefficient compared to aerobic respiration.

At some point during a hard run you will shift into anaerobic respiration and only make 2 ATP for each sugar molecule being used as fuel. You must then radically speed up anaerobic respiration so that the needed ATP continues for muscle contraction. You are now burning sugar 18X faster and producing lactic acid 18X faster than when the aerobic process is providing your ATP. Eventually you must stop and replenish the sugar and reprocess the lactic acid. By the way, lactic acid buildup is partially responsible for the pain you feel in the muscle during the next day.

You shift to anaerobic respiration when the muscles run out of oxygen.

In humans, anaerobic respiration happens for short periods of time and only in the skeletal muscles. During strenuous exercise this maintains metabolism when oxygen is temporarily depleted. Other organs of your body are incapable of anaerobic respiration, and their cells are damaged or die when oxygen is used up.

Aerobic Respiration

Aerobic respiration is the efficient process of extracting electron energy from the chemical bonds in organic molecules (food) and converting that energy into the most useful form of energy (called ATP) to run cell activities. This cell process uses oxygen and produces carbon dioxide. The complete equation is:

$$\text{Food} + \text{Oxygen} \longrightarrow \text{Carbon Dioxide} + \text{Water} + \text{ATP} + \text{Heat}$$

Aerobic respiration occurs inside the ***mitochondria***, which are cellular organelles in both plant and animal cells. The mitochondria have a remarkable structure that is somewhat like a factory. The high-energy electrons are stripped off the food molecule in the fluid matrix, and then the energy of those electrons is used to generate ATP energy along the cristae membranes.

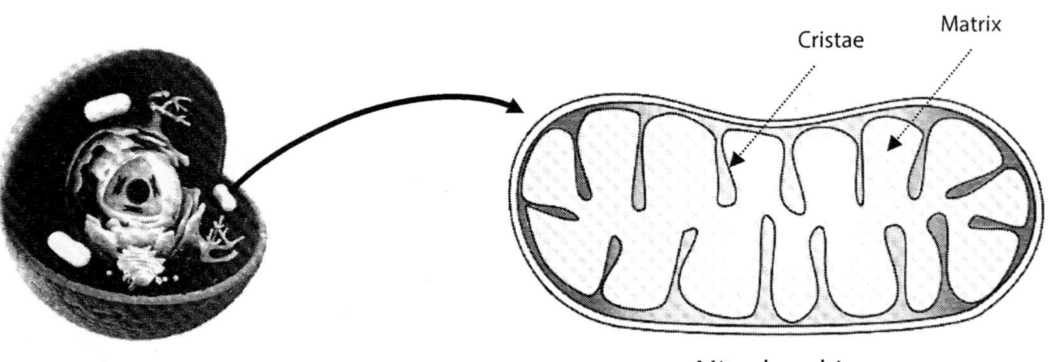

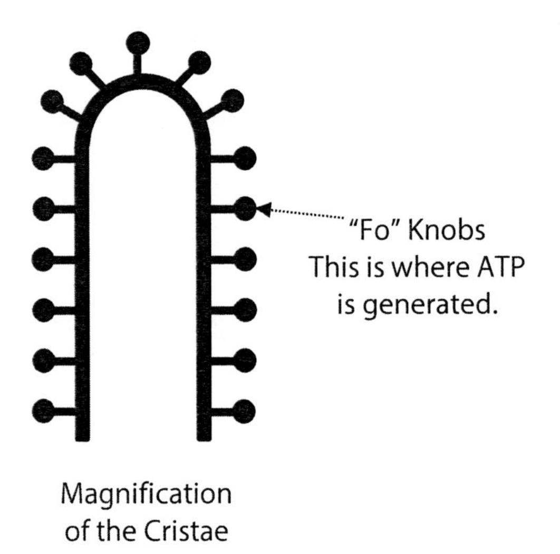

Figure 8.4. Structure of the mitochondria. This is the location of aerobic respiration.

? Question

1. Which is the high energy form – ADP or ATP?

2. Define aerobic respiration.

3. Where in our body does anaerobic respiration occur?

4. What substance builds up when we shift to anaerobic respiration?

5. Some of the cells of your body have many mitochondria and other cells have few mitochondria. Why would there be differences?

6. Complete the following table comparing aerobic and anaerobic respiration.

Table 8.1. Comparison of Aerobic and Anaerobic Respiration.

Comparison	Aerobic	Anaerobic
Is oxygen necessary?		
Which came first on the planet?		
What are the end products?		
How much ATP energy is generated?		
Where in the cell does it happen?		

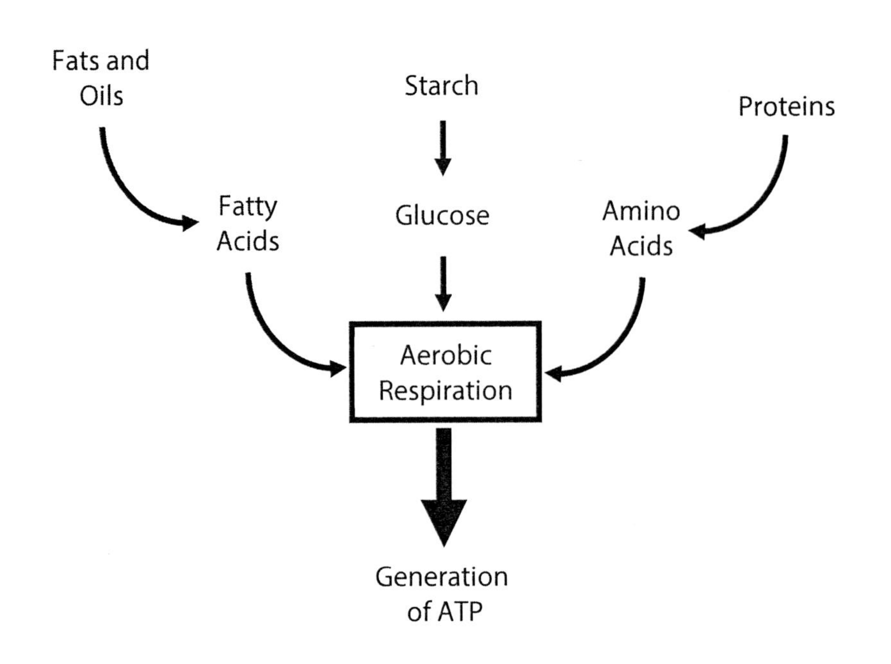

Figure 8.5. Carbohydrates, Fats, and Protein are fed into aerobic respiration. They generate ATP as the universal energy supply for cell processes.

Sugar is not the only source of energy.

Sugar is not the only type of food molecule that can be metabolized during cell respiration. Fats, proteins, and starch are other energy sources for the generation of ATP. The amount of ATP produced from these nutrients depends on the size of the molecule and the number of high-energy electrons that can be stripped off. Starches are easy to metabolize because they consist of glucose sugar molecules hooked together. Starches are quickly digested into monosaccharides in the intestine. Proteins also are broken into amino acids in the intestine and then modified into fuel by body cells. During the process of protein metabolism, nitrogen atoms are broken off the amino acid molecule. Ammonia is produced from that nitrogen, and the ammonia is then converted into urea and dumped into the urine. The part of the amino acid remaining after nitrogen removal can be used as a fuel by aerobic respiration. Amino acids generate about the same amount of ATP energy as does an equal weight of sugar.

Fat molecules have many more high-energy electrons than an equal weight of either sugar or protein. Protein and sugar provide about 4 Calories of energy per gram. Fat provides about 9 Calories of energy per gram. You can see why it's easy for those high-energy electrons to pile up!

? Question

1. What factor determines the amount of energy that can be extracted from a nutrient molecule?

2. Which nutrient provides the most ATP energy per molecule metabolized?

3. Urea is one of the substances that gives urine its characteristic smell. Urea in the urine means that you have been metabolizing which nutrient?

Exercise #2
What is Metabolism?

Historical Discovery Process

The earliest experiments on metabolism were performed about 300 years ago and involved both plants and animals. Very little was understood about chemistry, so most explanations depended on unseen forces and substances.

Description of Experiment	Results
Mouse in a sealed jar	Mouse dies in an hour or so.
Mouse in a sealed jar with a large plant in the light	Mouse lives just fine.
Mouse in a sealed jar with a large plant in the dark	Mouse dies in an hour or so.

Figure 8.6. Early experiments on metabolism.

"I'm outta' here guys!"

These early experiments revealed several facts. There is something in the air that animals need to live. Somehow plants "regenerated" the air that animals need, and light was necessary. Later experiments measured the changes in the weight of substances during metabolism, and it did not take long for the gaps to be filled in.

Mouse in a sealed jar with enough air but no food.	→	**Experimental Results:** Mouse loses weight. Mouse produces water. Air gets heavier.

The chemists soon learned that:

- Mouse Tissue can be represented by the formula for sugar - $C_6H_{12}O_6$.
- Light Air is oxygen - O_2.
- Water is H_2O.
- Heavy Air is carbon dioxide – CO_2.

The general equation for the metabolic process became:

Mouse Tissue + Light Air ⟶ Water + Heavy Air

? Question

1. Write the molecular formulas below the words in the general equation (above) for metabolism.

2. When radioactive isotopes of oxygen atoms are put into O_2 molecules and the mouse is allowed to metabolize in a chamber with that "labeled" oxygen, only the H_2O produced by the mouse is radioactive. Based on that evidence, draw dotted lines to show where each atom on the left side of the equation goes to on the right side of the equation.

3. Where does the CO_2 that you breathe out specifically come from during metabolism? Draw dotted lines to show this.

4. Where does the O_2 that you breathe in specifically go to during metabolism? Draw dotted line to show this.

Measuring Metabolic Rate

This is a very accurate method, but it is expensive and difficult.

Energy output is the term used to represent all of the energy required to maintain the metabolic processes and activities of an organism. The most accurate and direct way of determining the amount of energy used by the body during an activity is to measure the ***amount of heat given off.*** Physics tells us that heat is released whenever energy is transformed from one form into another (such as nutrient energy into physical work). A technical problem with using this method of analysis is that the person must be inside an insulated container surrounded by a known quantity of water. The temperature of the water increases as heat is released by the person. Although this procedure is very accurate, it is also expensive and difficult.

The usual method of measuring energy output is the ***oxygen consumption technique***. If the oxygen requirement of a resting person is known, then that value can be compared to the increased amount of oxygen used during a particular physical activity. This approach is easier and less expensive than the heat method. Most general studies of energy expenditure are based on oxygen consumption.

Figure 8.7. The oxygen consumption method for measuring energy output is much easier to do.

Factors Affecting Metabolic Rate

It is possible to estimate how much energy you expend during the day using tables and calculations. The basic calculations involve two parameters.

Energy Output = BMR + Daily Activity

- **Basal Metabolic Rate (BMR).** This is the fuel burning rate (O_2 consumption) required to maintain resting body needs. It is influenced by sex, age, and other factors.
- **Amount of Daily Activity.** This is all of the work and movement you do during the day beyond BMR.

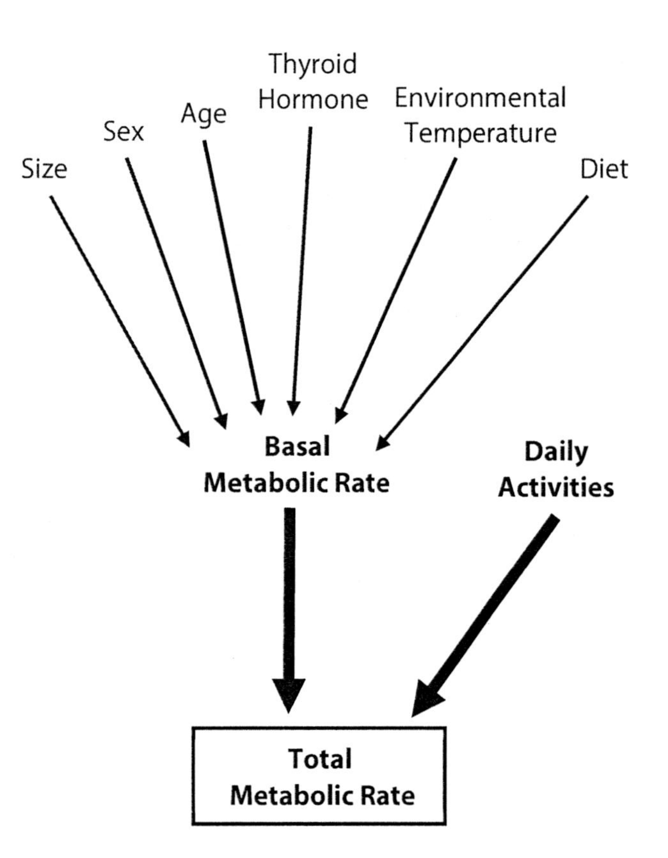

Figure 8.8. Factors that influence the metabolic rate. Basal Metabolic Rate plus Daily Activities equals Total Metabolic Rate.

The total metabolic rate is the product of BMR plus all daily activities. Physical activity can increase metabolism to 10X BMR. This is the dominant influence on metabolic rate. We will examine the details of metabolic rate during physical activities in another lab.

Important factors affecting BMR are shown in Figure 8.3. The effect of size on BMR is best described by surface area and not by weight. If a 100 lb person has a BMR of X, then the BMR of a 200 lb person won't be 2X. It will be X plus about 30% more. Smaller people metabolize faster per pound even though larger people eat more food. This can be somewhat frustrating for one of your larger clients to accept. They may naturally think that they need much more food than a smaller person. They don't.

Sex and age also impact BMR. The female is usually 5-10% lower than a same-size male, and BMR drops about 0.2% each year after the age of 2 or 3. Another determinant of BMR is the amount of thyroid hormone. This hormone directly influences the speed of carbohydrate and fat metabolism. Thyroid hormone can increase BMR to 2X normal, and it is part of the stress reaction. Another part of the stress reaction is the sympathetic nervous system which can increase BMR almost as much as thyroid hormone, but its effects last only an hour or two. Diet and environmental temperature have smaller effects on metabolic rate. The BMR can be increased 10-30% after eating. The stronger effect comes after meals that are higher in protein.

Finally, the influence of environmental temperature is significant, and we will study it next. Our metabolic rate increases in a cold environment in order to produce more heat for maintaining a normal constant body temperature.

? Question

1. Why would measuring the amount of heat given off by an organism be useful for estimating metabolic rate?

2. Why is the oxygen consumption method the usual way to estimate the metabolic rate?

3. Does a 200 lb person metabolize twice as much as a 100 lb person?

Exercise #3
Metabolic Rate and Environmental Temperature

Metabolic Rate of a Mouse

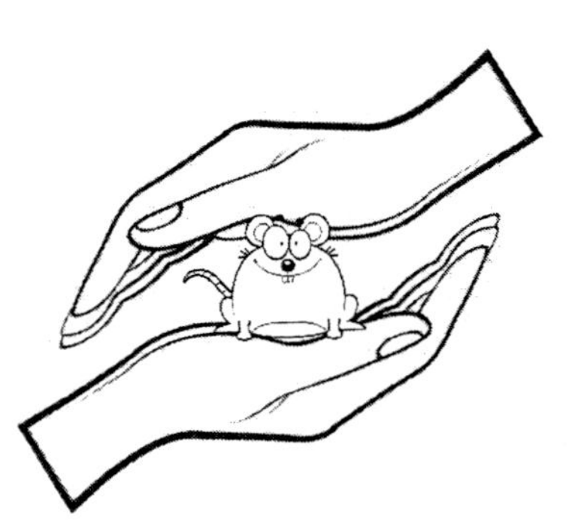

These are professional mice, and we treat them well.

Mice are ***endotherms***. That is, they get most of their heat from inside their own body (endo means inside). Cellular respiration generates the heat that keeps these animals warm. During this Exercise you will monitor the rate of cell respiration (also called metabolic rate) in a mouse. Then you will investigate the influence of environmental temperature on metabolic rate by comparing the mouse in a cold environment with what you observe at room temperature.

How to Handle Mice

- Mice should be picked up by their tail and immediately rested on your hand, and then marched into the Metabolic Cage.
- Do not grab them. Grabbing scares the hell out of them, and they may bite you or pee on you because of that fear.
- Also, don't play with the mice (on table tops, etc.) because there is a possibility of them getting loose on the floor. These are professional mice. They work several years for us, and we treat them very well. So, please be careful.

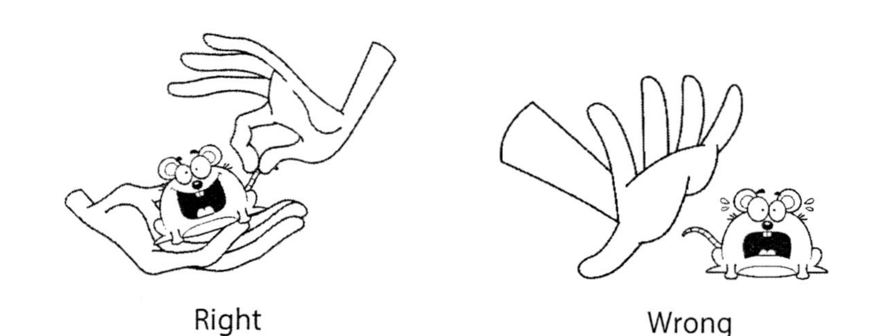

Experimental Design – Room Temperature

The basic question is: ***What effect does environmental temperature have on the metabolic rate of an endotherm (mouse)?*** Do this experiment at two temperatures: Room Temperature and Packed in Ice.

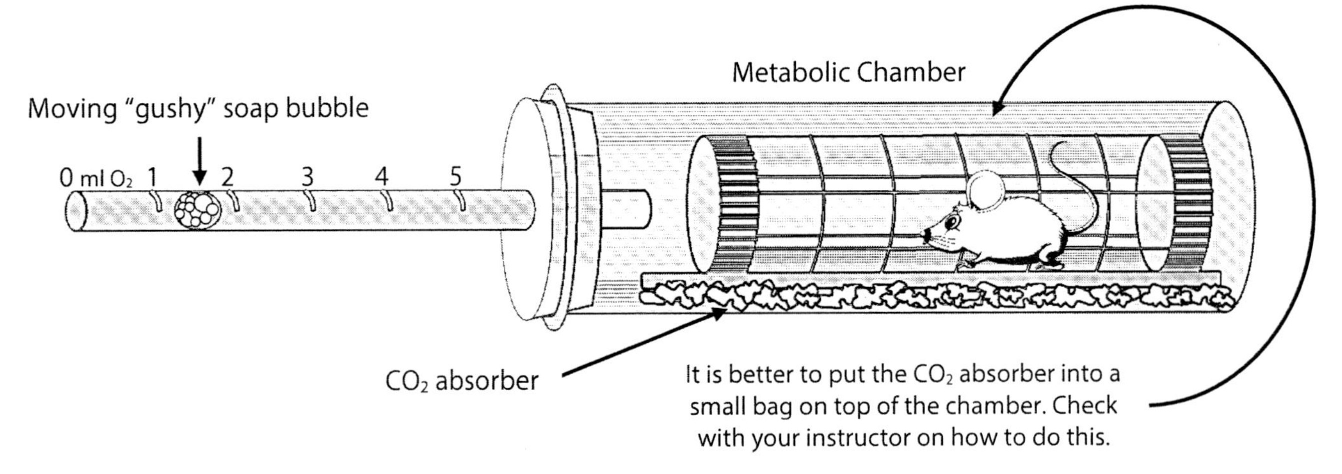

Figure 8.9. Design of the metabolic chamber. The soap bubble will move as oxygen is consumed by cell respiration in the mouse.

Procedure

- Weigh the wire cage part of the chamber: ______ grams.
- Go get your mouse, and put it into the wire cage. Then weigh the cage with the mouse in it.

  Weight of the mouse = _____ grams.
- If your class is using small bags of CO_2 absorber instead of loose material, then place one bag on top of the mouse cage so the mouse can't pee on it. Otherwise, put one tablespoon of CO_2 absorber (soda lime) into the trough at the bottom of the Metabolic Chamber.
- Wet the inside of the glass tube with soapy water. This will help prevent the "gushy" soap bubble from "popping" during the experiment.
- Put the caged mouse into the chamber and seal the cork tightly. Don't worry! The mouse won't suffocate. Leave the chamber alone for 10 minutes (sealed up with cork on and no soap bubble). This will equalize the temperatures inside and outside of the chamber.
- Use your finger to make a "gushy" soap bubble on the open end of the glass tube. Measure the time it takes (in seconds) for the bubble to move between the marks on the tube. This bubble movement indicates that 5 ml of O_2 have been consumed by the mouse.
- Perform three trials, and record the results.

Time for Mouse to Consume 5 ml of Oxygen (in seconds)	
Trial 1	
Trial 2	
Trial 3	

? Question

1. During cell respiration a mouse will consume O_2, and CO_2 will be produced in its place. If no CO_2 absorber had been used in your experiment, would you have seen a change in air volume?

2. If you use a CO_2 absorbing substance in the Metabolic Chamber, then what happens to the CO_2 that is produced during cell respiration?

3. Now, with the absorbing substance in the chamber, what happens to the air volume during your experiment as the O_2 is consumed during cell respiration?

Experimental Design - Packed In Ice

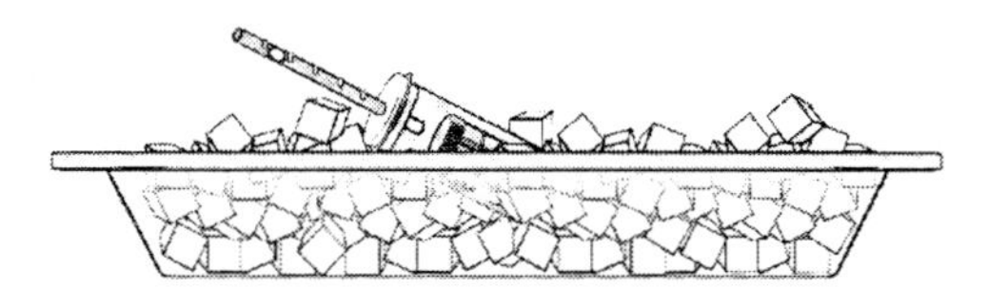

Procedure

- If ice is packed around a Metabolic Chamber like the type we are using, the temperature inside will stabilize at 5°C. This cold temperature will not harm the mouse as long as it is removed before 45 minutes. Our experiment will take less than 20 minutes.
- Let the "Packed in Ice" chamber equalize temperature for 10 minutes before applying the "gushy" soap bubble. (The chamber should be sealed during this 10 minute period.)
- After the 10 minutes of temperature equilibrium, apply a "gushy" soap bubble and perform three separate measurements of metabolic rate. Record the results.

Time for Mouse to Consume 5 ml of Oxygen	
Trial 1	
Trial 2	
Trial 3	

- Disassemble the chamber and carefully return your mouse to its home. Return the CO_2 absorber to its storage container, and dump feces into the special waste jar. Don't wash the apparatus unless you are told to do so. The chamber must be dry for the next lab class. Wash your hands!

Calculations of Metabolic Rate

You must convert the mouse's O_2 consumption to an hourly metabolic rate. Calculations 1 and 2 will make that conversion. This is accomplished by dividing the bubble time (in seconds) into 3,600 (the number of seconds in one hour). The resulting number is to be multiplied by 5 (5 ml of O_2 used in each trial).

Calculation 1

Calculate the average time of the three trials at room temperature.

5 ml O_2 consumed in ____ seconds. (Room Temperature - $\approx 20^0$C)

Calculate the average time of the three trials packed in ice.

5 ml O_2 consumed in ____ seconds. (Packed in ice - $\approx 5^0$C)

Calculation 2

Based on Calculation 1, how much O_2 would your mouse consume in one hour? (There are 3600 seconds in one hour.)

$$\frac{3600}{\text{Calculation 1}} \times 5 = \underline{\qquad} \text{ ml } O_2 \text{ consumed in one hour}$$

O_2 consumed in one hour at room temperature = _____ ml.

O_2 consumed in one hour packed in ice = _____ ml.

Calculation 3

In order for the metabolic rate of a mouse to be compared with a bigger or smaller animal, we must correct the calculations considering the mouse's weight. Use the following equation.

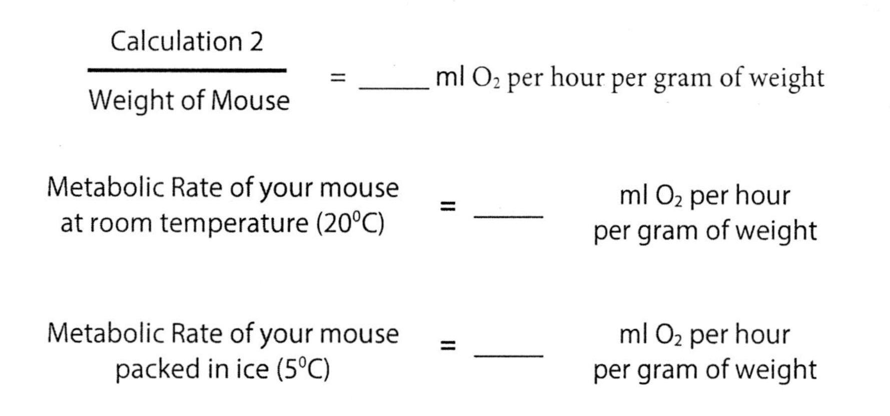

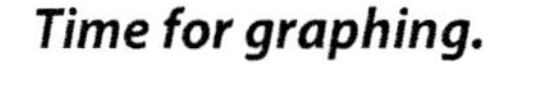
Time for graphing.

- Put a dot on the graph in Figure 8.10 for each of the metabolic rate values in your experiment.
- Draw a line between those two dots.
- Write the word "endotherm" on that line.
- Check with other lab groups to see how your calculations compare with theirs.

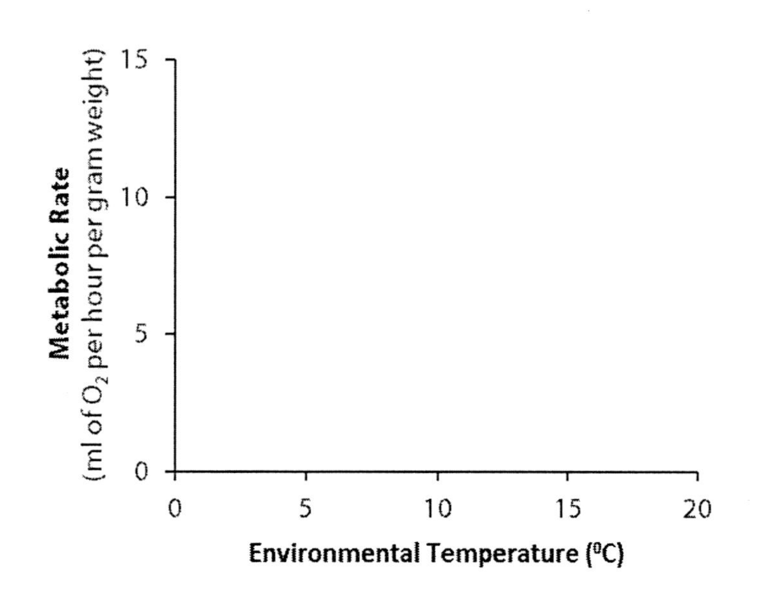

Figure 8.10. Effects of environmental temperature on the metabolic rate of an endotherm (mouse).

Exercise #4
Comparison of Endotherm and Ectotherm

A frog in the cold is cold!

An ***ectotherm*** (ecto means outside) gets its heat from the environment. The body temperature of an ectotherm is warm when the environment is warm, and the body is cooler when the environment is cold. The following results are taken from experiments that measured the metabolic rate of a frog (ectotherm) of about the same size as your mouse.

Table 8.2. Measurement of the Metabolic Rate of a Frog at Different Environmental Temperatures.

	Metabolic Rate Packed in Ice (5°C)	**Metabolic Rate** Room Temperature (20°C)
Frog #1	0.05	0.30 ml O2 / hour / gm
Frog #2	0.03	0.28
Frog #3	0.04	0.25

Procedure

- Calculate the average metabolic rate for the three frogs at each of the two temperatures.
- Put a dot on the graph in Figure 8.11 for each of the average values.
- Draw a line between those two dots.
- Write the word ectotherm on the line.

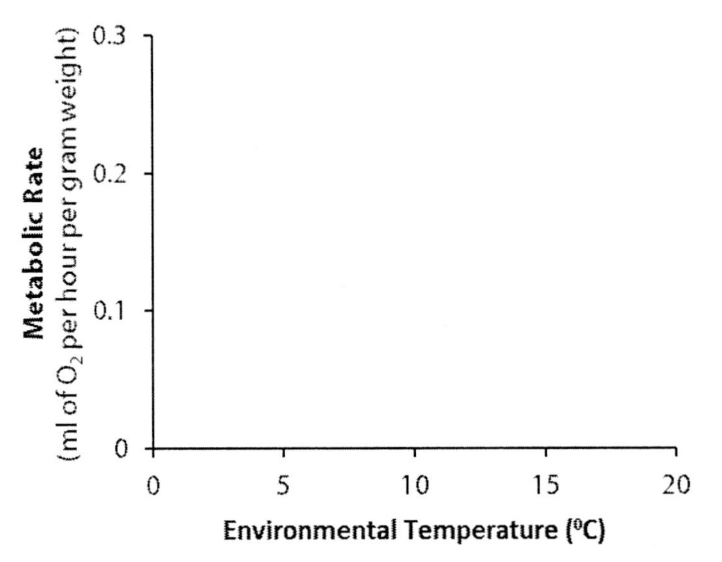

Figure 8.11. Effects of environmental temperature on the metabolic rate of an ectotherm (frog).

? Question

1. Which organism has the slower rate of respiration?

   endotherm or ectotherm

2. Which organism needs less food to survive?

   endotherm or ectotherm

3. How much food does the ectotherm need compared to the endotherm? Hint: Divide metabolic rates.

4. Which organism would do better if the amount of food is very limited, but the environment is fairly warm?

5. What kind of environment is described by question #4?

6. Which organism would do better in a cooler environments where the food is plentiful?

   endotherm or ectotherm

7. Will the organism in question #6 do fine in warmer environments if the food is plentiful?

   Why or why not?

Exercise #5
Food Demand for Humans

How much food does a student need to survive one hour of biology lab class?

We can borrow data from experimental research to help us estimate the amount of food that is required to support a human. Our calculations will be based on grams of sugar as the nutrient. Also, notice that the word Calorie is capitalized. When capitalized, this term represents 1000 times the value of a single calorie (not capitalized).

The Caloric demand for food varies greatly for a human depending on activity and environmental conditions. The energy demand might be as slow as 50 Cal per hour during sleep to as fast as 2,000 Cal per hour during extreme exercise. (That high rate of metabolism could be maintained for only about 2 minutes without total exhaustion.) An average student in biology lab class uses about 100 Calories per hour if they aren't walking around.

Information

- Assume a food demand of 100 Cal/hour for students.
- A human gets about 4 Cal of energy from 1 gram of sugar.

Procedure

- How many grams of sugar are required to "fuel" an average student during one hour of biology lab class?

  _____ grams of sugar used in one hour

- Weigh out that much sugar and show it to your instructor.

Use a weighing paper or tray for the sugar.

Exercise #6
Dietary Supplements for Increasing Metabolism

"I need something that will raise my energy level but not my self esteem – I'm already full of myself."

The Arguments

The Supplement Industry has developed alongside the fields of Medicine, Sport, and Nutrition. Its annual revenues in the United States are more than 30 billion dollars. Obviously, many people support its products. What can be said objectively about the health value of these supplements? And what scientific evidence is in support of the claims that they can safely change your metabolic rate?

Why is Medicine so slow and expensive?

Medicine's basic approach is to identify serious health problems in patients and to intervene with treatments that eliminate the cause or relieve symptoms. Typically the addition to the understanding of health and disease processes is slow and steady. Medicine is highly regulated, and its focus is on scientific experimentation. Medicine is slow and expensive. There can be a general public frustration of "Why can't they cure me faster and cheaper?"

A very effective marketing tool is to overwhelm the consumer with information from carefully selected research studies, and then add what seems to be reasonable conjecture.

Sport and Nutrition industries are not regulated like the medical fields. This greatly affects how money is spent. The money in these industries is spent on advertisement and promotion rather than on scientific experimentation. Reliability of Sport and Nutrition recommendations is more dependent on the evaluation and judgment of consumers. However, there are several professional organizations and publications that help to debate the validity of claims.

The Supplement Industry is almost completely independent from regulation. This industry does not have to prove that their product is effective. The only requirement is that they must not commit fraud while representing the benefits of their products. This is a legal definition – not physiological or scientific. They can use loosely controlled correlation studies and make claims based on conjecture from known processes. Therefore, you should read their claims very carefully. Examples of misleading statements include:

- Some studies have found ...
- Many experts are calling for ...
- This supplement makes it easier to ...
- A recent study found that ...
- You may already know that ..., but did you know that ...

A primary source of information and claims about supplements is available on the Internet. These sites flood the consumer with misleading information that is not corrected by any regulating agency. Only a few other

"I've got something special for you."

Internet sources (Quackwatch, NCAHF, etc.) attempt to counter the questionable claims. Most of the supplement claims rely on emotional appeal.

- A particular substance is in short supply in your diet – a supplement is needed.
- Take as insurance.
- RDA's (Recommended Daily Allowances) are too low.
- Natural is better than synthetic.
- This way is quick and easy compared to conventional medicine.
- Stress and modern life deplete normal sources.
- Special words are used: cleanse, detoxify, etc.
- Lots of personal antidotes and endorsements.
- Medicine is the enemy – don't trust doctors.

How about Metabolic Supplements?

People want to change their metabolic rate in order to feel better (peppy) or lose weight or to improve physical and mental performance. How much do they want these results? They use illegal substances and go against doctors' warnings, and they risk being discovered by overseeing Sports authorities. Although each metabolic supplement would have to be investigated separately to judge its claims, there are three indicators of whether a supplement might actually change metabolic physiology.

- Does the supplement contain a substance that has been tested by Medical and Academic researchers and shown to have physiological effects?
- Is the supplement being used in the commercial meat industry? If so, it will probably affect human physiology.
- Is the supplement a hormone in a form that actually gets into the blood? If so, it could affect human physiology to a significant degree.

None of these facts tells us that a supplement should be used. Changing physiology is a risky venture. There will be unexpected results and negative side effects. So, is it worth trying to change metabolic rate? The supplement industry says yes, but can we believe them?

What Can We Conclude?

Coffee may be the safe answer.

Metabolic supplements can be divided into four categories: (1) stimulants, (2) thyroid hormone or precursors to it, (3) special elixirs, and (4) an added food nutrient that speeds normal digestion or metabolism. Let's briefly consider some critiques of each category.

There are illegal stimulants that have powerful effects on metabolic rate, appetite, and other neurological mechanisms. These substances have been outlawed because of the addictive properties and damage to the body. Cocaine, amphetamines, and their derivatives head the list of these illegal drugs. The less powerful stimulants are usually legal, and their effects are not as immediately threatening to health. However, the long term effects can compete with those of the powerful stimulants. They usually increase metabolic rate a bit but are more significant in affecting mental alertness. They can be effective in Sport (both endurance and reaction time). Moderate use of some of the less powerful legal stimulants by the general public has not caused serious health concerns. But there are side effects like irritability, sleep disruption, heart arrhythmias, and diarrhea when using these supplements.

Most legal metabolic stimulants have very little effect on metabolic rate. The ones that have some effect include two general categories: (1) those that are from plant extracts or are synthesized copies of certain ingredients in plant extracts, and (2) those that contain small amounts of pharmaceutical stimulants. The best known in the first category include caffeine and all other similar acting stimulants (Guarana, Ginseng, Gingko Biloba, and many others). Ginko Biloba was recategorized from "safe" to "avoid" by the Center for Science in the Public Interest because it contains a toxic substance that induces liver cancer in laboratory mice. All herbal extracts present possible unknown risks. The remarkable fact is that so many consumers would rather pay handsomely for untested and unregulated plant extracts than for known caffeine or theophylline doses (various teas and coffee).

Or, maybe not.

The second category of stimulating supplements are those that contain various amounts of pharmaceutical drugs like ephedrine (MaHuang and Ephedra Sinica), sibutramine (slimming teas), and methylexaneamine (added to geranium oil and other products). The proper discussion of these stimulants would require a course on pharmacology. Consumers use these supplements as a way to avoid doctor and prescription costs, but they risk the side effects from mistaken self-diagnosis and lack of dosage and quality control.

Other metabolic supplements contain hormones or precursors to them. Steroids and growth hormones are a significant part of the Supplement Industry, but they do not affect metabolic rate much at all. The hormone supplement that does increase metabolic rate is thyroid hormone which is a

major controller of the metabolism of fats and carbohydrates. Its effects are profound, and it is a major ingredient in supplements that do stimulate metabolism. Supplement names include: metabolism booster, fat burner, thyroid helper, hyper shred, and many other similar variations. Most thyroid supplements contain thyroid extracts, and their recommended dose is nearly three times what a clinical patient usually starts with as treatment for hypothyroid. The prospective customer uses self-diagnosis with a questionnaire on the Internet, but only 10% of people over the age of 50 have a somewhat low thyroid hormone, and most of those receive something other than thyroid prescription when treated by the Medical professions. There are potent side effects of thyroid overdose including heart damage, anxiety, and thyroid disease itself. Some thyroid supplements contain iodine which is necessary for the natural synthesis of thyroid hormone in the body. But the amount of iodine needed to make normal levels of thyroid hormone is already supplied in a healthful diet, and there are negative side effects of excess iodine in the diet. All of these consequences are risked when using thyroid supplements.

There is another supplement category that could be more accurately called special metabolic elixirs. They are heavily marketed and have no supportive evidence except personal testimonies. The usual claim portrays them as treating nearly every ill of mankind, and they typically contain B vitamins, caffeine, or something else with a recognizable name. Any metabolic rate effect associated with them is accounted for by the placebo effect.

What about special food supplements that claim to improve digestion or increase metabolic rate? We know that a healthful diet is necessary for success in any health or sport endeavor. (That diet is described in the Diet Labs.) Special supplements have not been proven to surpass the benefits of the basic healthful diet. Of course, there are known modifications of diet (more of certain foods) that are important when a person has high levels of physical demand or is trying to lose weight.

We end of our discussion with a repeat of the known metabolic effects of exercise. The most powerful way to increase metabolic rate is to increase physical work. Some training methods are more effective than others, but the main factor is how much exertion you do. All of the illegal substances and most legal supplements have negative side effects, and no supplement effect on metabolic rate compares to a healthful diet combined with physical activity. Furthermore, physical exercise and good diet are free. Our best advice is to skip the supplements.

If farmers are using a particular supplement, it would probably have a physiological effect on you. But, what effect is another question.

? Question

1. Which industries have academic and governmental oversight and regulation?

   Medicine
   Sport
   Nutrition
   Supplement

2. Where is most of the money spent in the Medical Industry?

   Research and Testing
   or
   Advertising

3. Where is most of the money spent in the Supplement Industry?

   Research and Testing
   or
   Advertising

4. List three indicators that a supplement might actually change metabolic physiology.

5. List four general categories of metabolic supplements.

6. Which supplement type has the most powerful effect on metabolic rate?

7. What are the risks using those supplements?

8. What factor will increase metabolic rate many times more than any supplement?

Group Discussion

Discuss the following questions in your lab group.

1. Go on the Internet and search for "metabolic rate supplements". What are the three dominant themes?

2. In your lab group's opinion, what is the strongest argument against using supplements for increasing metabolic rate?

3. In your lab group's opinion, what is the strongest argument against the views presented in this Exercise?

LAB

9

Mitosis & Meiosis (Sameness and Variety)

Is it better to be the same as everyone else?
or
Is it better to be different?

We struggle with these questions in our personal lives. Would it surprise you to learn that all life, in terms of its reproductive strategy, has struggled with the same basic questions? It might also surprise you to learn that there is no one answer to reproduction, but two answers. The strategy of producing offspring that are genetically the same as the parent is called ***asexual reproduction***, and it is accomplished through a cell division process termed ***mitosis***. Asexual reproduction is the simplest and oldest form of reproduction, and it relies on a single parent. The strategy of producing offspring that express genetic variety is termed ***sexual reproduction***. It is accomplished through a cell division process called ***meiosis*** and a fusion process called ***fertilization***. Sexual reproduction is complex and it usually relies on two parents.

Many organisms have lost the means to reproduce asexually except for cell replacement or growth. But some less specialized species use both modes depending on the time of year. The one certainty is that organisms exist today only because they have incorporated both sameness and variety in their struggle to live and reproduce.

Exercise #1
DNA Replication

You don't reproduce - your DNA does!

Sixty years ago, the study of biochemistry confirmed the suspicion that a chemical called DNA reproduces — not the individual. And the sorting of DNA by the cell is what controls the sameness or variety of the next generation of organisms. This does not mean that the organism isn't central. The organism is made of cells that contain the DNA molecules. But the genetic unit that actually gets passed on to the next generation is the DNA. That's why the discussion of reproduction is centered on this unique molecule.

So, first let's take a look at the structure of DNA and how it reproduces itself. Then, we will see how an organism uses DNA to achieve sameness by asexual reproduction, or how it achieves variety by sexual reproduction.

DNA Structure

DNA is made up of many repeating subunits called ***nucleotides***. It is an extremely thin, long, ladder-like molecule that has two "rails" made of sugar and phosphate and many "rungs" made of special complementary bases. A nucleotide has three parts: a sugar, a phosphate, and one of four special bases. The DNA bases ("rungs of the ladder") are molecular units that combine only in two kinds of pairs. The base ***adenine*** (A) always will be paired with ***thymine*** (T), and ***cytosine*** (C) is always paired with ***guanine*** (G). Therefore, if you know one of the complementary bases, you can easily figure out the other side of the ladder rung.

A nucleotide is made up of a sugar, a phosphate, and one of four bases.

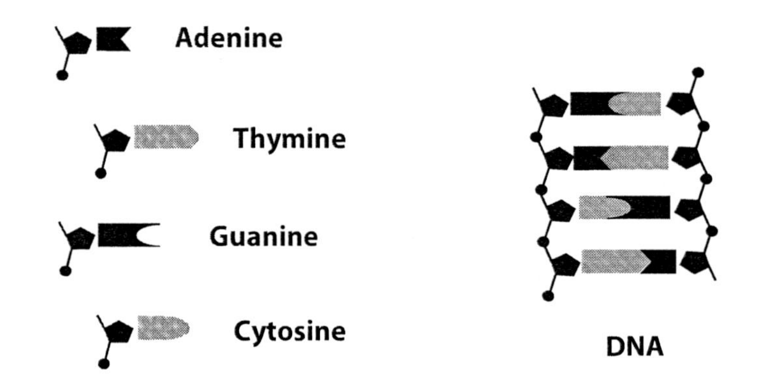

Figure 9.1. Complementary Base Pairing in DNA. There are four different nucleotides in DNA, and they pair in only one way within the "rungs" of the DNA molecule.

Complementary base pairing happens because each of the four nucleotides has its own very specific shape, and that shape determines the kind of attraction that nucleotides have for one another. Think of these nucleotides as being like magnets that will "stick" together if they can get close enough to each other. Adenine and thymine can get close enough, and guanine and cytosine can also. But, any other combination will not allow bonding between nucleotides.

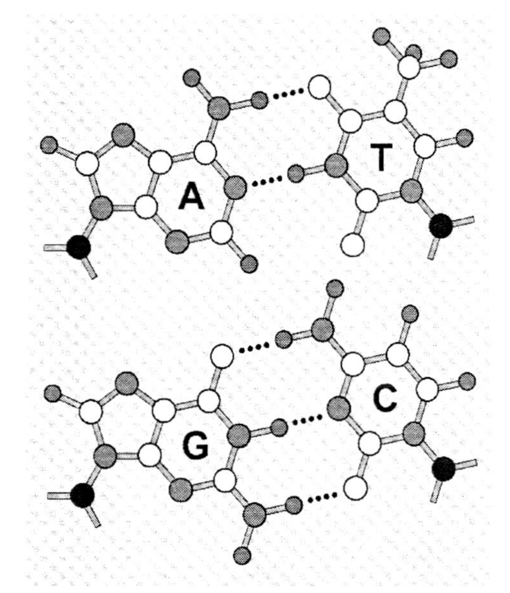

Figure 9.2. Bonding between complementary nucleotides. Bonding can occur because these nucleotides fit closely together.

? Question

1. Define complementary base pairing.

C
G
T
C
A
A
G
G
A
T
C
C
T

2. Guanine always pairs with __________.
3. Thymine always pairs with __________.
4. Fill in the complementary nucleotides in the DNA ladder to the left.

Reproduction of DNA

Before DNA can copy itself the cell must make lots of extra A, T, G, and C. Then the DNA unzips between the two bases and adds nucleotides to each side of the unzipped DNA molecule.

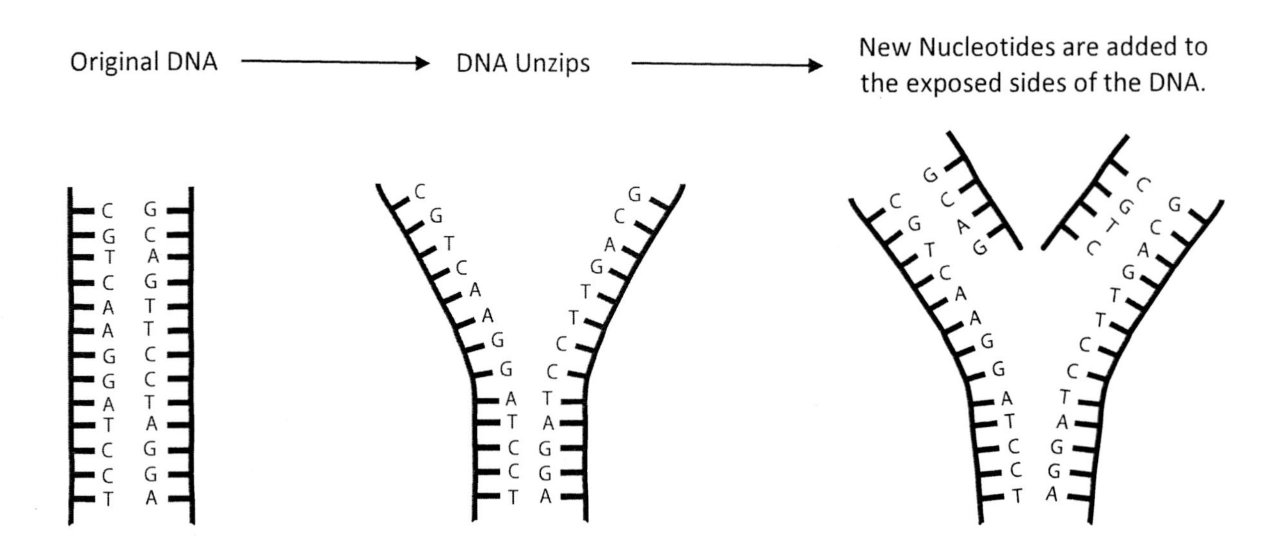

Figure 9.3. Reproduction of DNA. DNA is "unzipped" by a special enzyme, and floating nucleotides begin to specifically pair with each side of the exposed DNA ladder. Those nucleotides bond together and form two new DNA molecules.

? Question

1. What are the two "new" DNA molecules built from? Explain.

2. Are the two "new" DNA molecules identical to the "original" DNA molecule?

Procedure

- We have been illustrating DNA as a straight ladder, but actually it is twisted on itself like a spiral staircase. This shape is called a ***helix***. Go look at the model of DNA on the demonstration table.

Normally DNA exists as loose strands, called ***chromatin***, in the nucleus of a cell. This nuclear DNA sends a "message" (mRNA) to the ribosomes where protein and enzymes are synthesized. When stretched out, the length of a DNA molecule in a human cell can be 4 cm. However, the cell itself is a tiny fraction of that size. During cell reproduction the DNA must be able to move around, so the length of the DNA molecule is greatly shortened by tight coiling and compaction. In doing so, the DNA strands become wider and visible under a microscope. This visible form of DNA is called a ***chromosome***.

? Question

1. A chromosome is made up of tightly _______________ DNA.

2. Explain the reason for this shape.

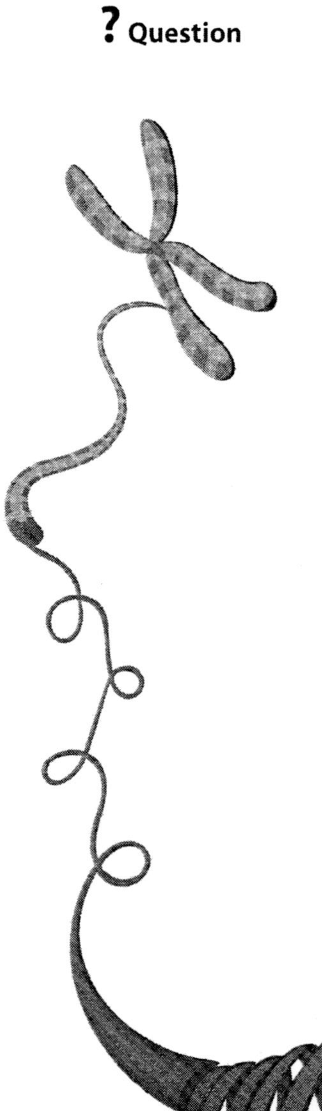

3. When not reproducing, DNA is found in the _______________ of a cell and exists in the form called _______________.

4. Visible DNA is called a _______________.

Exercise #2
Chromosome Sets

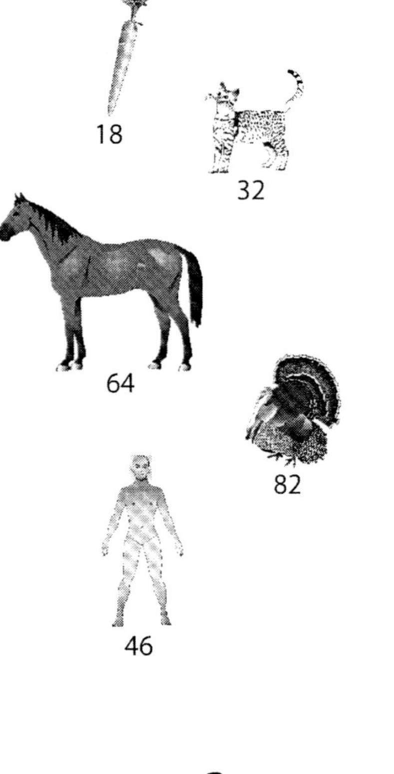

The One-Set Concept of Chromosomes: What is Haploid?

The number of chromosomes in a cell varies from species to species, but it is exactly the same among individual members of the same species. Each chromosome carries some of the genes for the species, and all of the chromosomes together complete the biochemical instructions for that species. Most species have either one or two chromosome sets. A ***set of chromosomes*** includes one copy of all necessary genes. Complex organisms have two sets of chromosomes as a "back up" mechanism to protect them against "broken" genes (bad mutations). If you have two copies of everything, then you probably will have at least one good copy of each gene.

In genetics the term for a single set of chromosomes is symbolized by the letter "***n***." Any cell that has only one set of chromosomes is termed ***haploid***. Haploid means that the cell has one of each kind of chromosome. The set concept will be used throughout the rest of this lab, so the following questions will aid you in understanding and recognizing sets. Remember that a ***set*** is a group of objects related in function and generally used together.

? Question

1. Pretend that the fingers of one hand represent chromosomes. (Count your thumb as a finger.) Hold up your hand.
   a. How many fingers (chromosomes) are on one hand? _____
   b. Are there different kinds of fingers on one hand? _____
   c. Is there more than one of each kind of finger on a hand? _____
   d. Judging by the definition of a set, you have _____ set(s) of fingers on one hand.

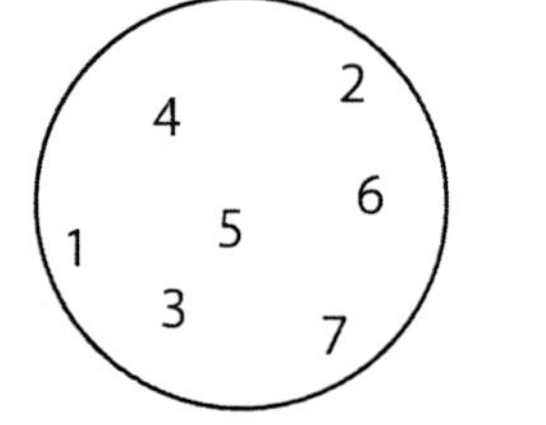

2. Pretend all the numbers within the circle represent chromosomes.
   a. How many numbers are there? _____
   b. Are there different kinds of numbers? _____
   c. Is there more than one of each kind of number? _____
   d. Judging by the definition, you have _____ set(s) of numbers.

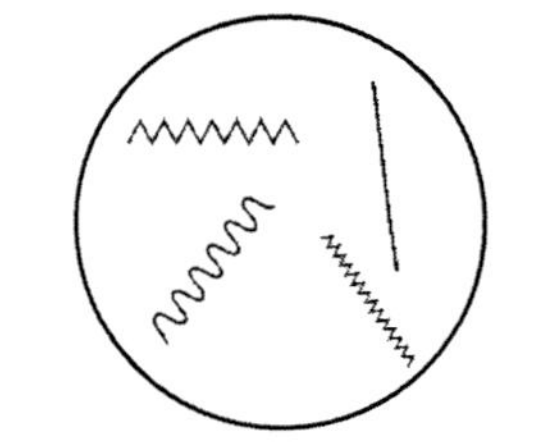

3. Pretend these lines represent chromosomes.
   a. How many lines are there?
   b. Are there different kinds of lines?
   c. Is there more than one of each kind of line?
   d. Judging by the set definition, you have _____ set(s) of lines.

Answers:

#1	#2	#3
a=5	a=7	a=4
b=yes	b=yes	b=yes
c=no	c=no	c=no
d=1	d=1	d=1

4. Explain the one-set concept of chromosomes.

The Two-Set Concept of Chromosomes: What is Diploid?

"Diploid" means two.

Simple organisms and the gametes of complex organisms are haploid. They have a single set of chromosomes. But complex organisms require two sets of chromosomes to survive the problems of mutant genes. We will discuss the details of this two-set requirement in a later lab "Mosses and Ferns". In genetics, the two-set condition is symbolized as "2***n***" and is called diploid. ***Diploid*** means that the cell has two of each kind of chromosome.

? Question

1. Hold up your hands. Pretend that the fingers of both of your hands represent chromosomes.
   a. How many fingers (chromosomes) do you have? _____
   b. Regarding both hands, do you have a duplication of each of the kinds of fingers? _____
   c. How many sets of fingers do you have? _____
   d. How many fingers are in a single set? _____

2. Now, pretend that all of the numbers within the circle represent chromosomes.
   a. How many total numbers are there? _____
   b. Is there a duplication of each kind of number? _____
   c. How many sets of numbers are there? _____
   d. How many numbers are in each set? _____

3. Pretend that the lines within the circle represent chromosomes.
   a. How many lines are there? _____
   b. Is there a duplication of each kind of line? _____
   c. How many sets of lines are there? _____

d. How many lines are there in each set? _____

Answers:

#1	#2	#3
a=10	a=14	a=8
b=yes	b=yes	b=yes
c=2	c=2	c=2
d=5	d=7	d=4

4. Explain the two-set concept of chromosomes.

5. Define haploid.

6. Define diploid.

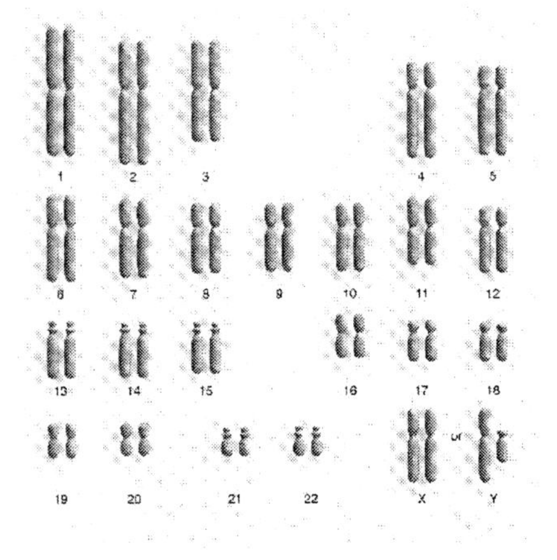

Humans have 23 pairs of homologous chromosomes.

Homologous Chromosomes: Maternal and Paternal

The word homologous means "the same", and the term ***homologous chromosomes*** refers to the pairs of chromosomes in a diploid cell that carry genes for the same traits. We will discuss more details about cell division in the next Exercise, but for now keep this fact in mind: You have two of each kind of chromosome, and those two are called a ***homologous pair***. Humans are diploid and have 46 chromosomes (two sets of 23). This means that we have 23 homologous pairs. And, each member of your 23 pairs came from one of your parents. In other words, one of your sets of chromosomes came from your mother, and we call them ***maternal chromosomes***. You get them from "Ma". Your other set of chromosomes came from your father, and we call them ***paternal chromosomes***.

Although both chromosomes of a particular homologous pair carry the same gene traits, these genes may be slightly different variations. For example, one might be the "blue eye" form, and the other might be the "brown eye" form. (More about this in the Genetics Lab.)

? Question

1. Pretend that the fingers of both of your hands are chromosomes. Hold up both your hands.
   a. How many individual fingers (chromosomes) do you have? _____
   b. How many homologous pairs are there? _____
   c. How many sets of fingers do you have? _____
   d. How many homologous pairs are in one set of fingers? _____

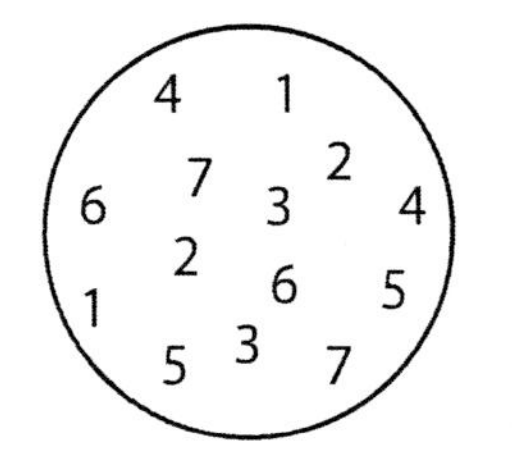

2. Pretend that the numbers in the circle are chromosomes.
   a. How many individual numbers are there? _____
   b. How many homologous pairs of numbers are there? _____
   c. How many sets of numbers are there? _____
   d. How many homologous pairs are in one set of numbers? _____

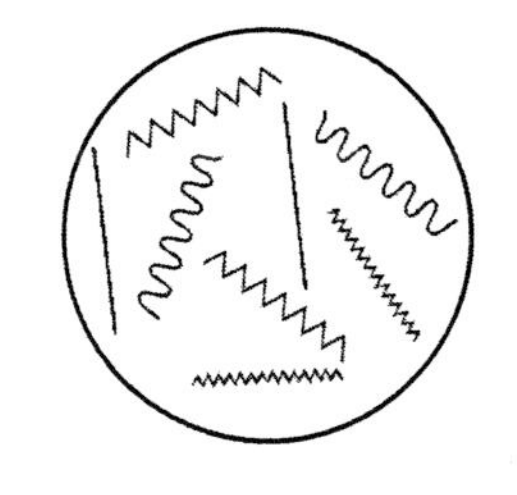

3. Pretend that all the lines within the circle are chromosomes.
   a. How many individual lines are there? _____
   b. How many homologous pairs of lines are there? _____
   c. How many sets of lines are there? _____
   d. How many homologous pairs of lines are in one set? _____

Answers:

#1	#2	#3
a=10	a=14	a=8
b=5	b=7	b=4
c=2	c=2	c=2
d=0	d=0	d=0

4. Explain the concept of homologous chromosomes.

5. What are maternal chromosomes? How many do you have?

6. What are paternal chromosomes? How many do you have?

Exercise #3
Asexual Reproduction of Cells - Mitosis

Mitosis produces more of the same.

Asexual reproduction of cells is called ***mitosis***. Immediately before this cell division process begins, the DNA duplicates itself creating two identical copies of every DNA molecule. The DNA copies move to opposite ends of the cell. Then the cell divides itself into two cells (each with exactly the same DNA as the original cell). The purpose of ***asexual reproduction by mitosis*** is to create new cells that are ***genetically identical*** to the original cell. Haploid mitosis does happen in some simpler organisms, but it will be covered in a more advanced biology course.

For a dividing cell to maintain its original species set number in the new cells, it must duplicate its genetic material prior to beginning mitosis. In addition, the genetic material must be divided in such a way that no new cell is either missing any DNA or has extra DNA than the original cell. Whatever amount of DNA the original cell has prior to mitosis, its offspring cells will have the same amount after the process is complete.

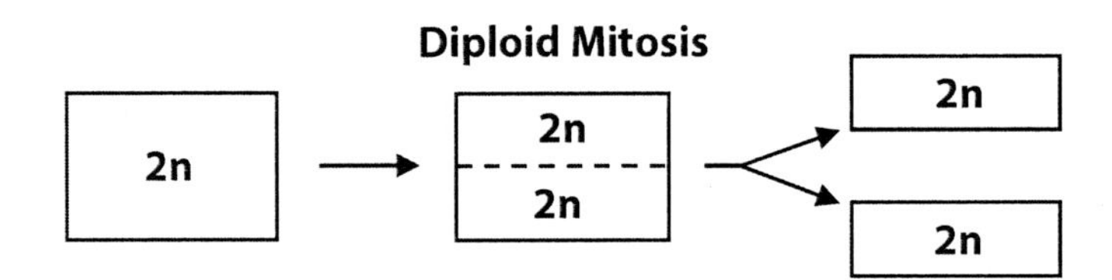

Figure 9.4. Mitosis (asexual reproduction of cells). Whatever amount of DNA the original cell has prior to mitosis, its offspring cells will have the same amount after the process is complete.

? Question

1. What would happen to the amount of DNA material in a cell if, when it reproduced, it did not duplicate the DNA first.

2. When the process of mitosis is used for organism reproduction, are the new offspring exact genetic duplicates of the parent organism?

3. If organisms use mitosis for reproduction, their offspring would exhibit (sameness or variety)?

4. If asexual reproduction produces identical offspring, then how does such a species "change" over time?

5. What would be the advantage of reproducing asexually?

6. What would be the disadvantage of reproducing asexually?

7. Starting with a diploid cell, draw the next three generations of that cell as it reproduces.

Exercise #4
Chromosome Movement during Mitosis

Chromosomes move during mitosis. They are like "suitcases" full of genes that replicate themselves, and the copies separate allowing two cells to be created from one. This movement of DNA material in the form of ***chromosomes*** (coiled DNA) has several "phases," which are described next.

Materials

- ✓ one red and one yellow crayon
- ✓ a package of chromosome beads. Each package should contain 8 chromosomes.

Procedure

- You will use snap-bead chromosomes as you work through the illustrations of the phases of mitosis.
- The human has 46 chromosomes (23 homologous pairs). We will follow the movements of 4 chromosomes (2 homologous pairs) to show what all the chromosomes are doing during mitosis. Start with 4 of the chromosomes from your package. This is how a cell would

look prior to duplicating its genetic material and undergoing the process of mitosis:

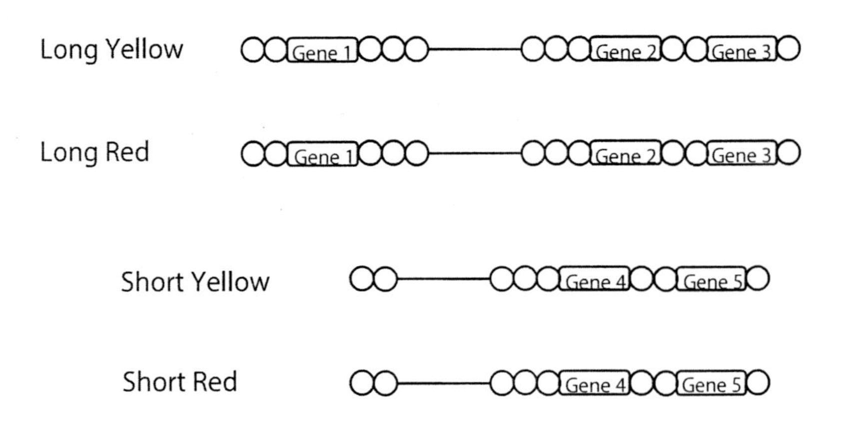

- The two short chromosomes represent one homologous pair, and the two long chromosomes represent another homologous pair. The red chromosomes are those from your mother (maternal), and the yellow chromosomes are from your father (paternal).

- Color the chromosome beads above with your crayons, and as you go through this Exercise, use the crayons to help you keep track of the chromosomes that came from your father and from your mother.

- The bead chromosomes are labeled A^1, A^2, B^1, and B^2 in the following discussion on the phases of mitosis.

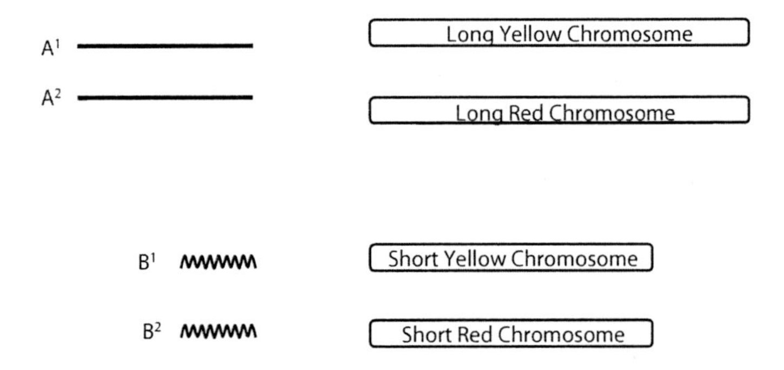

Phases of Mitosis

Biologists sometimes describe mitosis as having several phases. Ask your instructor if you are required to memorize the names of the phases. If so, remember the phrase: **P**ay **M**e **A**ny **T**ime. This will help you to remember the sequence of phases in mitosis. Pay attention to the different events as they occur in each phase, and mimic the phases by using your bead chromosomes. ***Interphase*** is usually considered to be the stage before mitosis actually begins. We include it as part of the mitosis discussion, but your textbook will say that mitosis begins with prophase.

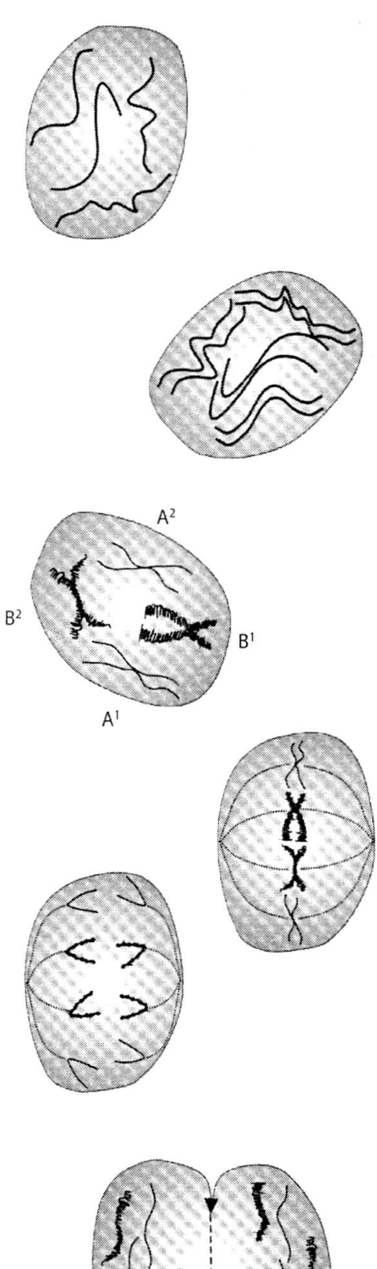

1. **Early Interphase:** We can not see the DNA during interphase because it is in long thin strands called ***chromatin***. In reality it is not coiled up yet and is not visible as chromosomes until prophase. However, it is best to label the DNA at this stage so that we can remember what the parent cell starts with.

2. **Later Interphase:** Each DNA molecule has duplicated itself. We have diagramed these "doubled" DNA molecules as though we could see them. Actually, DNA is still in the long, thread-like form. Duplicate your beads now, using the other four bead chromosomes from the package.

3. **Prophase:** This is when the DNA coils up and the chromosomes are now visible under the microscope. Each chromosome is doubled and consists of two absolutely identical "chromatids." A ***chromatid*** is the name for one of the duplicated DNA molecules that has coiled itself into a chromosome "package of genes" and is attached to the other chromatid.

4. **Metaphase:** The doubled chromosomes (each consisting of two chromatids) are lined up in random order along the midline of the cell. ***Spindle fibers*** have formed and are attached to the chromosomes and to opposite ends of the cell. You will have to imagine fine threads attached to your beads.

5. **Anaphase:** The spindle fibers pull the duplicated chromatids apart and move them to opposite ends (poles) of the cell.

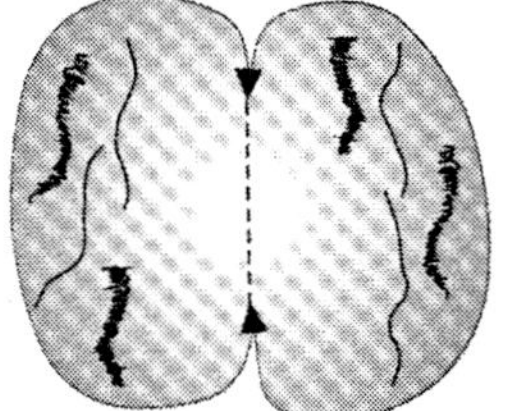

6. **Telophase:** Single chromosomes are at opposite ends of the cell, and the cell divides into two cells. Notice that you started with a cell having two sets of chromosomes, and you ended with two cells each having two sets of chromosomes.

? Question

1. What is the hint used to remember the sequence of stages in mitosis?
2. Describe the main event during Interphase.
3. Describe the main event during Prophase.
4. Describe the main event during Metaphase.
5. Describe the main event during Anaphase.
6. Describe the main event during Telophase.
7. Is the number of chromosomes at telophase identical to the number of chromosomes you started with in interphase prior to DNA duplication?
8. Are the new cells identical to the original cells?
9. What is the name for this cell division process?
10. What kind of reproduction is it?
11. Is mitosis going on in your body right now? What kind of cells are you producing by this process?
12. Name two processes during which your body must reproduce cells by mitosis.
13. What process do you think might not be happening as accurately when you age (get wrinkles, grey hair, lose your hair, etc.)?

Exercise #5
Onion Root Tip

It's time to review what you learned in Exercise #4. You will use a microscope to find the various phases of mitosis (cell division) in the root tip of an onion.

Materials

✓ compound microscope
✓ prepared slide of an onion root tip

Procedure

- Under low power, notice that the root tip is covered by a root cap (like a thimble over your finger). Behind the root cap is an area of square-shaped cells that are undergoing cell division.
- Next, look at this area under the high power (430x). If the cells are rectangular (not-square), then you are in the wrong place.
- Find every stage of mitosis.
- Draw a simple sketch of what you see at each phase of mitosis.

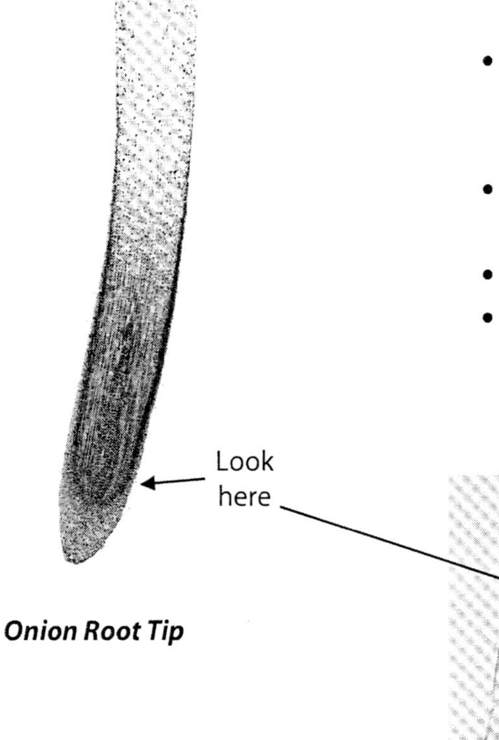

Onion Root Tip

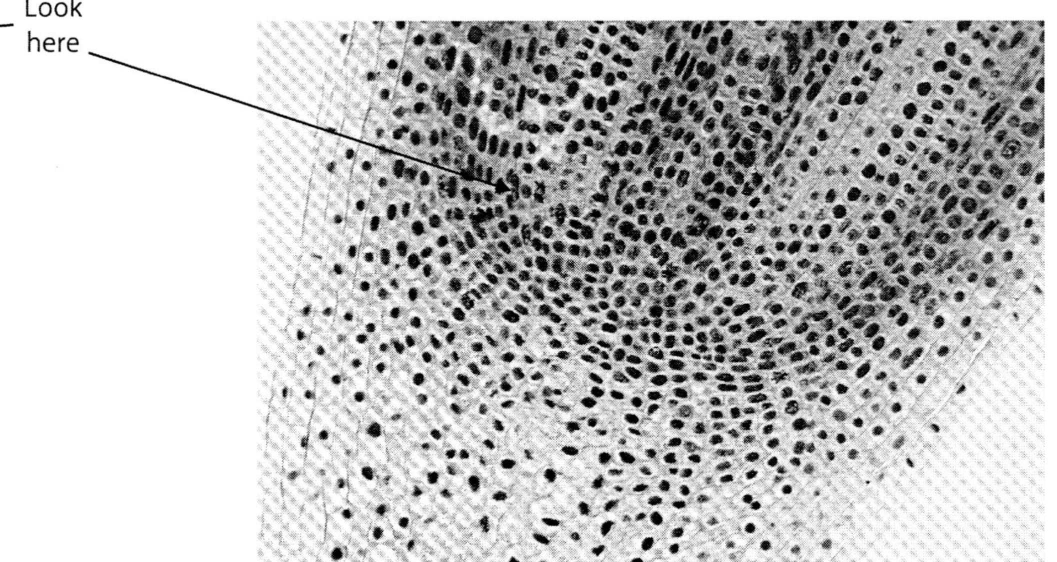

Figure 9.5. Higher magnification of the onion root tip.

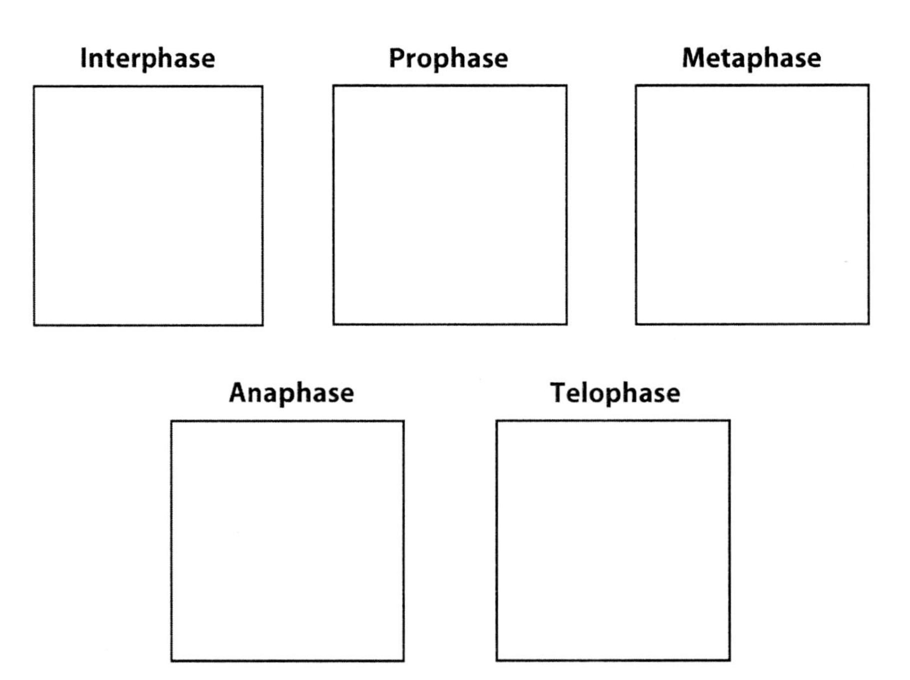

*Interphase will appear as a stained nucleus. You won't be able to see the actual DNA threads, only a general stained color.

- **Optional:** (Ask your instructor if you are to do this experiment.) You can estimate the relative amount of time that a cell spends in each phase of the cell cycle by counting all of the cells in the zone of cell division and recording how many of them are in each phase. Calculate the percent each stage is of the total. This is an indication of the relative time a cell spends in each stage of cell division.

Table 9.1. Estimating the Time Spent During Each Cell Division Stage. The cell is in Interphase most of the time of its life. The percent of cells in a particular stage is a reflection of how fast that stage is compared to the other stages. (Fewer examples means shorter time.)

Phase	# of Cells in Phase	% of Total Cells
Interphase		
Prophase		
Metaphase		
Anaphase		
Telophase		

Exercise #6
Sexual Reproduction

Sexual reproduction is the process of creating variety in the offspring of a species. It consists of two parts: ***meiosis*** and ***fertilization***.

Meiosis and fertilization produce variety.

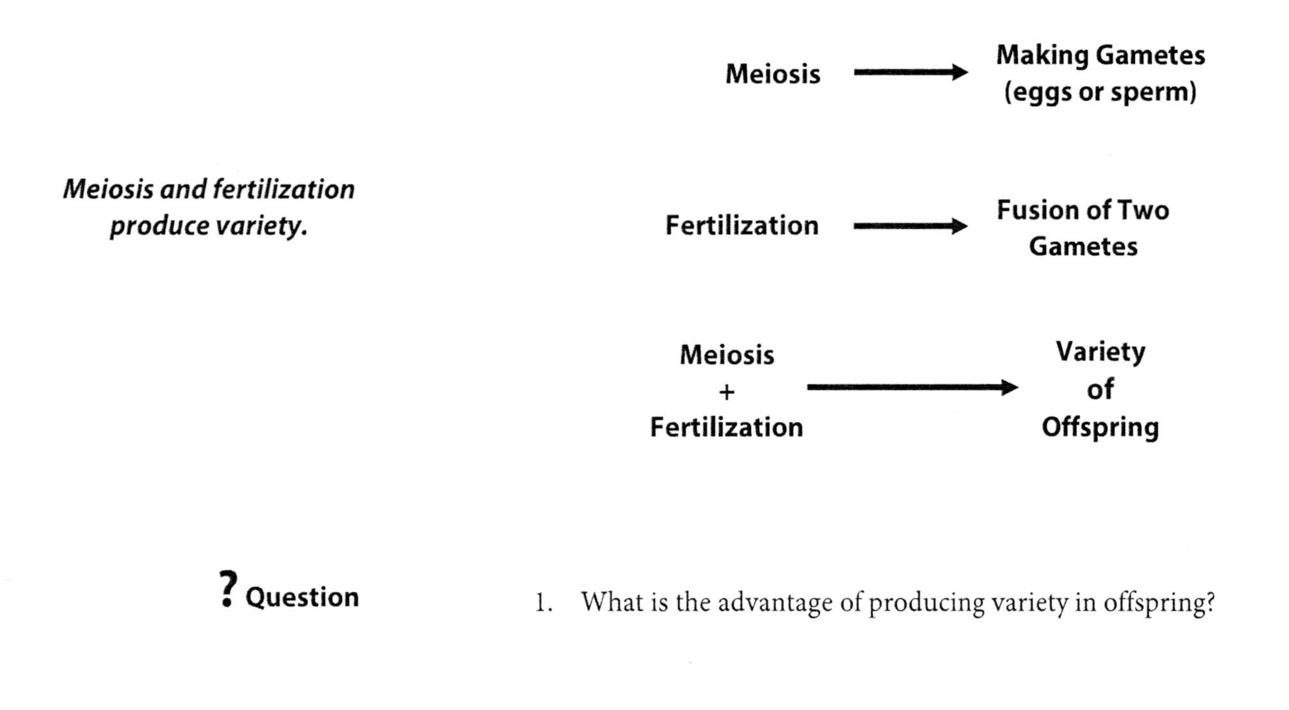

? Question

1. What is the advantage of producing variety in offspring?

2. What is the disadvantage of producing variety in offspring?

Chromosome Set Changes During Meiosis and Fertilization

Meiosis starts with a single diploid cell and ends with four haploid cells.

Meiosis starts with a single diploid cell and ends with four haploid cells. The original cell has two sets (2n) of chromosomes (DNA molecules) which are then duplicated. After DNA duplication, the original cell divides twice producing four cells—each with a single set (1n) of chromosomes. The four cells produced by meiosis are haploid. Also, their chromosomes have been mixed to produce genetic variety. More details of that mixing process are presented in Exercise #7.

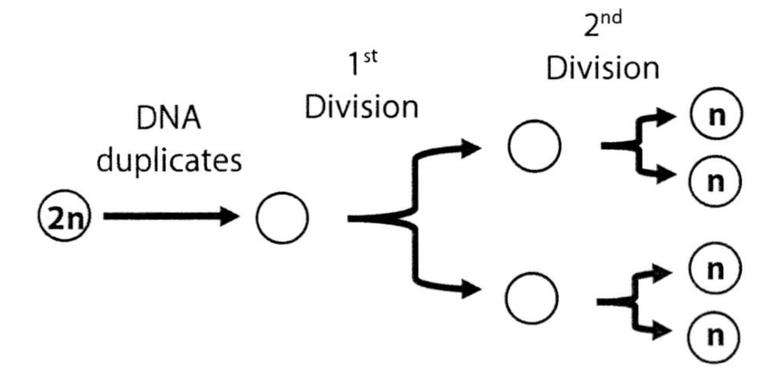

Fertilization is the fusion of two haploid gametes.

The product cells of meiosis are called ***gametes***, and these can fuse with gametes from another organism of the same species to begin the next generation. ***Fertilization*** is the fusion of two haploid gametes. This results in the chromosome set number returning to 2n. It allows the next generation to have the same "set" number of chromosomes as the parent generation.

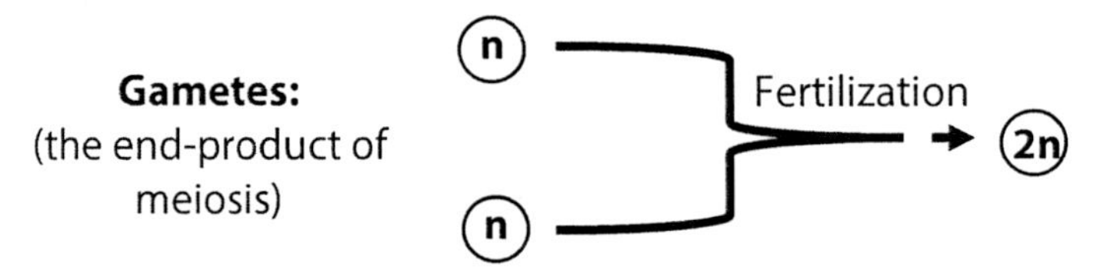

? Question

1. What type of cell division reduces the set number of chromosomes?

2. What process takes this reduced set number and returns it to match the original set number of the species?

3. How many sets of chromosomes are in a human sperm?

4. How many chromosomes are in a human sperm?

5. How many sets of chromosomes are in a human egg?

6. How many chromosomes are in a human egg?

7. When a human egg and sperm fuse, how many sets of chromosomes are there?

8. When a human egg and sperm fuse, how many chromosomes are there?

Exercise #7
The Events that Create Variety

Three events during meiosis and fertilization produce genetic variety in the species: crossing-over, independent assortment, and random fertilization.

Meiosis is the production of haploid cells from diploid cells. The essence of fertilization is the recombination of two haploid gametes to produce the next diploid generation. During meiosis and fertilization there are three events that create genetic variety in the next generation: ***crossing-over***, ***independent assortment***, and ***random fertilization***. None of these genetic variety events would be possible without a very special process during meiosis called ***synapsis***. This is the single most critical event that makes meiosis so different from mitosis.

Synapsis

Synapsis is defined as the "pairing up" process of homologous pairs early in meiosis. Let's illustrate the differences between mitosis and meiosis using the same chromosomes from our earlier mitosis diagrams.

Homologous chromosomes "pair-up" during meiosis. In mitosis they do not!

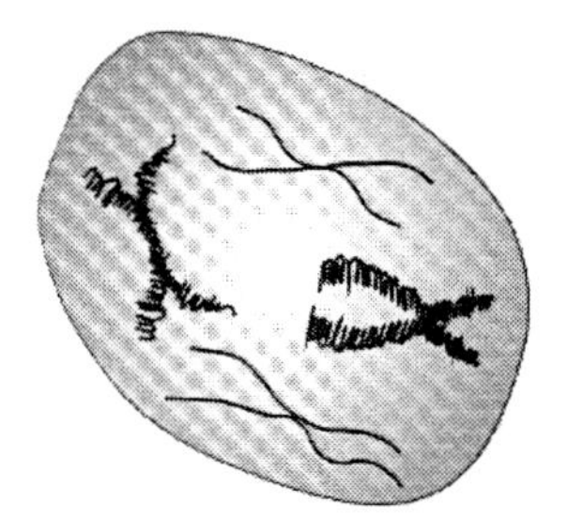

Prophase of Mitosis

Notice that the two members of the "A" homologous pair have duplicated themselves, and they are moved around the cell separately from each other.

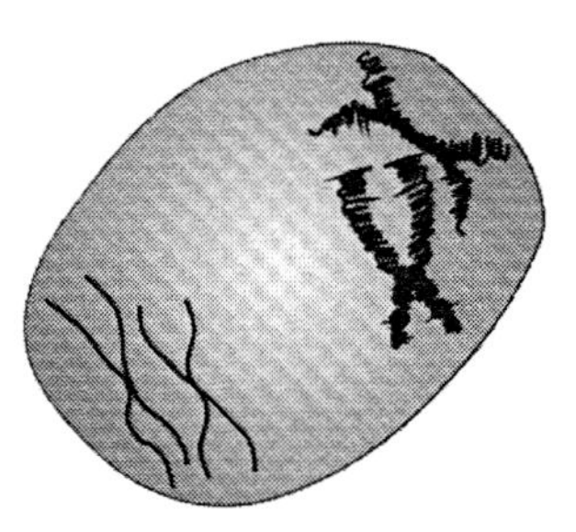

Prophase of Meiosis

Notice that both members of the "A" homologous pair have duplicated and are "paired-up" (***synapsis***). This paired grouping is called a ***tetrad*** (meaning four chromatids), and they are moved around the cell together.

Figure 9.6. The essential difference between meiosis and mitosis is synapsis and what happens because of it.

? Question

1. The two "A" chromosomes are concerned with the same traits and are called _______________ pairs.

2. Are the chromatids of the A^1 chromosomes identical or different?

3. Are the chromosomes A^1 and A^2 absolutely identical?

4. Are the "A" and "B" chromosomes homologous pairs?

5. In meiosis, do the "A" and "B" chromosomes pair up with each other (synapse)?

Crossing-Over

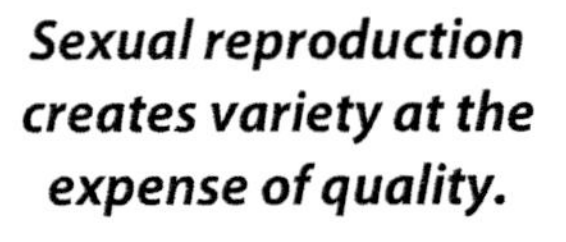

Sexual reproduction creates variety at the expense of quality.

As a result of synapsis, four chromatids stick together during the early stages of meiosis. These chromatids are so nearly identical (carry the same genes) that they will exchange equal pieces of chromatids when touching each other. ***Crossing-over*** is the exchange of DNA between the four chromatids in a ***tetrad***. This is an exchange between ***maternal*** and ***paternal*** chromosomes. When you make gametes (eggs or sperm), you mix up the chromosomes that you received from your mother and your father.

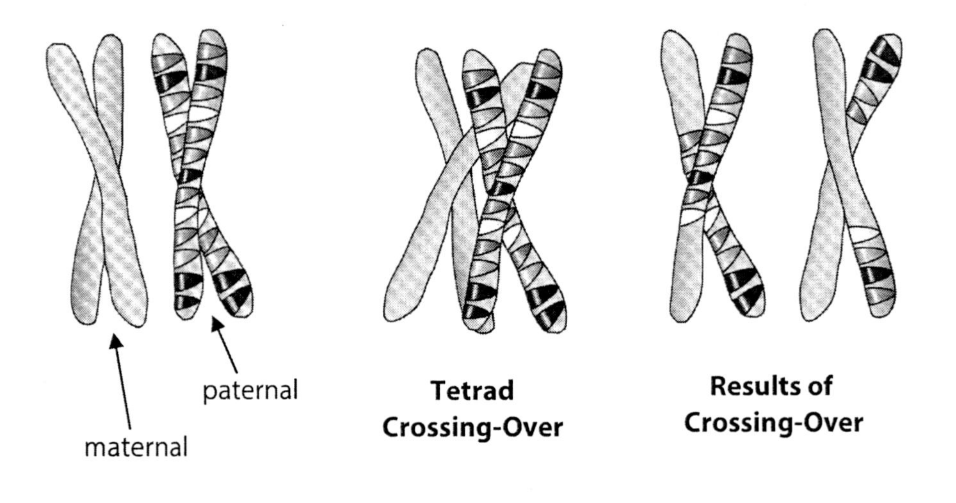

Figure 9.7. Crossing-over during meiosis. The maternal and paternal chromosomes of a homologous pair exchange equal pieces of chromatids.

Consider that a single chromosome carries a thousand or more genes. Many small cross-over exchanges are capable of creating hundreds of mixtures of chromosomes. You received one of each homologous chromosome pair (A^2) from your mother, and the other (A^1) from your father. The process of crossing-over makes "new mixes" of those chromosomes.

? Question

1. What event takes place during prophase of meiosis (as opposed to mitosis prophase) that makes it possible for crossing-over to occur?

2. Crossing-over happens between ______________ and ______________ chromosomes.

3. As a result of crossing-over, will the "A" chromosome that you pass on to your children be your mother's, your father's, or will it be a mixture of your mother's and father's?

Independent Assortment

During meiosis the tetrads move around the cell and divide in different ways than we saw in mitosis. The simplest description of the difference is:

- Tetrads line up in the middle of the cell (during metaphase).
- The original cell divides twice, separating the tetrads first into pairs of chromatids and then into single chromosomes. The result is four separate gametes.

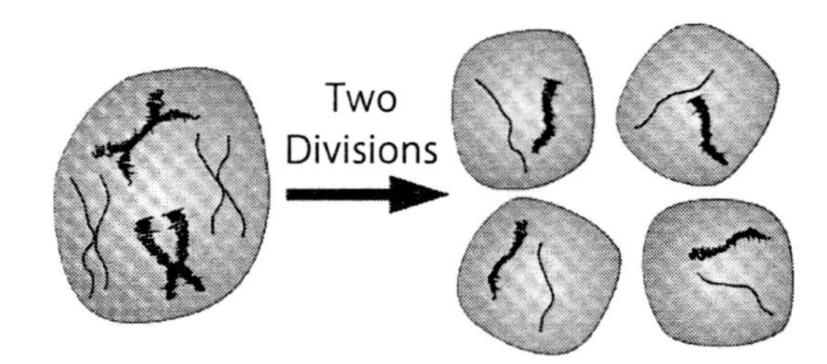

Each of these gametes contains only one of each kind of chromosome. They are haploid. However, there is a very important detail called ***independent assortment*** that occurs during the chromosome separations of meiosis. This happens because of the way that the tetrads line up in the middle of the cell during metaphase.

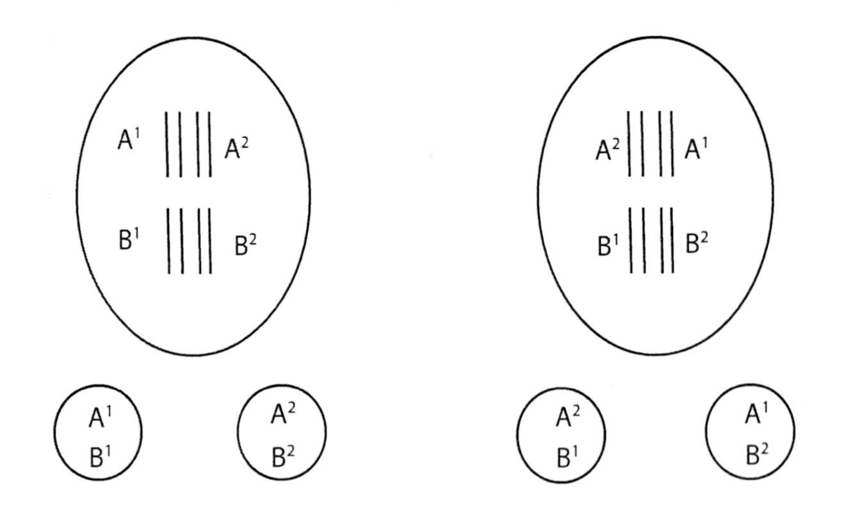

Figure 9.8. The particular way that a tetrad lines up during metaphase of meiosis will determine whether maternal and paternal chromosomes end up together or not. In this example, the cell on the left produces gametes with maternal together and paternal together, and the cell on the right produces gametes with maternal and paternal mixed.

In the cell on the left in Figure 9.8 the tetrads are drawn so that the A^1 and B^1 chromosomes are placed on the left side of the cell, and the A^2 and B^2 chromosomes are lined up on the right. If the tetrads separated equally as drawn, you would see only two kinds of gametes from this process (as shown). But, if the top tetrad had originally lined up during metaphase with the A^1 chromosomes on the right, then A^2 would have moved with B^1 into a gamete, and A^1 would have moved with B^2 into a gamete as pictured on the right in Figure 9.8.

This production of gametes containing different combinations of chromosomes is called ***independent assortment*** because one pair of homologous chromosomes is separated (segregated and assorted) into individual gametes independently of how another pair is separated. A^1 and A^2 have been separated independently of how B^1 and B^2 have been separated. Remember: Every gamete gets a complete haploid set of chromosomes with only one A chromosome and only one B chromosome.

? Question

1. How many genetically different gametes were produced in the independent assortment of these two homologous pairs?

2. The human contains 23 pairs of homologous chromosomes, all of which are independently assorted. What do you think the chance

would be that one of your gametes would contain either all of your mother's or all of your father's chromosomes?

Random Fertilization

Someone out there is more genetically like you than either of your parents.

Two mating individuals have the same kinds and same number of chromosomes, but those chromosomes are not exactly identical. Because individuals possess different variations of genes, the ***random fertilization*** from any two individuals will add even more genetic variety to the offspring. Fertilization recreates the diploid set number with the offspring receiving one chromosome of a homologous pair from one parent and the other chromosome of the pair from the other parent. Because there are so many sperm and eggs, each of which is genetically different, this is called random fertilization. No two offspring are the same, and none are identical to their parents.

To keep an example relatively simple, let's consider only one of the homologous pairs of the human—the "A" chromosome. (Actually, human chromosomes are referred to by numbers from 1 to 23.) We have labeled each of the A chromosomes differently because each is carrying different variations of genes. They are labeled from A^1 to A^4.

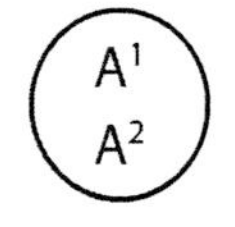

You

A^1 came from your father.

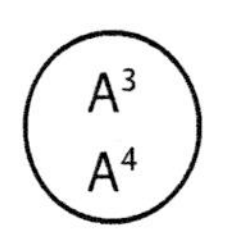

Your Spouse

A^3 came from your spouse's father.

? Question

1. Where did chromosome A2 come from?

2. Where did chromosome A4 come from?

3. What are your possible gametes?

4. What are your spouse's possible gametes?

5. Determine the four possible combinations of your gametes with your spouse's gametes. In other words, what genetic variety can we expect in your offspring?

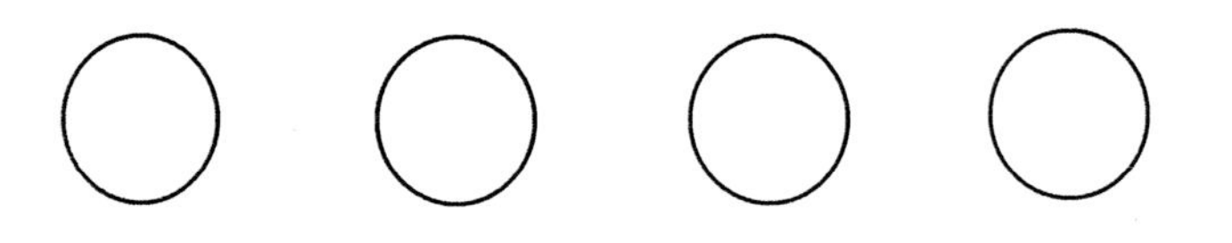

6. Are any of the offspring identical to either parent?

7. What are the three events during sexual reproduction that prevent identical children?

LAB

10

Genes and Protein Synthesis

"I would maintain that thanks are the highest form of thought, and that gratitude is happiness doubled by wonder."

- *G.K. Chesterton*

Dennis kicked the ground as he sat outside the math lecture hall. He was waiting for Jane, his best friend. She was a graduate student and understood science better than anyone else he knew. Dennis was an art student, and he needed her opinion right now. In his biology class he learned that small pieces of DNA, called genes, are responsible for producing traits in people. Genes control how amino acids assemble into proteins, and proteins are the fabric of life. They produce all the structures and functions in living things. This new explanation of life seemed to threaten some of the mystery he felt was certainly part of life. He had lived many adventures in his daydreams, and his boyhood fantasies first inspired him to draw and paint. And that led him to study art in college.

Jane burst from the classroom, walking briskly toward the chemistry building. "Jane!" he shouted, then ran to catch up with her. Jane glanced at her friend and said, "Dennis, you look upset. What's the matter?"

"Jane, how can I still keep the magic in my heart after learning how things really happen in nature?" Jane stopped, then laughed out loud. "But Dennis, I learned that from you! You are the one who notices every little thing. You show me the details I often miss." Dennis blushed with surprise. His face filled with a smile. "Jane, I'd like to take you to lunch." She smiled back. "Sure—after chemistry. But, why was that bothering you so much?"

"Oh," he said, touching her hand. "I guess I just forgot the details."

Exercise #1 **The Molecular Blueprint**
Exercise #2 **Genes and Enzymes**
Exercise #3 **Genes and Traits**

Exercise #1
The Molecular Blueprint

Proteins make up the fabric of cell structure, and they also form the enzymes that control biochemical reactions. The kinds of protein are what make one cell different from another cell. So, what makes proteins? Proteins are made of subunits called ***amino acids***. There are 20 different amino acids, and the sequence and arrangement of these amino acids are what makes one protein different from another. The ***nucleic acids*** DNA and RNA are the biochemical blueprint molecules for making proteins in the cell. These molecules are similar in structure, and both are examples of organic molecules that have some properties of life. DNA is the master blueprint molecule that is passed from generation to generation. RNA is a copy of the DNA blueprint, and it does the actual mechanics of building proteins.

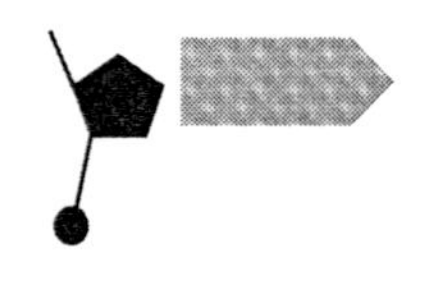

The nucleotide is the basic building block of both DNA and RNA. It consists of a sugar and phosphate connected to a specially shaped nucleotide base.

Complementary Base Pairing

During the lab on Mitosis and Meiosis you learned that DNA is an extremely long and thin ladder-like molecule that has two "rails" made of sugar and phosphate, and many "rungs" made of special ***complementary bases***. The DNA bases (rungs) are molecular units that combine only in two kinds of pairs. The base ***adenine*** (A) is always paired with ***thymine*** (T), and ***cytosine*** (C) is always paired with ***guanine*** (G). Therefore, if you know one of the complementary bases, you can easily figure out the other. The term ***nucleotide*** is the name for these repeating subunits in a DNA molecule.

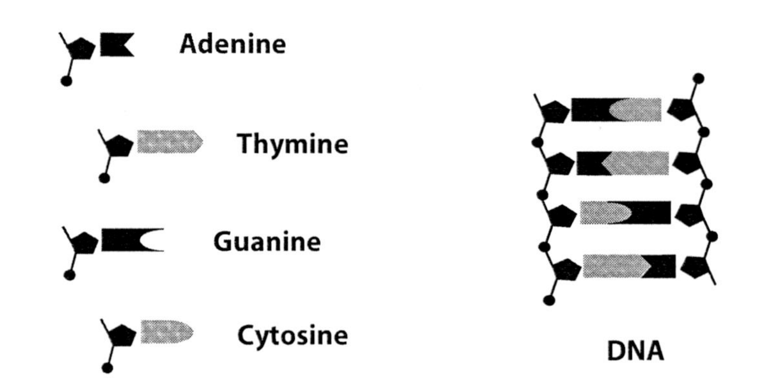

Figure 10.1. Complementary Base Pairing in DNA. There are four nucleotides in DNA, and they pair in only one way within the "rungs" of the DNA molecule.

RNA Structure

RNA is a single-stranded molecule made up of nucleotides linked together. Its structure is similar to DNA except that RNA has a different sugar in its four nucleotides. Also, RNA has one different nucleotide: RNA has ***uracil*** instead of thymine. RNA has no thymine!

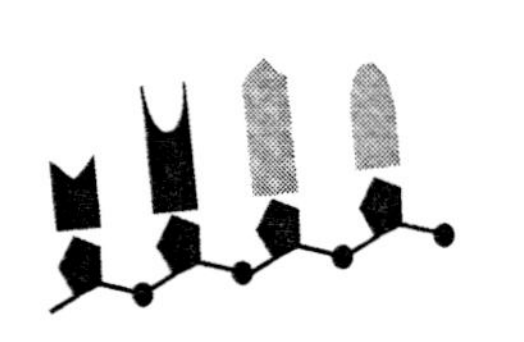

Figure 10.2. Differences between the structure of RNA and DNA. RNA is single-stranded, and it has the nucleotide uracil instead of thymine.

Laboratory experiments have shown that it is easy for DNA to replicate itself. If a DNA molecule is unzipped (by slightly heating or using an enzyme) and immersed into a tube containing a mixture of DNA nucleotides and special enzymes, it will make copies of itself until all the nucleotides are used up. This happens because of the ***specific nucleotide pairing*** (A with T, and G with C). Other experiments demonstrate that DNA can also make RNA. If a DNA strand is put into a tube of RNA nucleotides, it will make RNA copies until all the nucleotides are used up. This also happens because of the specific nucleotide pairing. The difference is that the U (uracil) of RNA pairs up with the A (adenine) in DNA. There is no T (thymine) nucleotide in RNA. Because uracil has a similar shape to thymine, it pairs up with the adenine nucleotide of the DNA molecule.

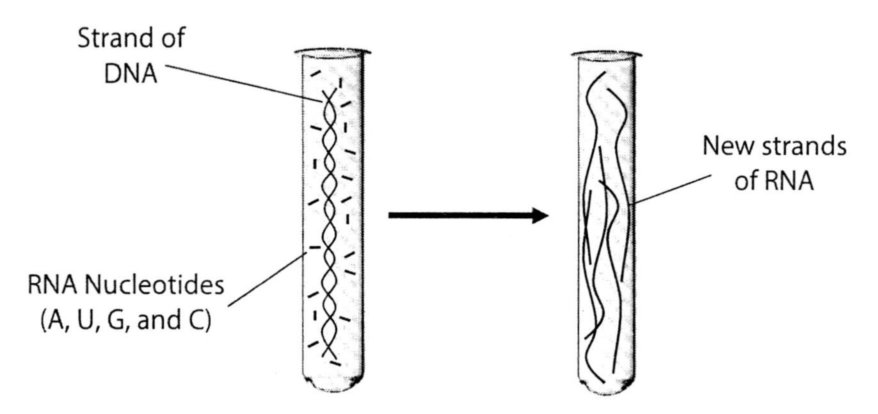

Figure 10.3. DNA makes RNA. The nucleotides of RNA are lined up along the nucleotides of DNA. RNA is a copy of the nucleotide "code" in the DNA.

? Question

1. List the four nucleotides of DNA.

2. List the four nucleotides of RNA.

3. What are the subunits of a protein?

4. Which molecule (RNA or DNA) is passed on from generation to generation?

5. Which molecule (RNA or DNA) does the actual mechanics of building proteins?

6. Which is a single-stranded molecule (RNA or DNA)?

7. An original strand of DNA has the following sequence of nucleotides:

DNA: C C A T C T G G A A C A C T A C T T A A A C A

RNA:

Fill in the corresponding nucleotides for the RNA strand.

Exercise #2
Metabolic Pathways, Enzymes, and Genes

The simplest explanation to start with was the "one-gene-one-enzyme" hypothesis.

By the 1950s it was well established that a gene is a portion of the DNA, and that each gene contains a "message" for making a particular protein. The first proteins investigated were enzymes involved in biochemical pathways of simple organisms. The idea was termed ***one-gene-one-enzyme hypothesis***. Further investigation revealed that the one-gene-one-enzyme hypothesis is an oversimplification. Some enzymes are synthesized under the direction of several genes, with each gene responsible for a different piece of the enzyme. In other cases, certain genes produce structural proteins needed by the cell.

Metabolic pathways are a series of cellular reactions necessary to convert a starting substrate into a finished product. There can be many steps in a pathway, but the next example has four steps controlling the production of different enzymes involved in the synthesis of a single product—melanin. Each step is catalyzed by its own special enzyme produced by a particular gene (or group of genes).

One way to understand human physiology is to discover all of our metabolic pathways.

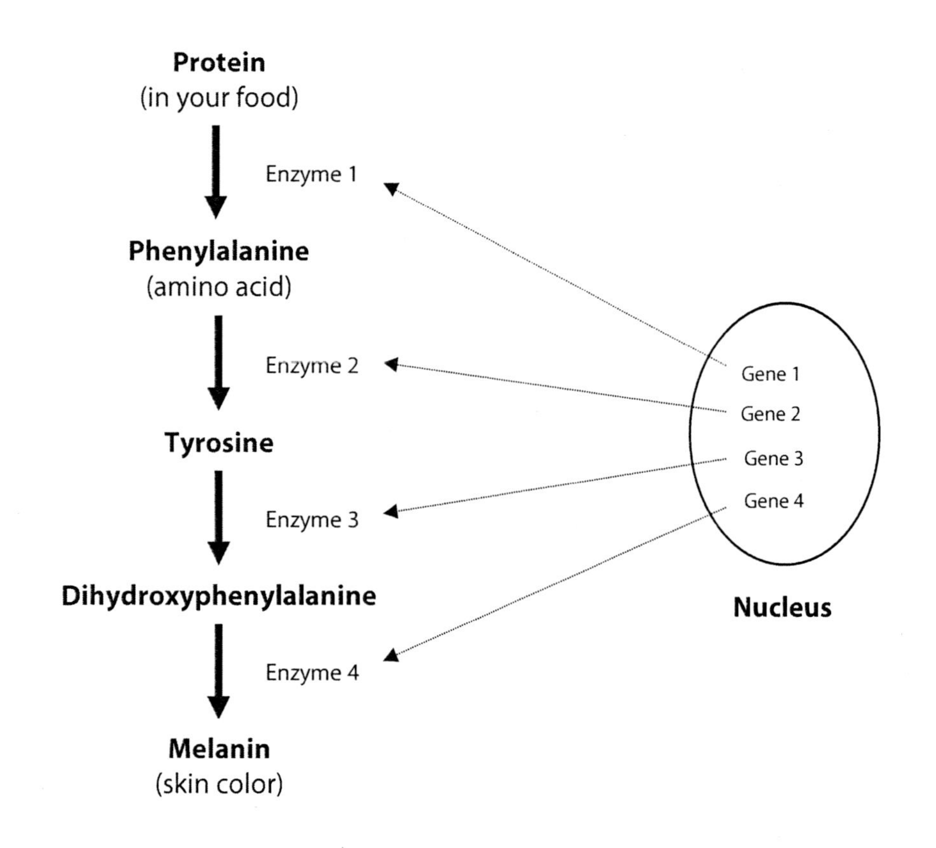

Figure 10.4. The role of genes and enzymes in metabolic pathways. This example shows four genes and enzymes involved in the production of melanin.

Genes Copied by mRNA

The message is transcribed from DNA language into RNA language.

If you wanted to make many copies of a valuable sculpture piece, one method of manufacture would be to use the "original" each time you made a copy. You would hesitate doing this because the original could be damaged and would soon wear out. A reasonable solution would be to make a mold of the original, and send that mold to the production line. When the mold is worn out, you make a new mold!

This analogy helps us to appreciate the cellular process whereby RNA copies genes and assembles proteins. RNA saves the wear and tear on the DNA original. Since a gene contains the "message" for building a particular protein, the RNA that copies the gene is called ***messenger RNA***, or simply ***mRNA***. It has the "message" to build a particular protein. The process of mRNA copying the gene is called ***transcription***. Each genetic message is "transcribed" from DNA language into RNA language.

DNA and RNA control the construction of enzymes.

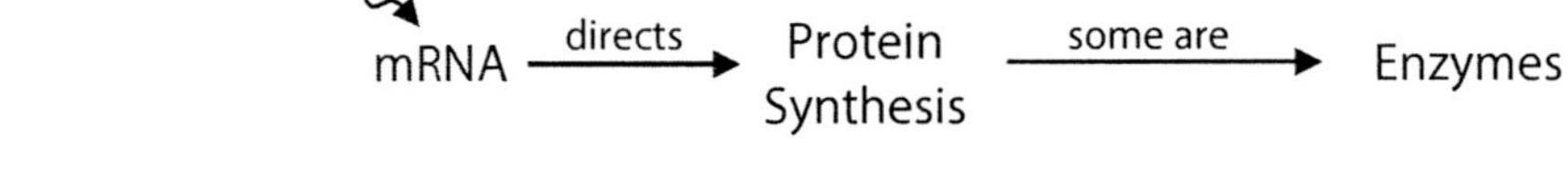

Figure 10.5. Mechanism of enzyme production.

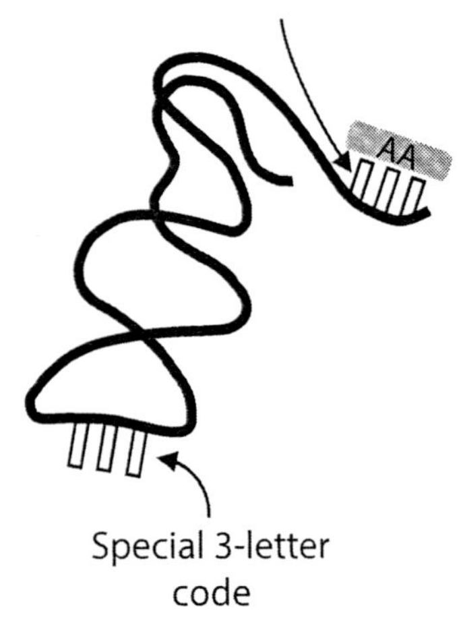

tRNA

Some of the genes in DNA are responsible for making another kind of RNA called ***transfer RNA***, or simply ***tRNA***. The tRNA is a twisted strand with two important reaction sites. One site attaches to a particular amino acid. Amino acids are the building blocks for all of life's proteins. There is a special tRNA for each of the 20 different amino acids, and each amino acid is "carried by" or "transferred" around the cell by its own special tRNA.

The second important reaction site is on the other end of the tRNA. It is a special 3-letter code (three nucleotides in a row) that is critical in protein synthesis. Your textbook calls this 3-letter code an ***anticodon***.

Codons

Each codon is responsible for lining up a particular amino acid.

Each mRNA molecule is a copy of a different gene. Some genes are as small as 30 nucleotides long, and other genes have a few hundred or even a few thousand nucleotides. The mRNA has a "start" end and a "stop" end. Between the two ends, there is a sequence of special 3-letter codes (three nucleotides in a row), called ***codons***.

Each codon of the mRNA is responsible for lining up a particular amino acid that is to be assembled into the protein. This is accomplished because a codon binds with only one kind of tRNA which carries only one kind of amino acid.

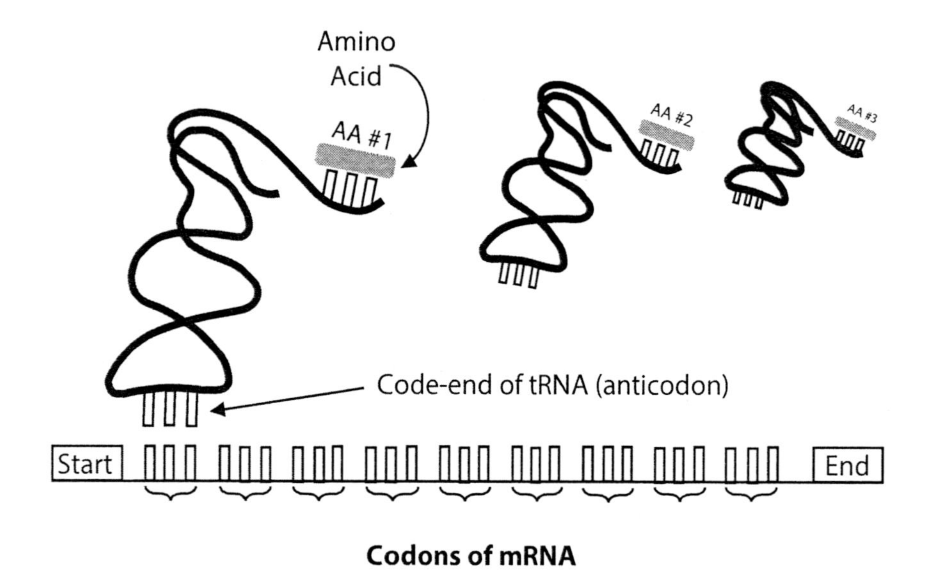

Figure 10.6. Building proteins one amino acid at a time. The role of the mRNA and tRNA is to arrange a sequence of amino acids in order to make a particular protein with an exact structure.

Table 10.1. mRNA Codons Responsible for Lining up Each of the 20 Amino Acids. You are to fill-in the "anticodon" column.

Amino Acid Codes for Protein Synthesis		
Amino Acid	**mRNA Codons***	**tRNA Anticodon**
Alanine	GCU	
Arginine	AGA	
Asparagine	AAU	
AsparticAcid	GAU	
Cysteine	UGU	
Glutamic Acid	GAA	
Glutamine	CAA	
Glycine	GGU	
Histidine	CAU	
Isoleucine	AUU	
Leucine	CUU	
Lysine	AAA	
Methionine	AUG	
Phenylalanine	UUU	
Proline	CCU	
Serine	UCU	
Treonine	ACU	
Tryptophan	UGG	
Tyrosine	UAU	
Valine	GUA	

* There are 64 codons. Some amino acids have several mRNA codons that code for them. However, there is no overlap of codes.

? Question

You should be able to fill in the 3-letter "code-end" of the tRNA molecules in the table above. Remember, in RNA A pairs with U, and G pairs with C. There is no thymine. Fill in the table.

1. What are the basic building blocks of proteins?

2. What is the one-gene-one-enzyme hypothesis?

3. Which molecule copies the DNA message for building a particular protein? What is this process called?

4. Where are tRNA and mRNA made?

5. How many different amino acids are there in living organisms?

6. Describe two special reaction sites on the tRNA.

7. How many nucleotides in a row make one codon?

Events at the Ribosome

The protein synthesis reactions at the ribosome are sometimes called ***translation***. The "message" in the mRNA is "translated" into the sequence of amino acids in a particular protein. Remember, when there is control over the line-up of amino acids, then an exact protein of any kind can be synthesized. Messenger RNA attaches to the ribosome and passes through it one codon at a time. As each of the codons passes through the ribosome, the appropriate tRNA carries in its own particular amino acid. Twenty different tRNA molecules carry the 20 different amino acids. See Table 10.1.

The mRNA codons assemble the amino acids. Amino acids are hooked together one at a time to make a protein molecule. The protein then breaks away from the ribosome, and the process of making another protein starts all over. It takes about one minute to make one protein molecule. A single mRNA can assemble many identical protein molecules until it either wears out or is "shut off" by the cell.

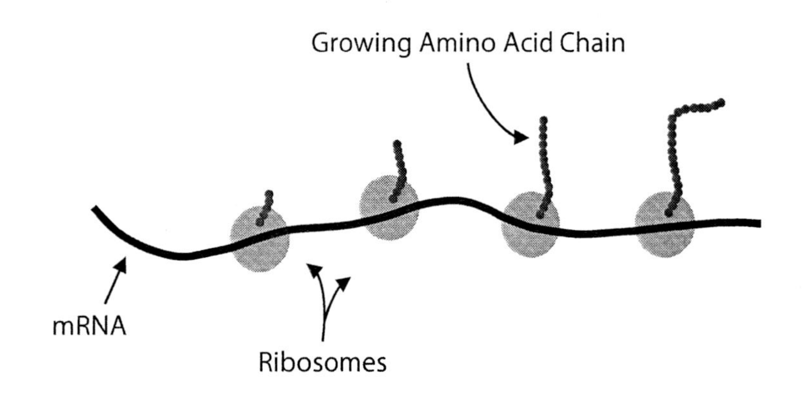

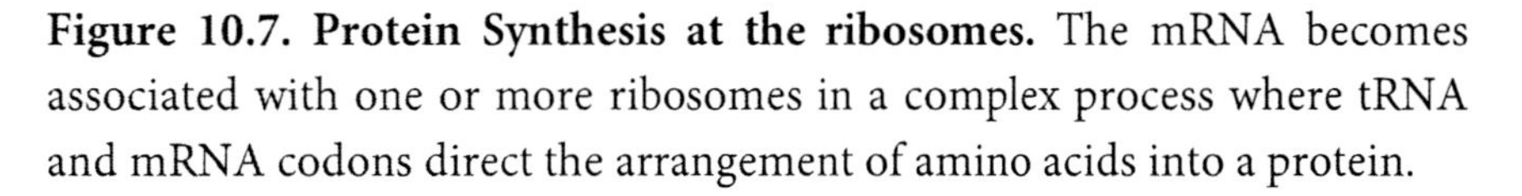
Figure 10.7. Protein Synthesis at the ribosomes. The mRNA becomes associated with one or more ribosomes in a complex process where tRNA and mRNA codons direct the arrangement of amino acids into a protein.

? Question

1. Define translation. Where does it happen?

2. Genes are sometimes changed by exposure to radiation. Radiation can cause one of the nucleotides in DNA to be replaced by another. When nucleotides are changed in the DNA, what happens to the codons in mRNA, and what effects result during protein synthesis? (Be specific.)

3. Practice your understanding of the events at the ribosome by writing a paragraph summarizing the important highlights.

Exercise #3
Genes and Traits

In any particular cell, some genes are on, and others are off.

Genetic traits are features (either structural or biochemical) of the organism that are produced by the action of one or more genes. If you have the gene, then you develop the trait. If you don't have the gene, then you don't develop the trait. All of the cells of your body have the same genes. However, in each cell certain genes are turned "on" and other genes are turned "off." How the cell functions is determined by which genes are on and which genes are off.

Science is not yet able to decipher all of the processes controlling whether genes are on or off in a particular cell. However, certain general mechanisms have been discovered. For example, in some cases a substance in the environment around the cell can activate a gene. Sometimes the proteins produced by one gene will activate or inhibit another gene. There are also master control genes that activate clusters of other genes.

During embryonic development, the genes of many cells are turned on or off in a particular sequence. This causes a cell to become specialized so it can function in part of the body (as bone or muscle or brain cells). Some of the genes that are turned on during embryonic development are essential only at that stage. These same genes are turned off in the adult person and would cause serious diseases like cancer if they were to accidentally turn on again.

Table 10.2. What Makes One Cell Different from Another Cell?

Which Genes are On?			
Type of Gene	**Skin Cell**	**Liver Cell**	**Intestine Cell**
Basic Metabolism Genes	On	On	On
Sugar Releasing Genes	Off	On	Off
Melanin Gene	On	Off	Off
Digestive Enzymes Genes	Off	Off	On

? Question

1. List the different ways that genes are controlled in the cell.

2. What is the relationship of the gene to the trait?

Traits

The complexity of genetic processes responsible for the different kinds of traits in an organism is beyond the scope of this exercise. Your textbook and lecture class will expand on this discussion. This brief review of gene-controlled traits will introduce you to the topic. The simplest example of genetic traits is the ***single-gene trait***. Gregor Mendel, who pioneered the science of modern genetics, discovered that pea plants with a "purple-making" gene produced purple flowers. This gene controls the synthesis of an enzyme that activates the biochemical pathway leading to the purple color. Other peas plants had a defective gene that would not synthesize the purple-making enzyme. These plants produced white flowers. They could not make the purple color.

Single-Gene Traits

The height of a person and their skin color are examples of ***multiple-gene traits*** because there are several genes contributing to the same trait. If skin color were a single gene trait, then there would be two colors of skin—brown and albino. Refer to the melanin pathway example in Exercise #2. If you have the gene for enzyme 3 in this pathway, then you make melanin (brown pigment). A defective gene produces non-functional enzyme 3, so no melanin is made (no pigment).

Multiple-Gene Traits

Now, consider the human example where there may be six duplicate skin color genes in each person. There are two possible choices for each color gene (melanin or no-melanin). People show a distribution of skin color from light to dark, depending on how many melanin or no-melanin genes they have. The size of a person depends on the amount of growth hormone they produce, and it is another multiple-gene trait similar to skin color.

Often the environment plays a crucial role in determining whether or not a particular gene has an opportunity to produce a trait. One example is a yellow pigment in certain plants that collects in the fat of rabbits eating those plants. It results in yellow-colored fat. In some rabbits there is a metabolic enzyme that breaks down the yellow pigment into a white by-product. Those rabbits have white fat. A gene controls the synthesis of this enzyme. Rabbits that have the active form of this gene produce white-colored fat. Rabbits who have the defective form of this gene produce yellow-colored fat. However, there is an environmental complication of this trait. (See the following questions.)

Environment-Controlled Traits

Yellow Pigment in Some Plants → Enzyme Made by Some Rabbits → White Fat in the Rabbits

? Question

1. If a rabbit that has the active form of the gene is fed plants with the yellow pigment, what color is its fat?

2. If a rabbit with the defective gene is fed plants with the yellow pigment, what color is its fat?

3. If a rabbit with the defective gene is not fed plants with the yellow pigment, what color is its fat?

4. Can you now see how the environment influences the expression of a trait?

5. What is different about a gene that creates albino color?

6. Describe the variation in the appearance among individuals in a population when a trait is single-gene controlled compared to a trait that is multiple-gene controlled.

Mutations

A ***mutation*** is a change in the sequence of nucleotides in a gene (DNA). Mutations can be caused by radiation, chemicals in the environment, or other spontaneous events that are surprisingly common during the life of an organism.

? Question

1. Which amino acid is supposed to be lined up at codon 6? (Hint: Refer to Table 10.1 in Exercise #2.)

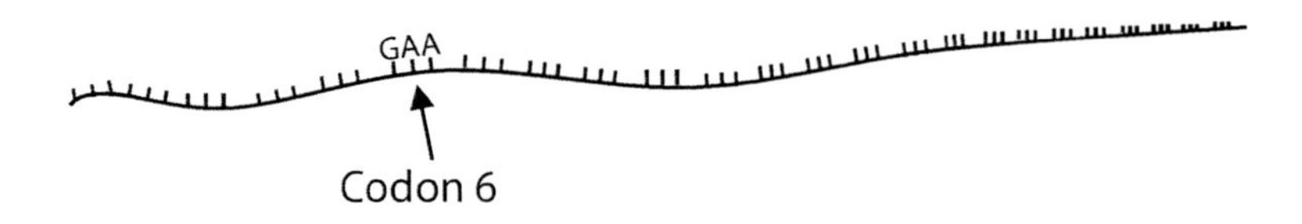

2. What would happen if a mutation in DNA changed codon 6 in the mRNA to GUA? (This one nucleotide substitution creates sickle-cell anemia.)

Mutations are the source of all change.

3. How does a mutation change the events at the ribosome?

Evolution

The very narrow definition of ***evolution*** is "the change in the frequency of alleles". ***Alleles*** are the variations of a gene in a population. Consider the following situation:

- Two colors of grasshoppers—brown and red—result from the two variations (alleles) of the color gene in this species. These grasshoppers live in two areas: an area with brown soil, and on an adjacent mountain with reddish soil.
- All the grasshoppers in both areas were accidentally killed by a fire.

- Ecologists re-introduced 1000 red and 1000 brown grasshoppers into each of the two soil areas. And, grasshoppers are a great food source for any bird that can catch them.

? Question

1. What do you think the ecologists discovered when they returned to the experimental sites five years later?

2. What factor controls the actual color of a grasshopper?

3. What determines the color frequency of grasshoppers in the two soil areas?

4. Is there only one direction of change during evolution? Explain your answer.

LAB

11

Genetics

A generation before Darwin's theory of evolution, an Augustinian priest, Gregor Mendel, discovered that hereditary particles are passed from parent to offspring during the reproductive process. These particles are called ***genes***, and the science of studying inheritance is called ***genetics***.

The investigation into the mechanics of inheritance—the mixing, the passing on, and the function of genes— is one of the greatest achievements in the 20th century. Genes can be traced backwards in time. When we do so, we find that all traits were new at some point in time, and that a gene's success correlates with adaptation, natural selection, and only partially understood chance events. However, genes do not last forever. Most have already gone extinct or been greatly changed. Understanding genetics has led to the prevention and treatment of several hereditary diseases. It has helped human beings to see their place in a family tree of life. In addition, genetics tells us that our genetic individuality is not the result of possessing a trait that no other individual has but is a result of a particular combination of genes. These genes came from our parents and their ancestors before them. Today's lab will explore some of the basic principles of genetics, introduce you to basic terminology, and help you apply genetic rules to some hypothetical inheritance questions.

Exercise #1	**Basic Terminology**
Exercise #2	**Some Rules of Genetics**
Exercise #3	**How to Solve Genetic Problems**
Exercise #4	**Genetic Problems**
Exercise #5	**Sex-Linked Traits**

Exercise #1
Basic Terminology

earlobe attachment

hair type

One Chromosome

In order to understand the mechanics of inheritance, you must understand the terminology used to describe this very complex process. A ***gene*** is a segment of the DNA molecule that is responsible for manufacturing a protein. That protein either becomes part of the organism's structure or it becomes an enzyme that controls biochemical events. A second fundamental concept is the chromosome. The ***chromosome*** is a vehicle for moving genes around the cell during cell division. It is like a suitcase – a very unusual one. A particular chromosome of a species always carries the same genes.

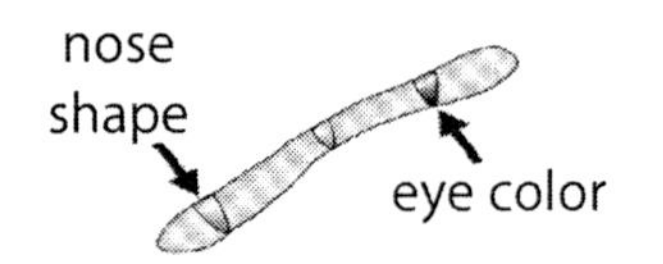

A Second Chromosome

Every organism has a certain number of chromosomes—the exact number depends on that species. Each chromosome carries many genes. DNA coils up into the form of a chromosome during cell division, and a gene is a distinct piece of that chromosome. Genes can be described by their exact location on a chromosome, and the process of locating genes is called ***mapping***. The location of a gene is its ***locus***, and geneticists go through great efforts to pinpoint these locations. Knowing where a gene is found on the chromosome is what allows scientists to start genetic research.

? Question

1. The particles that control inherited traits are called __________.

2. These particles are segments of__________, and are responsible for manufacturing a __________ that becomes either __________ or __________.

3. What is the basic purpose of a chromosome?

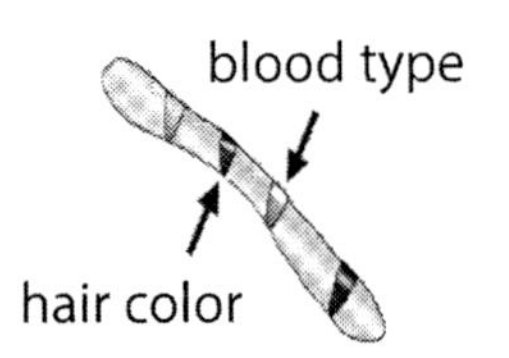

A Third Chromosome

4. Every living thing on the planet has the same number of chromosomes. (T or F) Explain your answer.

5. Every chromosome has the same identical genes as every other chromosome. (T or F) Explain your answer.

6. The place where you find a particular gene on a chromosome is called the __________.

Homologous Chromosomes

We have discussed homologous chromosomes before. This idea is essential to the understanding of genetics, so we will review it again.

- Very simple organisms have only one set of chromosomes and they are called haploid.
- More complex organisms have two sets of chromosomes and are called diploid.
- Haploid cells have one of each kind of chromosome and one of every kind of gene.
- Diploid cells have two of each kind of chromosome and two of every kind of gene.
- The two chromosomes of each kind in a diploid cell are called ***homologous chromosomes*** because they are carrying the same kind of traits (genes). Homo means "same."
- A human has 23 different kinds of chromosomes that are given numbers from 1 to 23. Because we are diploid organisms, we have two of each of the different kinds. So, we have 46 chromosomes in all, made up of 23 homologous pairs.

? Question

1. How many sets of DNA molecules or chromosomes does a diploid organism have?

2. How many sets of DNA molecules or chromosomes does a haploid organism have?

3. Humans are (haploid or diploid)?

4. Which of your cells are haploid?

5. How many homologous pairs of chromosomes does a human have?

6. Because chromosomes occur in pairs in a diploid organism, how many genes for one trait would a diploid organism possess?

7. How many genes for a trait would a haploid organism (or one of your gametes) possess?

Alleles: The Various Forms of a Gene

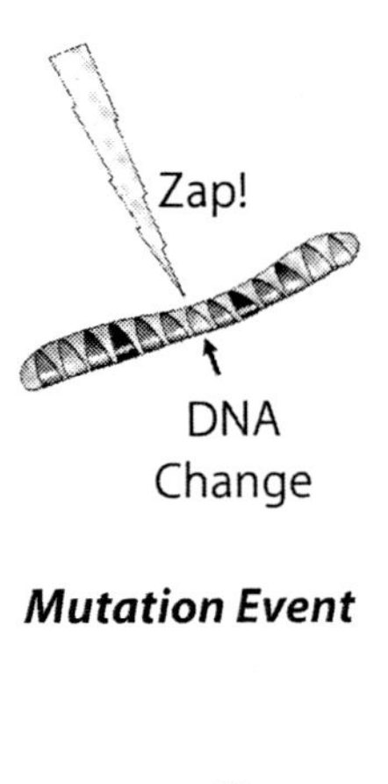

Mutation Event

Humans are diploid, and they have two copies of every kind of gene. One of the purposes of genetics is to figure out which form (variation) of these two genes you have and what expression of those genes you can expect. The alternate forms of a particular gene are called ***alleles***. For example, there are three alternate forms (three alleles) for blood type: A, B, and O.

The reason all species have various alleles (forms of genes) is that ***mutation*** events change the structure of genes. A gene can be mutated (changed) by radiation, by chemicals in the environment, or by other spontaneous events that are surprisingly common. There may have been a time when all the genes for eye color were identical and resulted in brown eyes. But over time, mutations occurred and changed the DNA of this eye color gene, creating a new "alleles" for the eye color trait. Perhaps this new allele was for blue eyes. Alleles are always for the same trait and are located at the exact same spot on homologous chromosomes. They have the same ***locus***. This is how we know that they are truly alleles of each other and not different genes. Remember: Alleles are variations of the same gene.

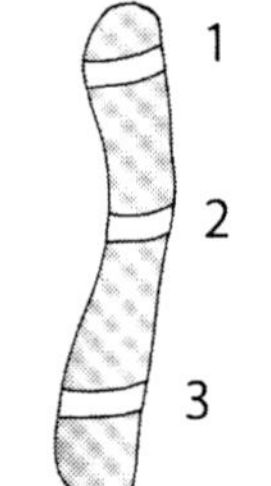

Chromosome #3

? Question

1. What is an allele?

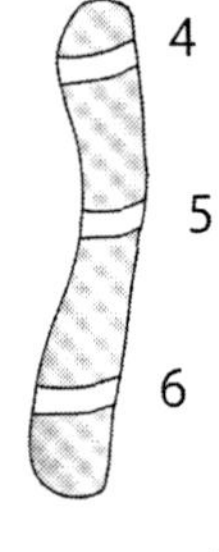

Chromosome #7

2. Where are alleles located?

3. What process creates the various alleles in a species? Explain how.

4. Which of the genes (1 through 9) on the chromosomes to the left are alleles of each other?

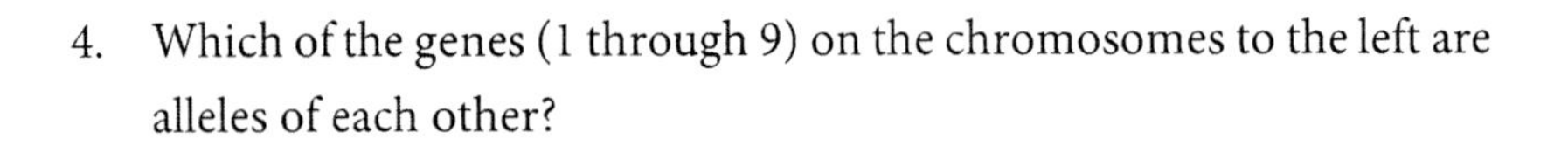

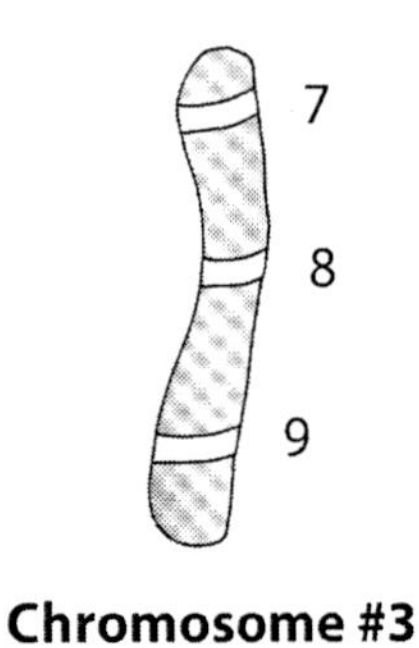

Chromosome #3

Genotype

Genotype = Your Genes

A ***genotype*** is a description of the alleles that you have for a particular trait. Even though two genes can look different, they are alleles of the same gene if they are at the same locus on homologous chromosomes.

Considering the two chromosomes on the left, draw and label the three combinations of eye color alleles that are possible.

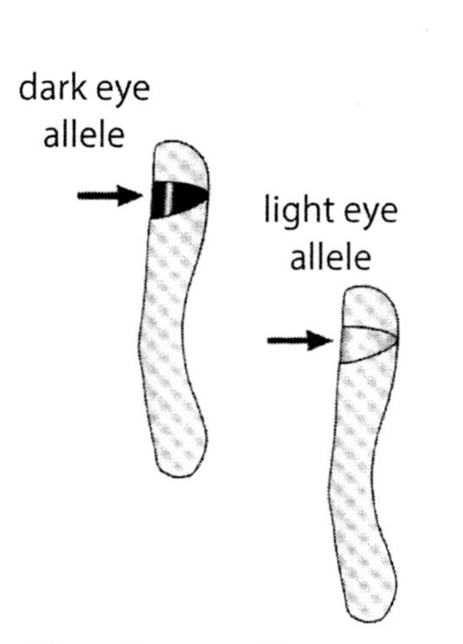

Homologous Chromosomes

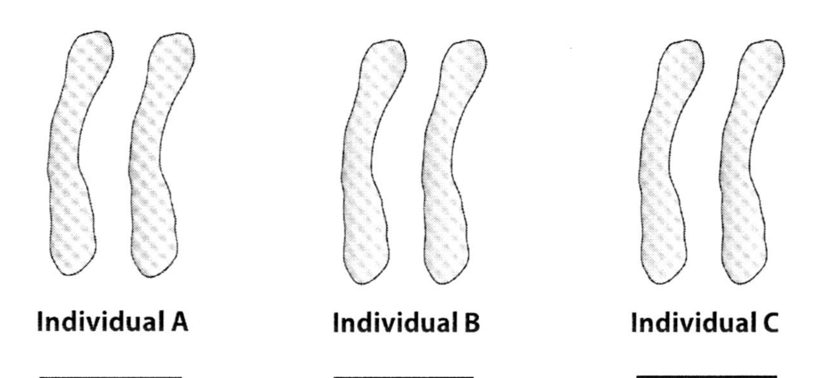

If an organism has two identical alleles, we say it is ***homozygous*** for that trait (meaning the "same" two alleles). If an organism has two different alleles, we say it is ***heterozygous*** for that trait (meaning "different" alleles). Go back to the diagram of the three individuals above and label each as to whether it is homozygous or heterozygous.

Phenotype

Phenotype = What you look like.

The physical expression of the alleles—what an organism looks like— is called the ***phenotype***. Because there are different possible combinations of alleles (genotypes), there are also different possible phenotypes for a trait that can be expressed in a population.

- The phenotype is the description of the physical expression of a trait (brown eyes), whereas the genotype is the description of the exact combination of alleles (for example; 1 allele for brown eyes + 1 allele for blue eyes).
- The genotype results from the combination of genes you inherited from your parents.
- The phenotype results from the physical expression of the genotype, and it may be influenced by the organism's environment. In some cases there are only two phenotypes for a trait, and in other cases there are more than two phenotypes for a trait.

Draw Table 11.1 on the whiteboard, and record your phenotype for each of the six traits. After everyone in the class has recorded their phenotypes, write the class totals in Table 11.1. Your instructor will tell you the genotypes of these traits at the end of the next Exercise.

Table 11.1. Frequency of Various Traits in Lab Class.

Trait	Class Phenotype Totals		
Eye Color	*Dark =	Light =	
Earlobes	Attached =	Unattached =	
PTC Paper	Can Taste =	Cannot Taste =	
Hairline	Widow's Peak =	Straight Forehead =	
Hair Type	Straight =	Wavy =	Curly =
Fingers	Five =	Six =	
Little Finger	Bent =	Straight =	
Tongue	Roller =	Non-Roller =	
Long Palmer Muscle	Present =	Absent =	

* These examples are oversimplifications of reality. When human physical traits are investigated and completely understood, there never is only one gene pair that controls what we see expressed.

Exercise #2
Some Rules of Genetics

Nine times out of ten, in the arts as in life, there is actually nothing to be discovered; there is only error to be exposed.

—H. L. Mencken American editor and critic (1880–1956)

Rule of the Gene

You must have the gene to pass it on.

The parent must possess the gene in order to pass it on. The source of all genes in the offspring is the parents. Always look to the parents to figure out what genes the sperm or egg can possibly carry. A parent does not have all of the alleles found within a reproducing population of a species.

? Question

1. How many different alleles for a single trait can a homozygous parent pass on?

2. How many different alleles for a single trait can a heterozygous parent pass on?

Rule of Segregation

Only one gene of the two alleles that you possess is put into each gamete that you make. Alleles are located on homologous chromosomes, and since homologous chromosomes are segregated during meiosis, the genes are also segregated. Numerous gametes are formed during gamete production, and if the alleles are different (heterozygous), 50% of the gametes will carry one allele and 50% of the gametes will carry the other. When alleles are the same (homozygous), 100% of the gametes will carry the same allele.

? Question

1. A parent possesses two copies of each gene. When this parent passes on its alleles for a gene, how many does it contribute to each of the offspring?
2. How many copies of a gene does the other parent contribute to each offspring?
3. How many copies of each gene for the trait does each offspring receive?

Rule of Dominant and Recessive Alleles

Some alleles control the phenotype even if they are paired with a different allele. If two different alleles are together in an organism, and only one phenotype is expressed, then the allele that is expressed is called ***dominant***. The other allele that is "hidden" is called ***recessive***. One example of a dominant allele is the dark-eye allele that will create the dark-eye phenotype in an individual even if the allele for light eyes is also present.

? Question

1. Can the individual carry an allele that is not expressed? Explain.
2. What word is used to describe the genotype condition in which two different alleles occur together in the same organism?
3. What word is used to describe the genotype condition in which two of the same alleles occur together in the same organism?

Since dominance and recessiveness have intricate biochemical explanations, the only easy way of determining dominance is to cross two individuals that are homozygous (pure) for the two different phenotypes. This produces the heterozygous condition. Whichever phenotype is exclusively expressed is said to be the ***dominant phenotype***.

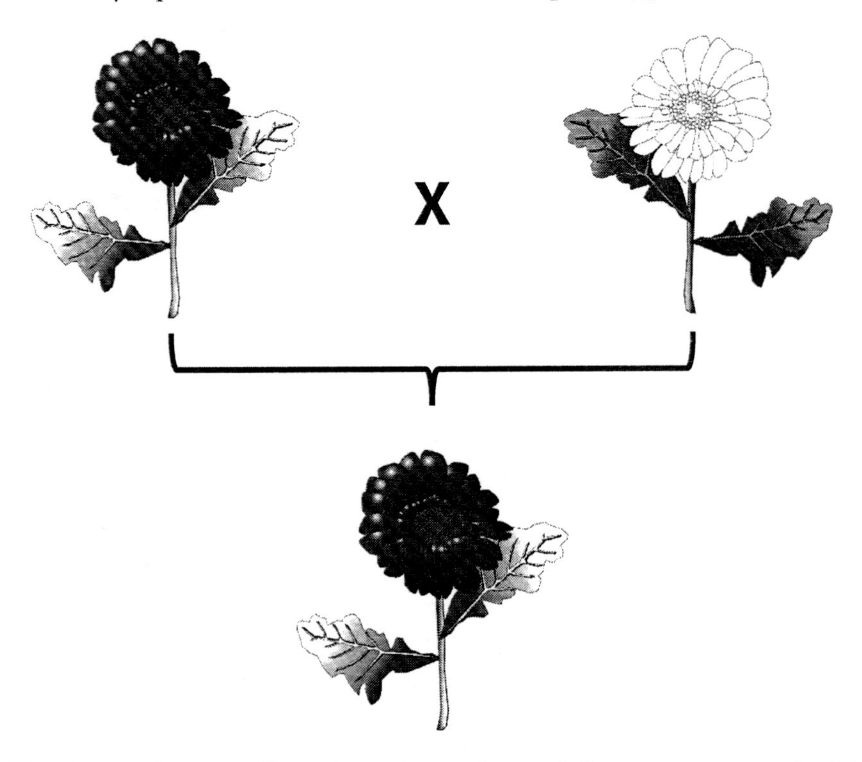

Figure 11.1. In some flowering plants when red flowers are crossed with white flowers, the result is all red flowers. This is an example of complete dominance.

? Question

1. A homozygous blue-eyed mouse with short whiskers mates with a homozygous brown-eyed mouse with long whiskers. All of their offspring have brown eyes and short whiskers. Which alleles are dominant?

2. A homozygous five-clawed cat is crossed with a homozygous six-clawed cat and all of the kittens have six claws. Which allele is dominant?

3. In humans, the five-fingered condition is recessive to the six-fingered condition. Yet, most people have five fingers. Explain how this can happen.

Dominant does not mean most common.

4. Ask your instructor which alleles are dominant in the class Phenotype Chart (Table 11.1). It is a common mistake to assume that the allele found most frequently is always the dominant allele. The success of an allele is usually determined by how successful the trait is in response to natural selection.

Rule of Incomplete Dominance

When two different pure-breeding strains are crossed, and their offspring show a blending of phenotypes, then neither allele is dominant. Incomplete Dominance is easily recognized whenever there is a phenotype somewhere between two extremes. Including the parents, there are three phenotypes (red, white, and pink) being expressed in these flowers instead of only two, and that third phenotype is intermediate between the other two. This heterozygous condition is called ***incomplete dominance***.

Figure 11.2. In some flowering plants when red flowers are crossed with white flowers, the result is all pink offspring. This is an example of incomplete dominance.

? Question

1. On the chart you did earlier, which of the three hair types (wavy, curly, or straight) represents incomplete dominance—the blended heterozygous condition?

2. You cross a herd of red cattle with white cattle and all of the calves appear to be roan (reddish white). Is this an example of incomplete dominance? How do you know?

3. You cross a blue flowering pea plant with a white flowering pea plant and all of the offspring are blue flowered. Is this an example of incomplete dominance? How do you know?

Exercise #3
How to Solve Genetic Problems

Using Letters for Alleles

For convenience, the genes of an allele pair are usually symbolized by a letter from the alphabet. A ***Capital letter is used for the dominant trait*** and ***a small letter for the recessive trait***. When we want to describe the genotype of an organism, we use letters to represent the alleles inherited from the parents. For example, free earlobes is a dominant allele and attached earlobes is recessive. You would use a capital "**F**" to indicate the dominant allele and a small "**f** " to indicate the recessive allele in describing an individual.

? Question

1. Write the three genotypes for earlobe attachment as it applies to the following individuals.

   a. Heterozygous _____
   b. Homozygous Dominant _____
   c. Homozygous Recessive _____

2. When it comes to symbolizing incomplete dominance with letters, it is best to use the letter "**C**" for one allele and "**C'** " for the other allele. List the three possible genotypes for hair type.

   a. Curly _____ b. Wavy _____ c. Straight _____

Why not use a small letter "**c**" for the heterozygous genotype in this case?

Using the Punnett Square

The ***Punnett Square*** is a simple method of predicting the probable outcome of genetic crosses.

Procedure

- Draw a square like this.

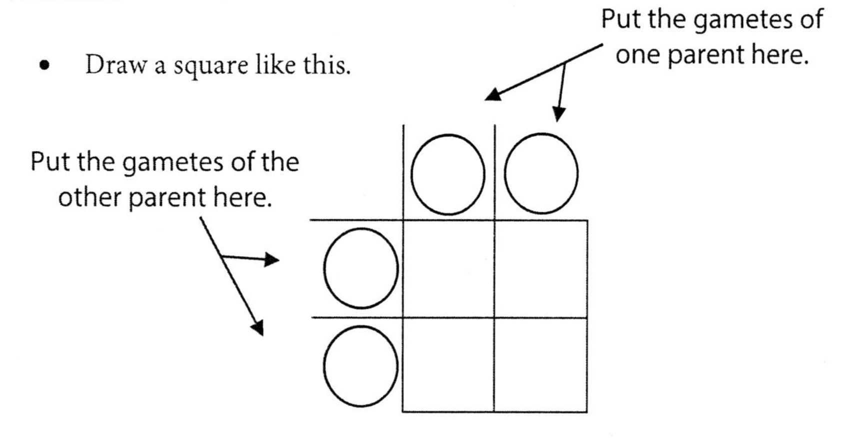

- Determine the kinds of gametes that are made by each of the parents in this cross (Ff x ff), and put those gametes into the parent boxes of the Punnett Square.

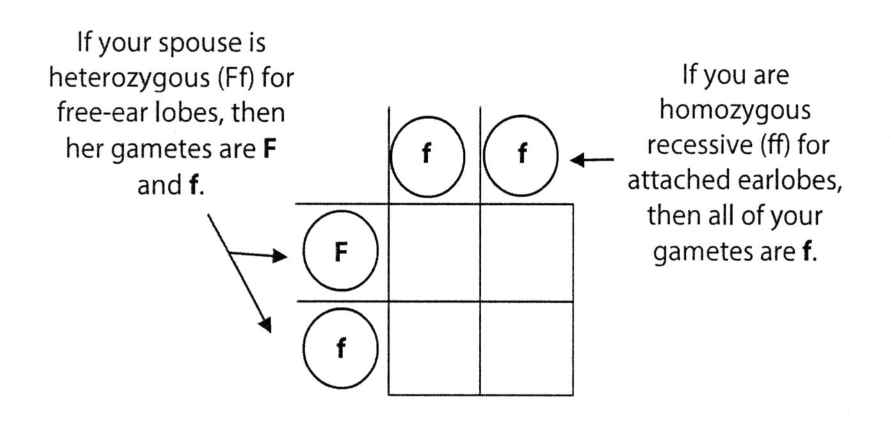

- Fill in the offspring boxes of the Punnett Square.

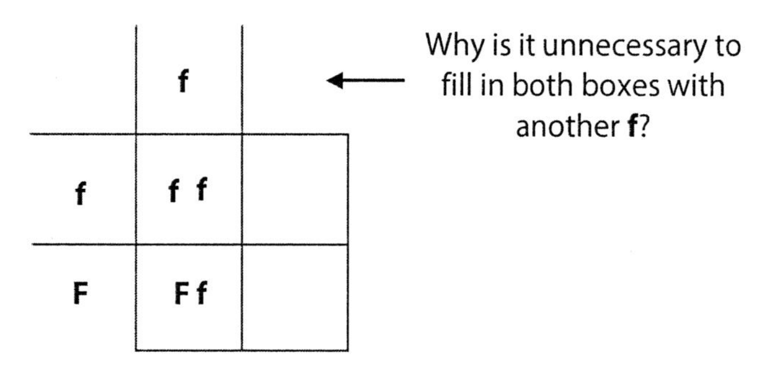

In this example there are only two possible offspring genotypes. The Punnett Square tells us to expect about 50% ff and 50% Ff. Sometimes the Punnett Square is more complex than this example, and you must figure out more than one trait at a time. Nevertheless, you use the same basic method.

- Make up your own genotype example and work out the crosses.

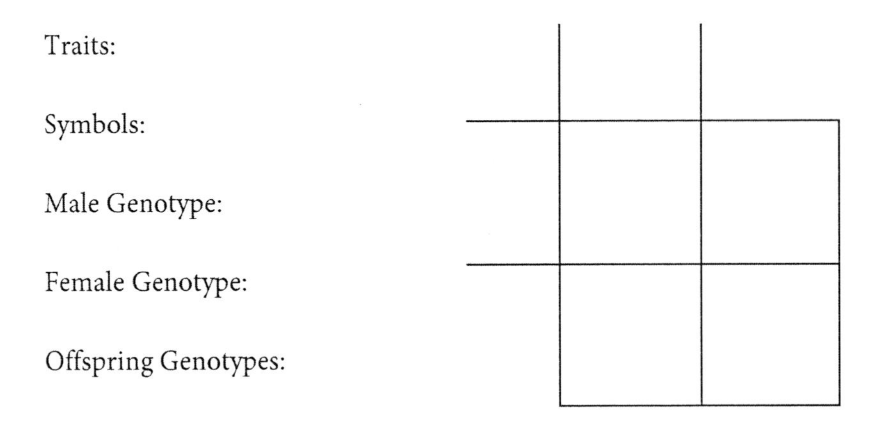

Exercise #4
Genetic Problems

Cases of Complete Dominance

Gregor Mendel grew different varieties of pea plants in his garden. When he crossed yellow-seed plants with green-seed plants, he always got yellow pea seeds.

1. What is the dominant allele?

2. What is the genotype of all green-seed plants?

3. Use the Punnett Square to show Mendel's cross.

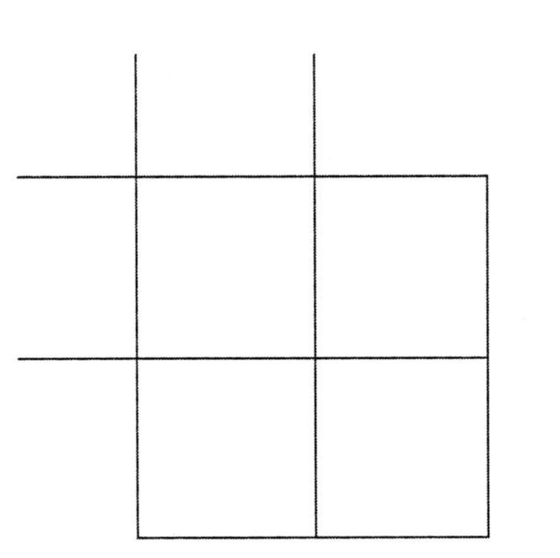

4. Do the parent yellow-seed plants have the same genotype as the offspring yellow-seed pea plant?
   Parent:__________ Offspring:__________

5. What genetic fact do you know about any yellow-seed pea plant?

6. If yellow-seed pea plants are dominant to green-seed pea plants, why are there mostly green pea seeds in nature?

A dark-eyed man has children with a light-eyed woman, and they have ten dark-eyed children.

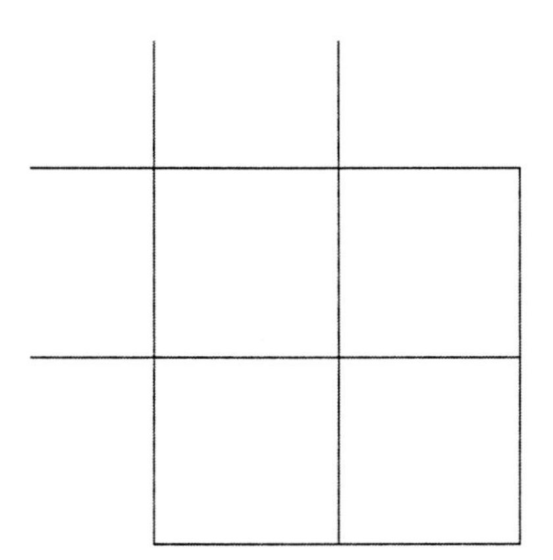

1. What is the dominant allele?

2. What is the genotype of all light-eyed people?

3. What are the genotypes of the two parents?

4. What is the genotype difference between the dark-eyed parent and the dark-eyed offspring?
   Parent:__________ Offspring:__________

5. When two heterozygous dark-eyed people (Dd) are crossed, what is the phenotype ratio of dark-eyed offspring to light-eyed offspring? Use the Punnet Square to show your answer.

Eye color is due to multiple alleles and more than one gene pair. The numerous phenotypes are determined by genes that control both the amount and the distribution of a dark pigment called ***melanin***. Except for albinos, everyone has some eye pigmentation. The actual eye color is determined mainly by the location of melanin in the iris of the eye. There aren't different colored eye pigments.

Blue Eyes: No melanin in the front part of the iris. The color is due to minimal amounts of melanin in the rear of the iris with the clear front portion scattering light reflected off the melanin. This scattering is greatest in the blue spectrum giving the iris its blue color.

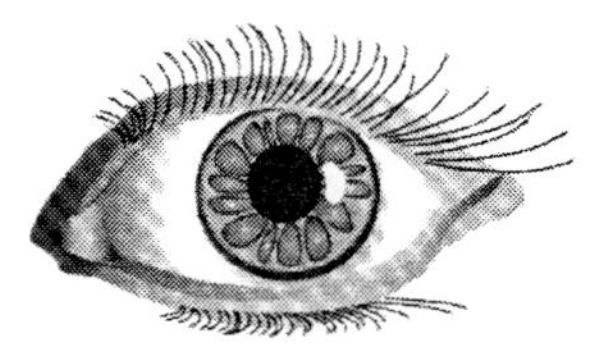

All eye colors are determined by where the brown pigments are located in the iris.

Grey Eyes: The same as blue, but with a slight amount of melanin in the front of the iris which tones down, or greys, the blue reflected from behind.

Green Eyes: A bit more melanin particles scattered in the front part of the iris creates a yellow appearance. Blended with the light blue from the rear of the iris, it produces an overall green color.

Hazel Eyes: Even more melanin particles in the front of the iris gives a slight brown color, and dilute melanin particles scattered throughout the iris add some yellow.

Brown Eyes: Melanin particles are in the front part of the iris and throughout the iris. The amount of melanin varies, leading to gradations of brown color in the eye.

Black Eyes: Large amounts of melanin in front and throughout the iris.

Test Cross to Check Genotype

Floppy Ears is a recessive trait.

If an organism shows the dominant phenotype, then one of its genes has to be the dominant allele, but you cannot be sure of the identity of the other allele unless you do a ***test cross*** to see if the dominant parent will breed pure. Imagine that you are in the rabbit breeding business. You know that straight ears on a rabbit is a dominant allele and floppy ears is recessive. You purchase a male straight eared rabbit. How do you figure out if your male rabbit is homozygous or heterozygous for straight ears?

Is this Straight Eared Rabbit homozygous or heterozygous?

1. Which genotype of female should you breed him to?

2. If a proper test cross is used, what phenotypes of rabbits would you see if your male rabbit is heterozygous dominant?

3. What ear phenotypes would you see if your male dog is homozygous dominant?

4. Complete the Punnett Square to show the test cross that would convince someone that your rabbit is homozygous for straight ears.

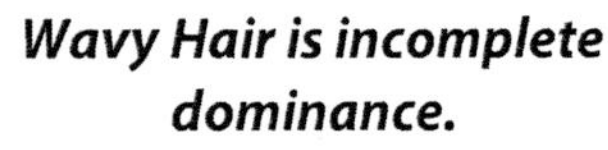

Wavy Hair is incomplete dominance.

Cases of Incomplete Dominance

When a straight-haired mouse is crossed with a curly-haired mouse, the result is always wavy hair. If two wavy-haired mice cross:

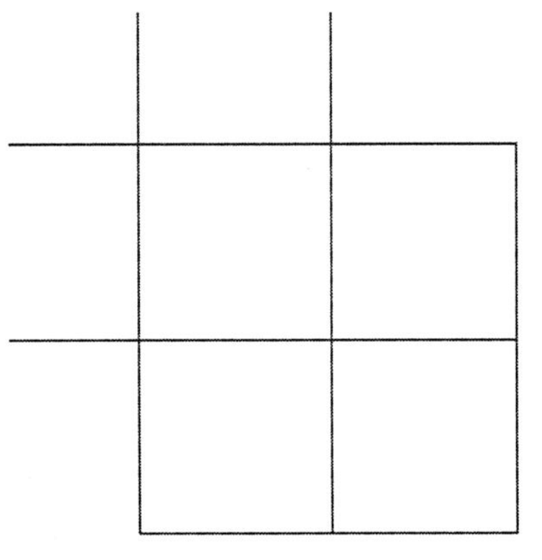

1. What are the genotypes of the two wavy-haired mice?

2. Draw the Punnett Square of a cross between two wavy-haired mice, and show the probable genotypes of their offspring.

3. What is the expected phenotype ratio of the offspring?

   ___% ___% ___%

4. What is the expected genotype ratio of the offspring?

Red orchids with straight petals are crossed with white orchids that have curly petals. The results are all pink orchids with wavy petals.

1. What are the genotypes of the two parent orchid plants? Remember: You are dealing with two different traits.

	(color)	(shape)
First parent:	_____ _____	_____ _____
Second parent:	_____ _____	_____ _____

2. What is the genotype of the offspring orchids?

Offspring:	(color)	(shape)
	_____ _____	_____ _____

Exercise #5
Sex-Linked Traits

The Y chromosome is missing these genes.

X Chromosome

Sex Determination

Humans have 23 homologous pairs of chromosomes. Twenty-two of these pairs are named using the numbers 1 through 22. The 23rd pair is individually labeled with the letters "X" and "Y". These labels distinguish them as the ***sex chromosomes***. They are not exactly the same. The Y chromosome has the "male making" genes, and the X does not. Also, the Y chromosome is missing some of the genes that are carried on the X chromosome. The female has two X chromosomes, and the male has one X and one Y chromosome.

During meiosis in the male two types of sperm are produced: those carrying the X and those carrying the Y chromosome. Females produce eggs carrying only the X chromosome. If a Y chromosome is present in the cells of an embryo, then the child becomes a male. If the Y is not present, the child becomes a female. It is the presence or absence of the Y chromosome that determines the sex of a child. This means that a male child receives a Y chromosome from his father and an X chromosome from his mother. A female child receives an X chromosome from her father and the other X chromosome from her mother.

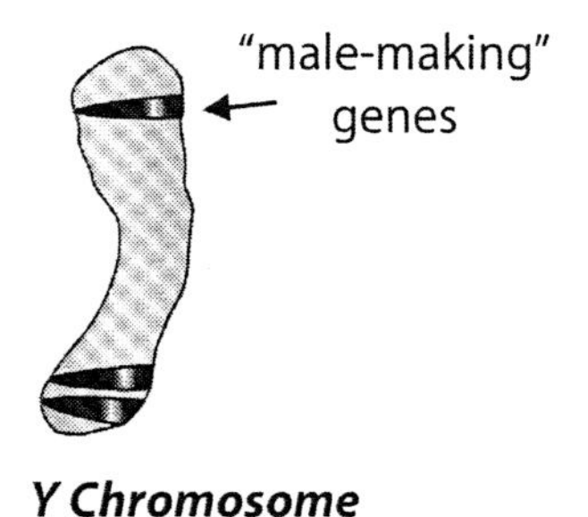

Y Chromosome

? Question

Draw a Punnett Square to show a cross of X and Y chromosomes in the fertilization of male and female gametes. The offspring boxes should reveal why we have about a 50% male to 50% female ratio within the human population.

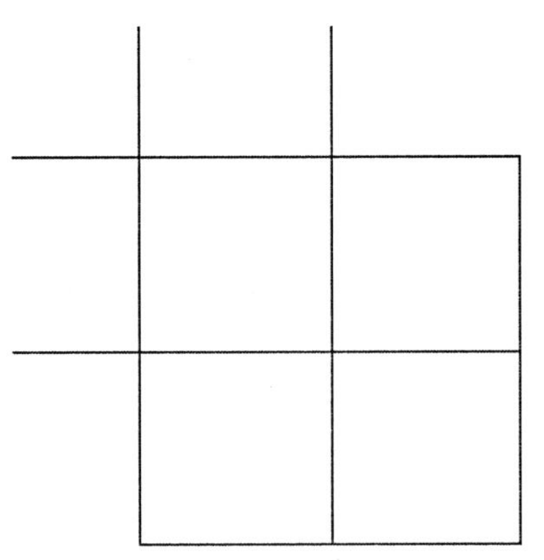

Sex-Linkage

"Sex-linked" means carried on the X chromosome.

The X and Y chromosomes are not exactly identical (Y is missing some genes that are on the X), and sometimes there are unequal frequencies of phenotypes in male and female offspring. If any phenotype is distributed unequally between male and female offspring and those differences are due to X and Y chromosome differences, then we call those traits ***sex-linked***. Actually, "sex-linked" means that the gene is carried on the X chromosome and not on the Y chromosome. It could be more accurate to call them ***X-linked***. It is easier to understand sex-linkage by looking at the sex chromosomes. See Figure 11.3.

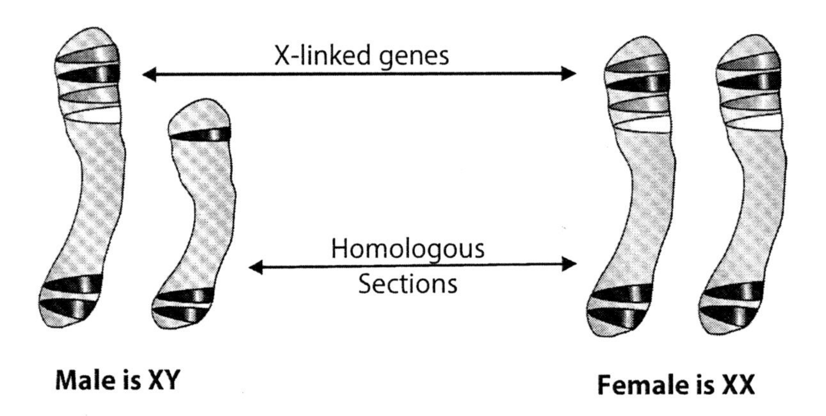

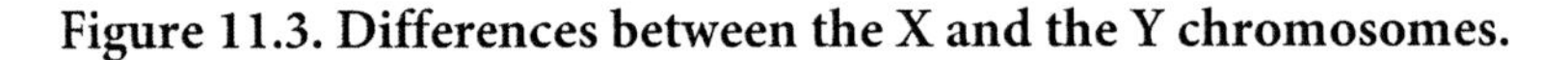

Figure 11.3. Differences between the X and the Y chromosomes.

There is a homologous section of the X and Y chromosomes that is the same, and these genes are inherited just like the genes on any other homologous chromosome pair. Notice that the Y chromosome is very short. It lacks some of the genes that are carried on the X chromosome. The X-linked section on the X chromosome carries genes that are missing from the Y chromosome.

? Question

1. How many copies of an X-linked gene does a male have?
2. Will a male be able to give his X-linked genes to his daughter? Explain.
3. Will a male be able to give his X-linked genes to his son? Why or Why not?
4. How many copies of an X-linked gene does a female have?
5. A male child gets X-linked genes from which of his parents?
6. A female child gets X-linked genes from which of her parents?
7. If a father is carrying an X-linked allele, then how many of his sons will get that allele?
8. How many of his daughters will get that allele?
9. If a mother has a defective X-linked allele on one of her chromosomes and the other chromosome is normal, then how many of her sons will get that defective allele?
10. Will any of her daughters get the defective allele? How many?
11. If we found that none of the daughters actually showed the defective phenotype, how could we explain it?

Tips for Solving Sex-Linked Genetic Problems

Follow the X, and follow the Y.

There is a sex-linked gene on the X chromosome that causes a disorder called ***hemophilia***, where the blood fails to clot properly when a person is injured. This disorder is recessive and can be symbolized by the small letter "n." Normal blood clotting is dominant and can be symbolized by the capital letter "N." In solving sex-linked cases we not only use letters to symbolize the genes, but we also include the X and Y chromosomes and follow the sex chromosomes into the next generation.

- Using these symbols we can indicate a female who is heterozygous for clotting as $X^N X^n$.
- A homozygous female for normal clotting would be $X^N X^N$.
- A hemophilic male would be X^n Y.

We would diagram the Punnett Square showing the cross between a heterozygous female and a normal clotting male like this:

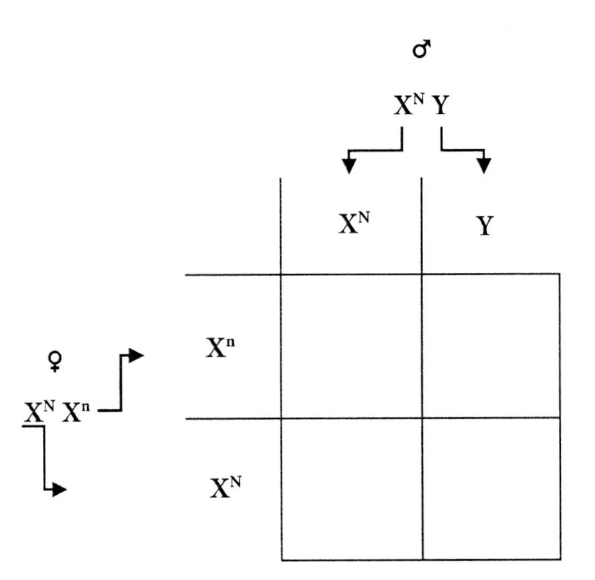

? Question

1. Complete the Punnett Square above showing the offspring.
2. What is the genotype for the female parent?
3. What is the genotype for the male parent?
4. What are the genotypes for their offspring?

5. What are the chances that any child will be a hemophiliac?

6. Is it the father or the mother that passes the hemophilia gene to the male child?

7. Failure to distinguish between red and green colors is caused by a recessive allele and is a sex-linked gene carried on the X chromosome. A red-green color-blind male marries a normal female. Of their six children (four boys and two girls), all have normal vision.

   What is the most probable genotype of the mother?

8. Will any of their four male children pass this disorder on? Explain.

9. Draw a Punnett Square of this cross to prove your answers.

10. A normal-vision female gave birth to a color-blind daughter. Her husband has normal vision. He claims that the child is not his. Does the genetic information suggest someone else is the child's father? Explain and prove your answer using a Punnett Square.

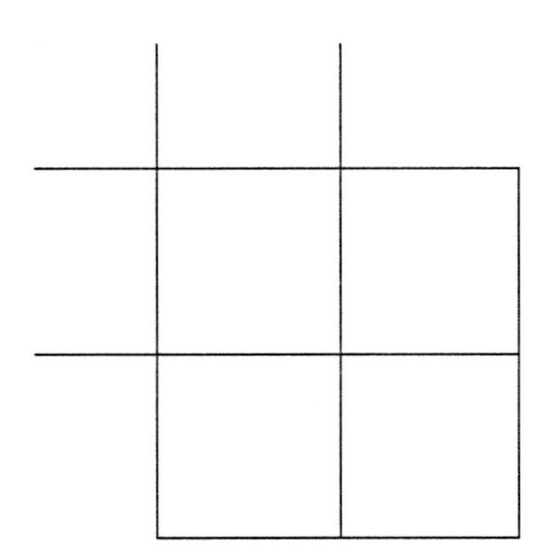

LAB

12

Biotechnology

Biotechnology uses organisms and their biological processes to create solutions for medical, agricultural, and commercial problems, as well as many other social and environmental applications. This research area is advancing rapidly, and its applications are expanding contemporary imagination. The advances can be compared to the invention of the computer and the blossoming information age. Biotechnology is fueled by discoveries of the biochemical processes that are controlled by proteins (enzymes), RNA, and DNA. It seems as if we have discovered one big secret of life—all that it makes, and all that it could make.

Has biotechnology uncorked the bottle and let out the Genie? For what do we wish? Which wishes will improve the world or human condition? How will we avoid making a foolhardy wish? This lab introduces a small part of the biotechnology world, focusing on DNA technologies and how they are valuable to us.

Exercise #1 DNA Fingerprinting Tools
Exercise #2 DNA Fingerprinting Simulation
Exercise #3 DNA Isolation from Human Cheek Cells
Exercise #4 Recombinant DNA

Exercise #1
DNA Fingerprinting Tools

DNA fingerprinting (more accurately called DNA profiling) is a procedure used to identify specific characteristics in DNA molecules. This procedure allows us to distinguish the DNA of people or any other organism. DNA fingerprinting has been applied to criminal investigations, paternity cases, genetic-relationship questions, identification of inherited disorders, and the personal DNA identification of individual humans.

The classic use of fingerprints taken from the fingers is for identification purposes only, and it lacks the ability to show genetic relationship between people. Identical twins have different hand fingerprints because those prints are partly determined by non-genetic conditions during embryonic development. The use of classic fingerprints focuses on whorls and intersections of ridges on the fingers (which are not inherited), whereas DNA fingerprinting identifies actual sequences of nucleotides in the DNA (which is inherited). DNA fingerprinting uses many biotechnology tools, and a discussion of several of those tools will give you a basic understanding of the process.

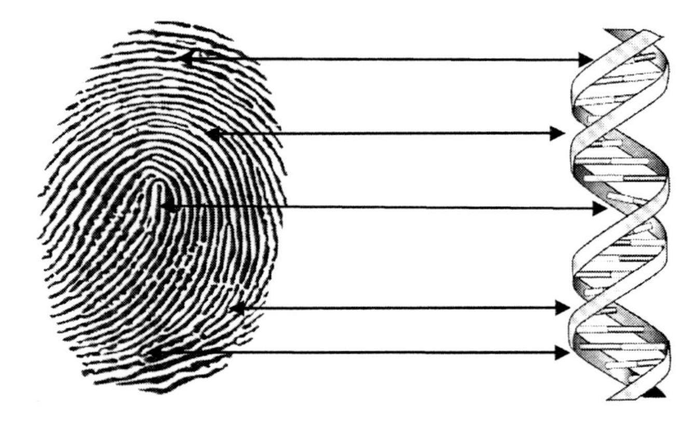

Figure 12.1. Comparison of Classic "Fingerprint markers" vs "DNA Fingerprint markers".

PCR—How to Clone DNA

One molecule of DNA can be replicated into as many copies as you want.

PCR (***polymerase chain reaction***) is the method used to make millions of copies of DNA from a small sample. It is sometimes called "molecular photocopying". The DNA is "unzipped" by heating the sample in a mixture of DNA nucleotides and a special enzyme called ***Taq DNA polymerase***. A small amount of "***primer***" (short piece of RNA that starts the chain reaction) is also added to the mixture. The DNA molecule copies itself under these conditions (A pairs with T; G pairs with C).

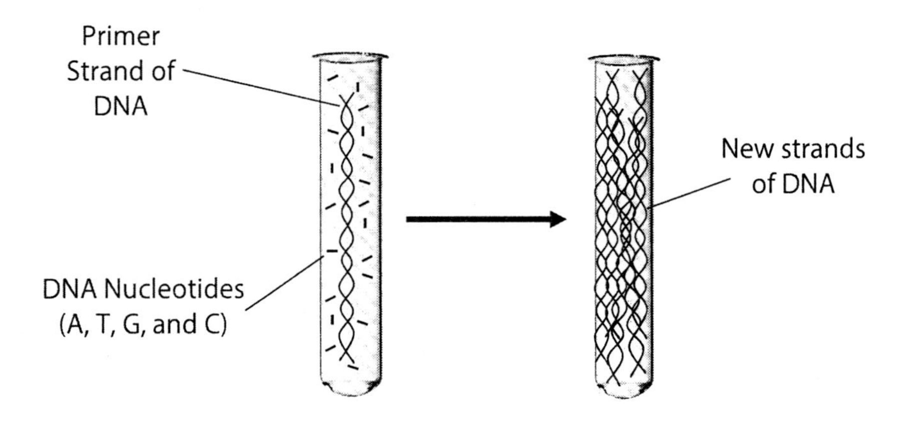

Figure 12.2. PCR (polymerase chain reaction).

The special enzyme used in PCR (Taq DNA polymerase) was discovered and extracted from thermal-pool bacteria like those in Yellowstone National Park. The bacteria's name is *Thermus aquaticus*, hence the name Taq polymerase. This enzyme is capable of functioning at the high temperatures during the laboratory PCR process. Most enzymes are destroyed by high heat, and use of this enzyme is another advancement in biotechnology.

There are a couple of tricks to this procedure. The starting sample is heated to about 95°C which causes the DNA molecule to unzip into two complementary halves. When the DNA is cooled to about 50°C, each half begins making a new complementary copy of itself from the mixture of nucleotides. A second heating of the PCR chamber separates the two new DNA molecules, and another replication occurs when the sample is cooled for the second time. An automated machine can repeat these temperature change cycles (30–40 times) to produce millions of exact copies of the original DNA.

Other versions of laboratory copying of DNA start with specifically designed primers that attach to a particular section of DNA and make multiple copies of that section. This approach is used in forensic investigations. The PCR technique creates plenty of sample for the DNA technician to do the many specific DNA tests.

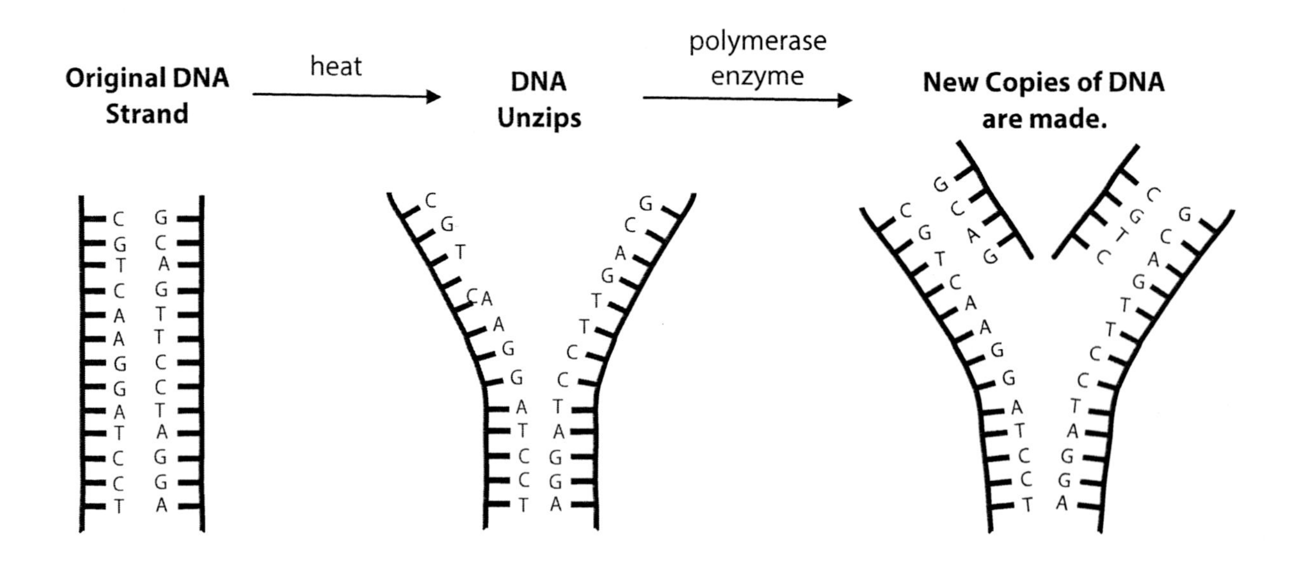

Figure 12.3. Details of PCR. Now, two DNA molecules exist where there was only one before.

? Question

1. What is a simple definition of biotechnology?

2. In which example is the "fingerprint" identical for identical twins?
   Classic hand fingerprints or DNA fingerprints

3. What does PCR produce?

4. What are the ingredients for the PCR process?

5. Where did biotechnology find the special DNA polymerase enzyme used in the PCR process?

6. What is a PCR primer?

Restriction Enzymes

When long molecules of DNA are mixed with ***restriction enzymes***, the DNA is cut into shorter threads. Restriction enzymes are a natural defense that bacteria use to destroy ("restrict") viruses that invade them. At first biotechnology used the natural enzymes that were discovered. But soon synthetic versions were being created in the lab. Lab-made versions have two important parts – a binding section and a cutting section. Several thousand have been studied and many are available commercially. They are designed to cut DNA at very specific nucleotide sequences.

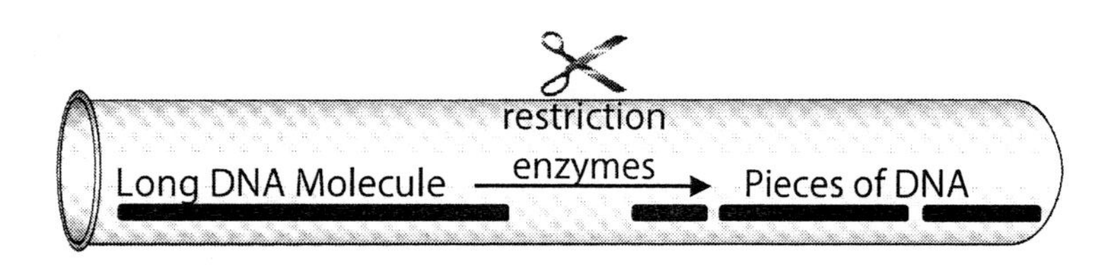

Figure 12.4. Restriction enzymes cut long pieces of DNA into smaller pieces. The DNA is cut wherever there is a "recognition site" which is determined by your genetics.

Restriction enzymes act like "scissors" cutting the DNA at specific nucleotide sequences called ***recognition sites***. Recognition sites are determined by your particular genetics. Let's see if you get the idea of how this biotechnology tool is used.

Research is what I am doing when I don't know what I'm doing.

-Wernher von Braun
1912-1977
German rocket scientist

Procedure

- Start with three examples of restriction enzymes (BamHI, EcoRI, and HindIII). Each enzyme cuts at its own recognition site. A recognition site could be as simple as CCCGGG where the restriction enzyme cuts the DNA between the G and C. There are hundreds of restriction enzymes that have been developed by biotechnology, and each has its own technical name. We have used a simple shape to represent the recognition site for each of the three cutting enzymes below. The particular restriction enzyme cuts only at its recognition site in the long thread of DNA.

Table 12.1. Examples of Restriction Enzymes and Recognition Sites Used in Biotechnology.

Restriction Enzymes (These cut DNA at a very particular spot.)	**Recognition Sites** (This is the very particular spot in the DNA where the restriction enzyme cuts.)
BamHI ✂	▲
EcoRI ✂	●
HindIII ✂	■

- Your job is to analyze the DNA of three subjects. Their DNA was mixed with the three restriction enzymes ("scissors"). The long threads of DNA from each subject are cut at the recognition sites. Determine the length of each resulting piece of DNA from the subjects after cutting has occurred.

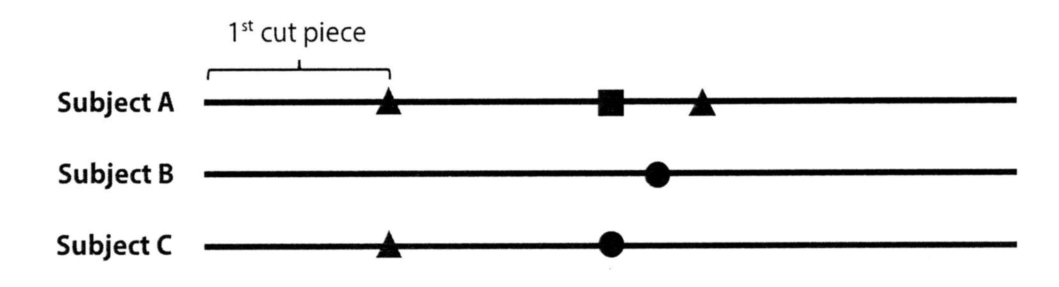

- Measure the length of each piece with a cm ruler, and make a thick line mark on the graph to indicate each of the pieces. We have done this for the first piece in Subject A. Finish measuring and making marks for all the DNA cut pieces of Subjects A, B, and C. There will be nine pieces.

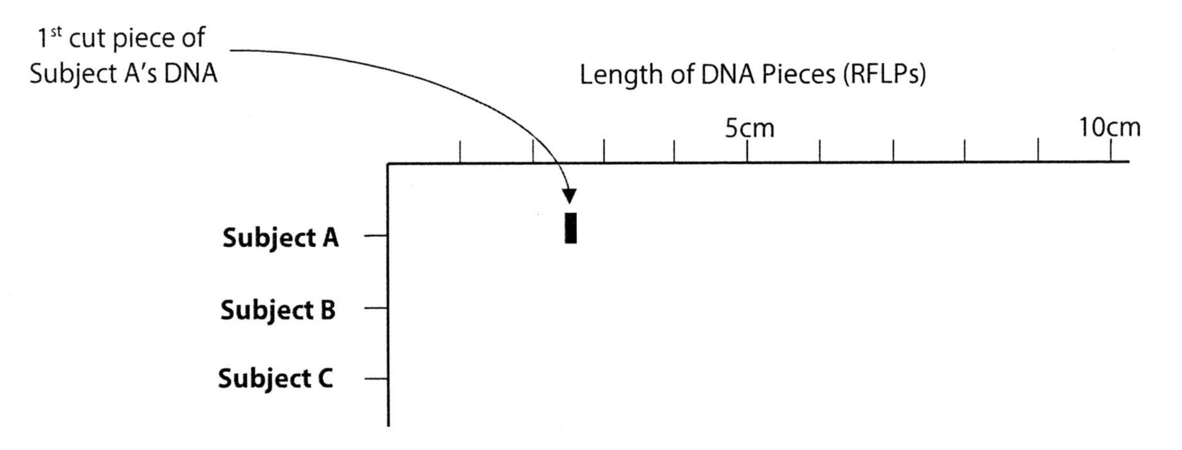

RFLP's and STR's

There are many techniques for analyzing DNA. They range from a complete decoding to an analysis of sections of the DNA. The "complete nucleotide sequence" approach is slow, expensive, and yields way more information than we can currently understand. The "targeting" methods are more manageable and focused in application. You have just used a focused approach in comparing the DNA of Subjects A, B, and C. The scientific name for these individual pieces of DNA cut up by the restriction enzymes is ***RFLPs*** (Restriction Fragment Length Polymorphism). RFLPs are separated from each other during a process called ***gel electrophoresis***, which is the biotechnology tool described next. The length and number of RFLPs is dependent on the restriction enzymes used and the unique DNA code in each person.

The forensic use of DNA fingerprinting has particular criteria for selecting restriction enzymes to analyze DNA crime scene samples. Researchers discovered that everyone has repeated sequences of nucleotides in their DNA called Short Tandem Repeats (STRs). These DNA segments are not part of our actual functioning genes but are unique to each of us. For example, one person may have a STR of ACACACACAC, and another person may have a STR of ACACACACACACACACACAC at the same place in their DNA. Crime investigation uses 13 STRs to compare DNA, and the international directory for analyzing the DNA samples is called CODIS (Combined DNA Index System).

Gel Electrophoresis

Gel electrophoresis is a precise method for separating pieces of DNA (called RFLPs or STRs) that have been cut by restriction enzymes. It is the most familiar part of the DNA fingerprinting process because it produces a visual gel record similar to the graph in the previous discussion.

- PCR amplifies the original DNA sample.
- Restriction enzymes cut the DNA in smaller pieces.
- DNA pieces are separated from each other by gel electrophoresis.

The basic procedure for gel electrophoresis is next. You will use the detailed directions in Exercise #2 for actually doing the experiment.

1. **Prepare an agarose gel sheet in a casting tray.** Agarose is made from seaweed. This gel-like sheet provides the medium that the RFLPs move through. The structure of this gel allows different size pieces of DNA to separate from each other. Think of the gel as a

matrix with many small particles suspended in it. The particles form an "obstacle- course" with uniform space between the particles. Bigger pieces of DNA move slower through the gel than smaller pieces do.

2. **Cover the gel sheet with a buffer solution.** This stabilizes the pH and serves as a conductor of electricity between the gel and the electrodes during the electrophoresis.
3. **Place the samples to be compared into small depressions "wells" at one end of the gel sheet.**
4. **Apply an electrical charge to each end of the gel.** DNA pieces (which have a negative charge) begin to move away from the negative electrode and towards the positive electrode.
5. **Wait for the "DNA fingerprint" to develop**. This might take 30 minutes or so.

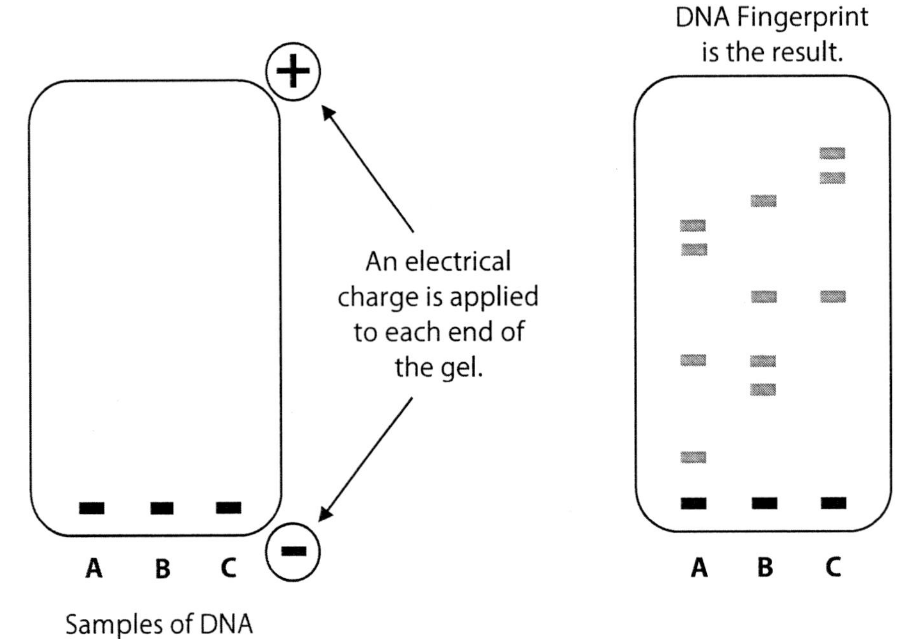

Figure 12.5. Gel Electrophoresis. The basic technique can be modified to improve the separation of DNA fragments. RFLPs of identical size but different genetic coding will separate into multiple bands in the gel.

? Question

1. What are restriction enzymes?

2. Where did researchers first discover restriction enzymes, and what does the word "restrict" refer to?

3. What are recognition sites?

4. What are RFLPs and STRs?

5. Why not decode the entire DNA of the person or organism being investigated?

6. How many STRs are used for comparison in forensic investigations?

7. What is the purpose of the gel?

8. What is the purpose of the buffer?

9. Where do you put the DNA samples?

10. When the electrical charge is applied to the gel, what happens?

11. Why are there different "bands" when the process is finished?

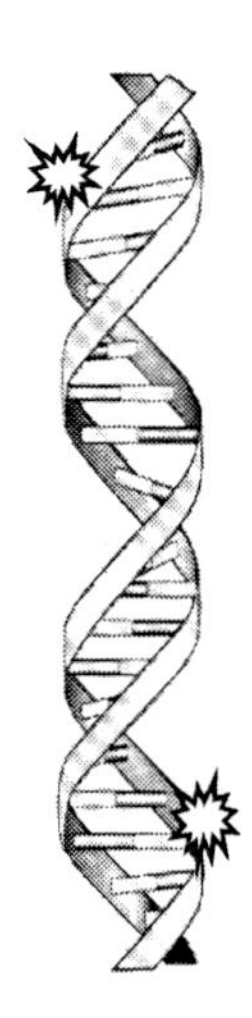

Probes target certain genetic markers in your DNA, if you have those markers.

Radioactive and Fluorescent Probes

Radioactive or fluorescent ***probes*** are special molecules designed to combine only with a particular sequence of nucleotides in the DNA sample. The "targeted" sequences are called ***genetic markers***. When the marker has combined with a specific probe, that DNA is either radioactive or fluoresces when a special UV light is shined on it. Researchers produce targeting probes to combine with a particular gene or with a specific gene variation responsible for disease. There are many useful applications of this biotechnology tool.

The basic procedure for using DNA probes is:

- Design an appropriate probe to target the genetic marker.
- Add the probe to the DNA sample you want to test, and incubate at the correct temperature for the probe.
- Wash the sample. This removes the probe if it has not combined with a targeted gene in the DNA.
- Test the DNA sample to see if it is radioactive or fluoresces. This will indicate if the probe is attached to the target in the DNA.

? Question

1. What is a genetic marker?

2. Refer to your lecture textbook (or the Internet) to find one clinical use for a DNA probe.

3. Refer to your lecture textbook (or the Internet) to find one forensic use for a DNA probe.

Exercise #2
DNA Fingerprinting Simulation

This exercise will simulate DNA fingerprinting used to identify the perpetrator in a murder investigation. There are three suspects: Suspect #1 (S_1), Suspect #2 (S_2), and Suspect #3 (S_3), and there is a DNA sample from the crime scene (CS). Your task is to discover whose DNA matches the forensic evidence collected at the crime scene. You will work in two groups at each table. Each group will perform all the steps.

Examine the DNA fingerprinting apparatus, and notice that there is a main chamber and a small gel tray. The gel tray is notched so that it fits into the main chamber in only one way. The "comb-like" piece hangs vertically, resting in notches of the gel tray. This comb may have two sides — one side with more "teeth" than the other side. (More about that difference later.) The main chamber has a lid with two wires (red and black) projecting from the top. The wires will be plugged into the power supply.

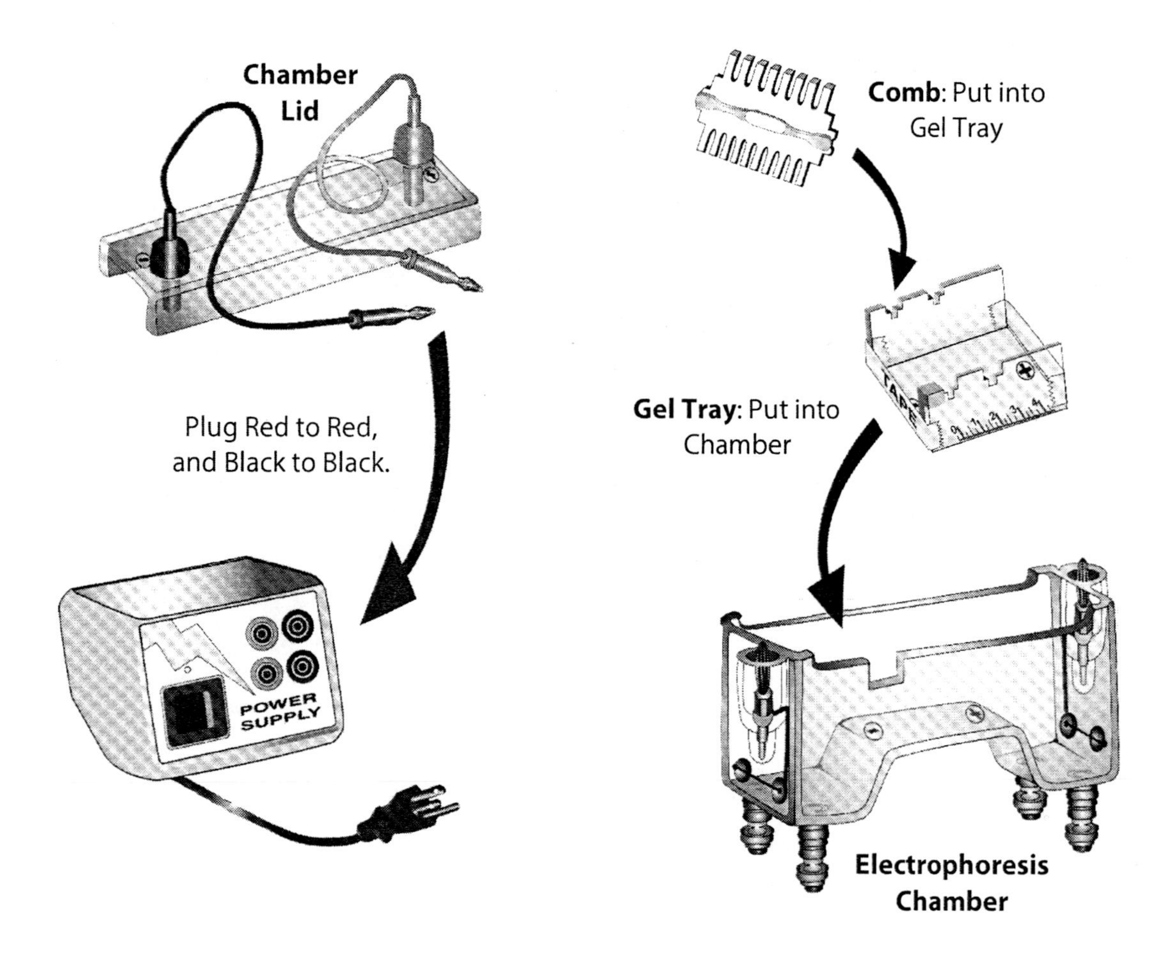

Figure 12.6. DNA Fingerprinting Apparatus.

Making the Gel

Be sure that you understand the basics of the DNA fingerprinting apparatus before you start making the gel. Ask your instructor if you have any questions.

Materials

- ✓ Masking tape
- ✓ 125-ml flask (for mixing agar)
- ✓ Weighing paper or weigh boat
- ✓ TBE buffer
- ✓ Graduated cylinder (for measuring fluids)
- ✓ Paper toweling (for top of agar flask)
- ✓ Weighing scale
- ✓ Microwave oven

Procedure

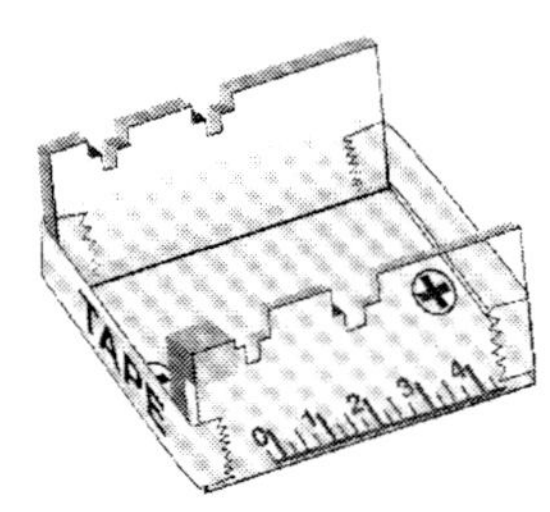

Carefully tape the ends of the Gel Tray.

- Close off the open ends of the gel tray with masking tape or with rubber dams, whichever is provided. You must do a good job of sealing the ends or the gel pour will leak out. Check with your instructor to be sure you are doing it correctly.
- Next, insert the gel comb into the notches at the end of the tray. Insert the comb so that the side with 10 teeth faces downward into the tray. The comb will make 10 depressions (wells) in the gel. You need only four wells for your samples, but you also need a space between each sample and a space along the side of the gel. This assures that each sample won't mix with adjacent samples during the electrophoresis. (The dyes used in this simulation spread more than actual DNA pieces during electrophoresis.)

Comb fits into the first notch in the gel tray.

Gel Comb
"10-teeth"
side down

- One group of students will work with your instructor to prepare agar solutions for everyone in the lab. Now, prepare the gel solution. First, determine how many groups are in the class. You need that many 125-ml flasks. Carefully measure 40 ml of TBE buffer and pour it into each flask.
- Predetermine the weight of a piece of weighing paper or weigh the boat and add 0.35 g of agar. This amount of agar is for each flask. Add the agar to each flask and swirl it until it dissolves. Give an agar flask to each of the student groups.
- Each lab group will heat their own agar gel solution. Put the prepared paper toweling over the top of your flask; this prevents boil-over. It also protects your fingers when you remove the hot flask. Set the microwave oven for 15 seconds, place the flask inside, and heat. The solution is not ready until it is clear. You will need to repeat 2 or 3 of these 15-second heatings. Stop the oven if the agar starts to boil. Grab the flask with the paper toweling and move it to a safe place. Remove carefully. The flask is hot!

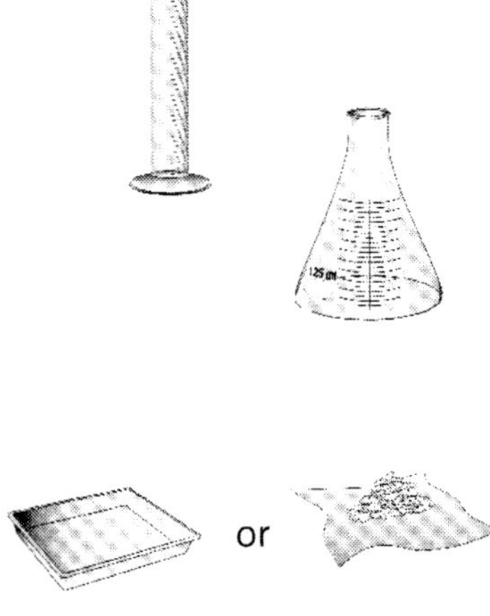

Pouring the Gel

- When the flask has cooled to 55 °C, it is ready to pour. A quick way to check for the proper temperature is to carefully touch the bottom of the flask to the back side of your hand. If it is painful, it is still too hot to pour. If it is hot, but not painful, it is ready to pour.
- Carefully pour the cooled agar into the small gel tray (with the comb in place). You only fill the small gel tray — not the main chamber. Ask your instructor if you get confused. It will take about 15 –20 minutes for the gel solution to harden.

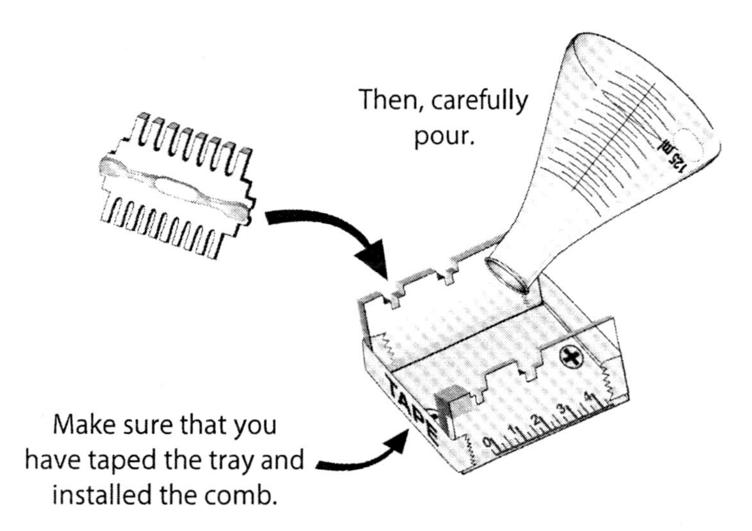

- After the gel has hardened, fill the electrophoresis chamber with 300 ml of TBE buffer. Use the graduated cylinder to measure the proper amount.
- Carefully remove the tape from the ends of the gel tray, and insert the tray into the main chamber so that the "well end" of the tray is oriented toward the negative (black) end of the main chamber (match the notches of the tray and chamber).
- The gel should be completely submerged. Next, carefully remove the comb. You are now ready to inject the samples.

Injecting DNA Samples into the Wells

Materials

- ✓ Micropipette
- ✓ Micropipette tip
- ✓ Samples from suspects and crime scene
- ✓ Cup of water (for rinsing micropipette tip)

Procedure

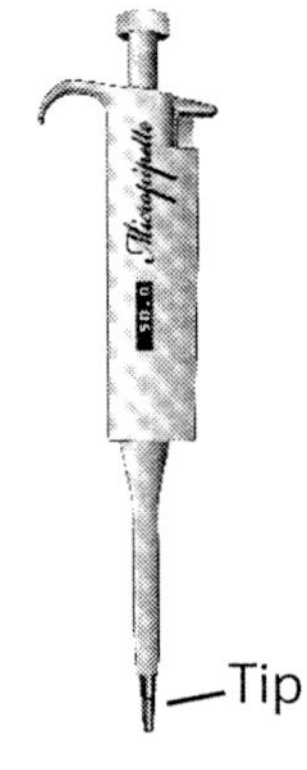

Micropipette

- Read all of these procedures before starting the sample injections.
- You are going to inject a small sample from each of the three suspects and one from the crime scene. These samples are to be injected into the wells of the gel tray. This is tricky, and you will need some practice with the micropipette until you feel confident of your coordination and accuracy. Use some of the rinse water to practice your hand coordination. Next, you will use actual samples to fill the wells in the gel tray.
- The micropipette holds 20 μ (micro-liters). Inject this amount of sample into its own particular well (CS in one well, S_1 in another well, etc.). Dispense each sample into its well about half-way down. Don't pierce the bottom of the agar gel. You must clean the tip between each sample so that you don't contaminate the next well. Use the cup of practice water to rinse the tip.
- There are four samples: Crime Scene (CS); Suspect 1 (S_1); Suspect 2 (S_2); and Suspect 3 (S_3). Leave one empty well between each sample. Also, leave an empty well on each side of the gel tray. Empty wells keep the dyes (DNA fragments) separate from each other. Mark the order of the sample on your Results Chart.

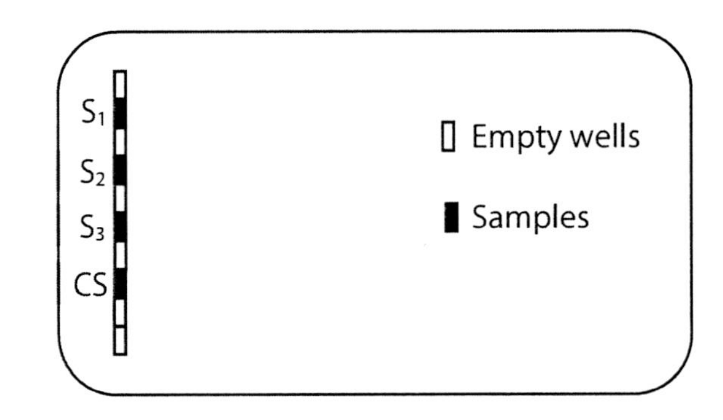

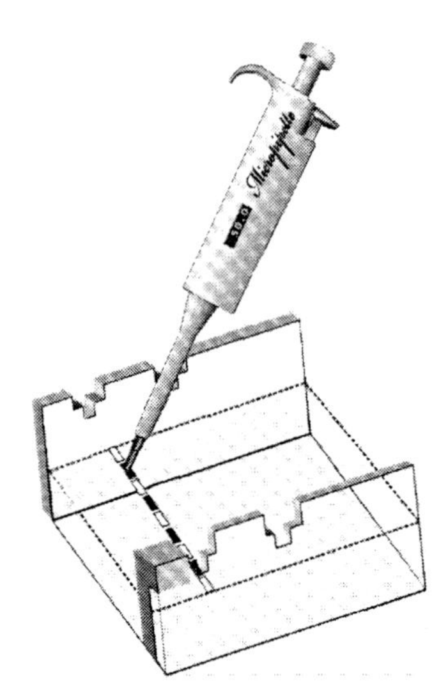

The secret is slow and steady.

- Now, it's time to inject the samples. Steady your pipetting hand with your other hand, and don't poke through the gel sheet. If any sample begins to "float out" of a well, use a disposable pipette to remove the spreading sample from the buffer. If you don't, there could be some color contamination of your results. Ask for help from your instructor.

Start the Electrophoresis Process

Procedure

- Your instructor will demonstrate the way the DNA fingerprinting apparatus is connected to the power supply and turned on. It will take 20+ minutes at 100V for the fingerprinting to be completed.
- When the samples have separated into bands, you can turn off the power. Then remove the gel tray and slide the gel from the tray into a weigh boat filled with water. This will make it easier to see the separations. The gel will remain stable in the water until you can show it to your instructor.
- Show your results to your instructor. Use colored pencils to record your results. Be sure each sample is labeled.

If the DNA fits, you must not acquit.

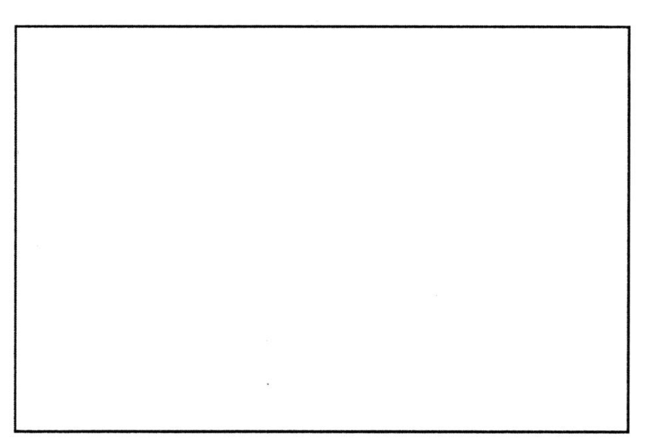

Which suspect matched the crime scene evidence?

- Put the gel into the trash can (not in the sink). Rinse the gel tray, comb, and 125-ml flask. Do not throw out plastic pipettes, tips, or buffer. Leave the buffer in the chamber for the next class, but do remove any floating bits of gel from the chamber.

Exercise #3
DNA Isolation from Human Cheek Cells

Everywhere you go, you leave a piece of yourself behind – skin, hair, etc.

DNA occurs inside the nucleus of every cell of your body (except red blood cells). If the DNA in a single cell nucleus were uncoiled, it would stretch about 2– 3 meters in length. You can actually see this "stringy-looking" molecule if you collect a small sample of cells and isolate the DNA. There is a simple technique for isolating DNA from some of your own cheek cells. The basics of this procedure are:

- **Collect cheek cells.** Rinse your mouth, then add more cells by gently scraping the inside of your cheek with a toothpick.
- **Break open the cheek cells.** Mix cheek cells with a solution of SDS (sodium dodecyl sulfate). It is a detergent that dissolves the fatty cell membrane. This releases DNA from the cells.
- **Add some salt.** Salt makes the DNA molecules less soluble in water by neutralizing some of their electrical charges.
- **Pour a little alcohol over the top of your sample.** DNA is not soluble in alcohol. If alcohol is carefully poured on top of a DNA solution, then a special "interface surface" forms between the alcohol and water. Chemists use tricks like this to start a precipitation process in a solution. The DNA will begin to form "strings" along the interface surface between the alcohol and the water.

Collecting Cheek Cells

Materials

✓ 1 disposable drinking cup
✓ 5 ml of drinking water
✓ 1 toothpick

Safety Precautions: Be sure to use a clean drinking cup, drinking water, and toothpick.

Procedure

- Put 5 ml of clean water into a disposable drinking cup.
- Swish the water vigorously in your mouth for 1 minute, and spit it back into the drinking cup. This will collect many cheek cells.

Researchers have extracted and analyzed small pieces of DNA removed from a 30,000 year old Neanderthal bone.

- Next, use the "blunt end" of a toothpick to gently scrape the inside of your cheeks several times. Don't dig in, but rub firmly to pick up cells. Dab the toothpick in the drinking cup to release more cells. Repeat the scraping and dabbing four times. You now have enough cheek cells to start the next procedure.

Isolation of DNA

Materials

- ✓ 1 regular-size plastic test tube
- ✓ 1 disposable pipette (not the micropipette)
- ✓ Saturated salt solution (NaCl)
- ✓ 10% SDS detergent
- ✓ 1 piece of parafilm
- ✓ Ice cold 95% isopropyl alcohol (get this just before you need it)
- ✓ Test tube rack
- ✓ 1 plastic microfuge tube

Safety Precautions: Alcohol is flammable. Keep it away from flame.

Procedure

- Look for the 1-ml graduation mark on the plastic pipette. Put 3 ml of your cheek cell suspension into a disposable plastic test tube.
- Add 1 ml of the 10% SDS detergent solution to the test tube.
- Add 1 ml of the saturated NaCl solution to the test tube.
- Stretch a small piece of plastic parafilm over the top of the test tube, and gently turn the tube upside down five times. Avoid making soap bubbles. Wait about 5 minutes and repeat the gentle inversion five more times.
- Get the bottle of ice cold isopropyl alcohol. (The alcohol is either in the freezer or in an ice bath in the room.) The next bit is a little tricky. Hold your test tube sample at a 45° angle while you carefully trickle 5 ml of ice cold alcohol down the inner side of the test tube. This slowly covers the surface of your sample with alcohol.
- Carefully place the sample into the test tube rack, and wait 10 minutes. Gently tap the side of the test tube several times during the waiting time. You should be able to see some white strands start to form between the alcohol and cheek cell sample. This is DNA (and some RNA also). Show your instructor.

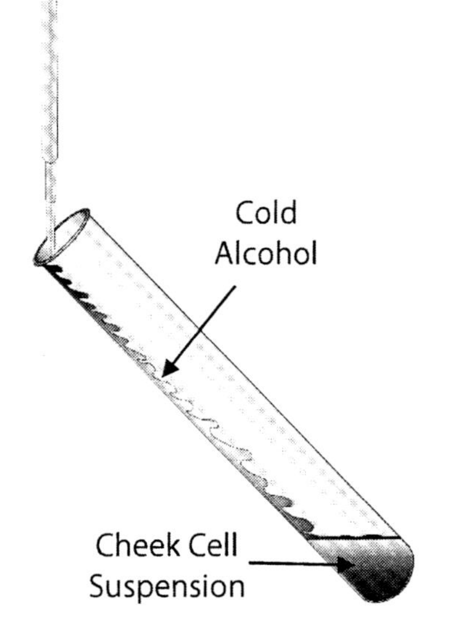

- To preserve the sample, use a rinsed plastic pipette to gently pull up the DNA strands from the test tube. Transfer them to the microfuge tube. You should be able to see the floating strands of DNA. The tube is yours to keep.

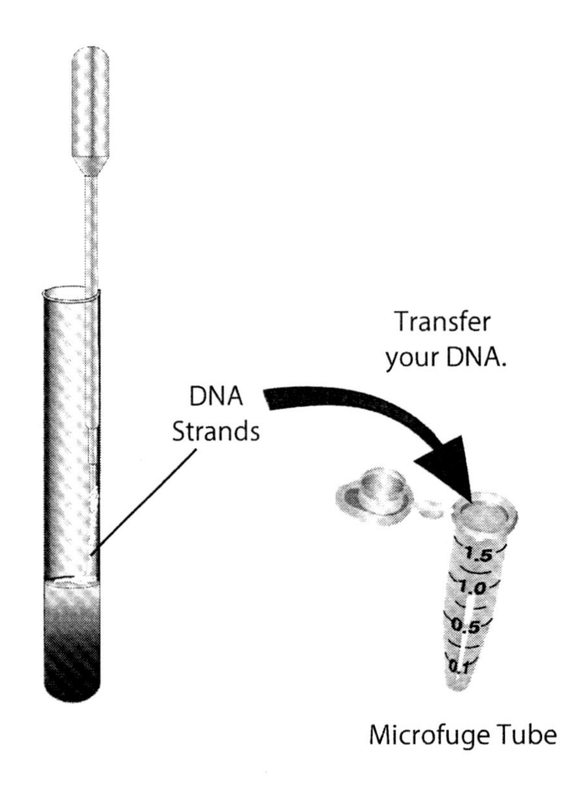

- Throw away the drinking cup, toothpick, plastic test tube, and plastic pipette. The liquids are safe to wash down the sink.
- There you have it —human DNA!

? Question

1. What is the purpose of adding the 10% SDS detergent solution to your cheek cell sample?

2. What is the purpose of pouring a small amount of alcohol on top of your cheek cell sample?

Exercise #4
Recombinant DNA

Theoretically, almost anything is possible with recombinant DNA.

Recombinant DNA, often called genetic engineering, is the process of combining a gene from one species with the DNA of another species. One example of this technology involves splicing a human insulin gene into a bacterial chromosome. The resulting "genetically engineered" bacteria are able to synthesize human insulin for diabetics. Animal genes can be inserted into plant species, and plant genes can be inserted into animal species. Theoretically, almost anything is possible. A brief description of the procedures will give you a simple understanding of the process.

- The first step is to select a gene to transplant into another species, and there are many considerations in this step.
  - What is the end purpose?
  - Which species have this gene?
  - Which species could receive this gene and accomplish the end purpose?
- The selected gene must be cut out of the donor organism's DNA. This is a very complicated and expensive lab process. Restriction enzymes must be designed by the biochemists to do the exact cutting needed.
- Many copies of the donor gene are made through the PCR process.
- The recipient DNA is cut and mixed with copies of the donor gene. Special "attaching enzymes" (example, DNA ligase) attach the free ends of the donor gene to the free ends of the recipient DNA. Eggs can be the target of recombinant DNA, or the genes can be inserted into the first cell of embryonic development.
- The last step is to wait and see if the DNA recombination has been successful.

What we call Man's power over Nature turns out to be a power exercised by some men over other men with Nature as its instrument.

- C.S. Lewis

Your instructor will demonstrate the results of one recombinant project at the college. The gene for "glowing in the dark" from a jellyfish (or firefly) was recombined with the DNA of bacteria. This gene has the code for making a protein called "green fluorescent protein," which fluoresces when UV light is shined on it. The bacteria were grown on an agar medium, and some of them should demonstrate the new "glowing" gene from the jellyfish.

Search on the Internet for "GFP Bunny".

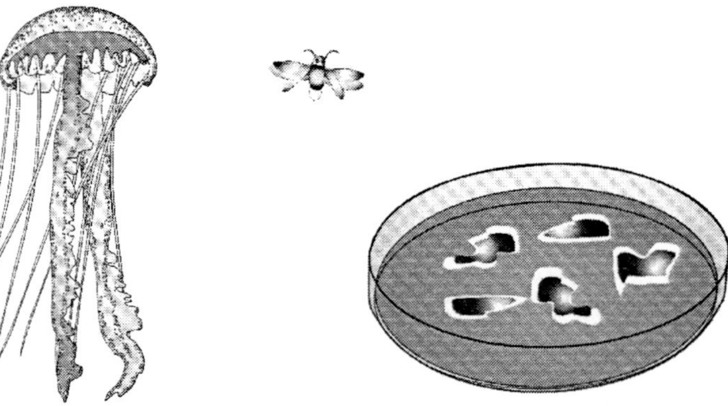

Figure 12.7. Another example of genetic engineering. The gene for GFP (obtained from either jellyfish or firefly) is recombined with the DNA of bacteria.

The 20th Century was about plant and animal hybridization.

The 21th Century will be about recombinant DNA.

Go see the demonstration of the "glowing gene" now. **Safety Precautions:** *Always wear UV safety glasses when viewing samples under UV light.* If your lab class meets during the daylight, take one of the plates outside into the sunshine and see what happens. The "glowing gene" from a firefly is not as fluorescent as the gene from a jellyfish, so if your college used a firefly gene for recombination, you may not see any difference in sunlight. This gene has been inserted into other organisms, too. Search on the Internet for "GFP Bunny".

? Question

1. Describe your observations. What happened when you took the bacteria into sunlight?

2. What is the name of the gene involved in this recombination project?

3. Which species was the "donor" for this gene?

4. Which species was the "genetically engineered" recipient of the gene?

Recombinant DNA holds an incredible future for us. Agricultural applications include plants with insect resistance, fungal and bacterial resistance, increased protein content, more sweetness, and many other traits that did not previously exist in those plant species. There are many hopeful possibilities for plant "re-engineering", but some people fear that we might produce ecological damage by creating new uncontrollable species. Those fears have led to strict rules for the development of genetically engineered plants.

Genetic engineering projects have also produced drugs for the treatment of diabetes, growth disorders, hemophilia, leukemia, some cancers, ulcers, anemia, heart attacks, emphysema, and other human ailments. The Human Genome Project is well underway to identify the location of every human gene on each chromosome. This project will discover the exact sequence of nucleotides in all human genes. And the genomes of other important non-human research species will be compared to the human genome. Moral and ethical considerations are being considered as science advances this project. There are many questions for Biotechnology, all leading to many unknown and debatable answers. What is done and how it is done are challenges for the next generation.

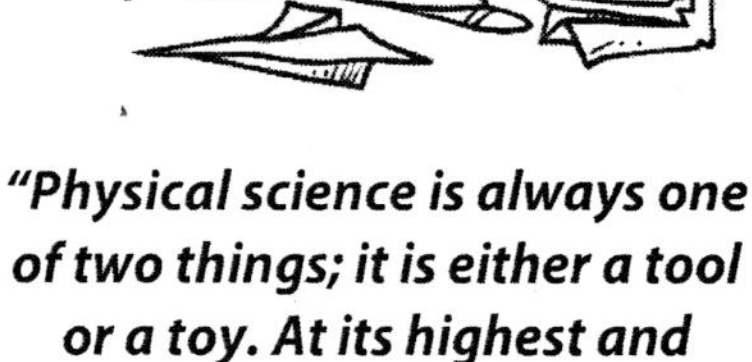

"Physical science is always one of two things; it is either a tool or a toy. At its highest and noblest, of course, it is a toy."

-G.K Chesterton
Greatest Journalist of 20th Century

"I do not know what I may appear to the world; but to myself I seem to have been only a boy playing on the sea-shore, and diverting myself in now and then finding a smoother pebble or a prettier shell than ordinary, whilst the great ocean of truth lay all undiscovered before me."

-Sir Isaac Newton (1642-1727)
English mathematician and philosopher

LAB

13

Microbes

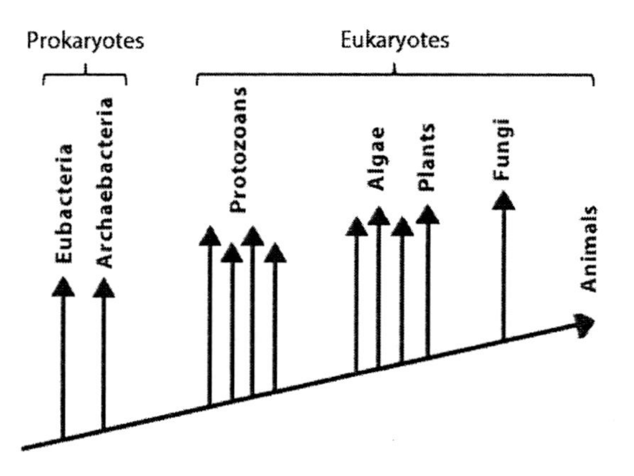

One-celled organisms probably developed on Earth sometime before 3.5 billion years ago. The vast majority of these creatures did not fossilize, so we have a very unclear picture of what happened during early evolution. Therefore, we depend on the few fossils that have been found and the characteristics of modern-day organisms to reconstruct a story of early life. That story begins with the simplest cell type called ***prokaryote*** ("before the nucleus"). Prokaryotic cells are very small and do not have a nucleus or other cell organelles. Those traits appeared after 2 billion more years with a second type of cell called ***eukaryote*** ("new nucleus"). Eukaryotic cells are larger and have a nucleus and other specialized cell organelles.

Three major lines (Domains) of life have been proposed by taxonomists. One group is called ***Domain Archaebacteria*** ("ancient bacteria"), and a second group is called ***Domain Eubacteria*** ("new bacteria"). Domain Archaebacteria includes bacteria that live in very exotic environments like hot springs and thermal vents in the deep ocean. Domain Eubacteria includes the "common bacteria" we encounter in our daily lives. The third domain is ***Eukarya*** (the eukaryotes), and it includes all other organisms. It may seem surprising that eukaryotes are put into a single domain, but the uniting feature is the similar structure of their cells.

Exercise #1	**Bacteria in Our Environment**
Exercise #2	**Some Basic Characteristics of Organisms**
Exercise #3	**Bacteria**
Exercise #4	**Cyanobacteria and the True Algae**
Exercise #5	**Protozoans**
Exercise #6	**Bread Mold and Mushrooms**
Exercise #7	**Comparison of Eubacteria and Eukarya**
Exercise #8	**Lichens**

Exercise #1
Bacteria in Our Environment

There is no way to clean away all of the bacteria.

This Exercise must be completed during the lab prior to this lab so that there will be enough time for the bacterial colonies to grow and be seen. Today your class will pose one or two questions related to bacteria that live with you on this campus. Then you will design a sampling procedure that answers your questions.

Procedure

- Divide the class into three large groups. Each group has 5 minutes to discuss interesting bacteria questions that you think are possible to answer by sampling with agar plates.
- Decide which question is the most interesting to your group. Each group will present its favorite question to the entire lab class.
- As a class you will decide the final experimental questions. You will have sterile agar plates for growing bacteria. As a class, decide how you will use these plates in your experimental design.

Agar Plates

Special glassware, called a ***Petri dish***, is used for growing bacteria. The top of the Petri dish loosely covers the bottom half. This permits oxygen gas in the air to enter the plate without allowing contamination by spores and other microbes from the environment. Agar is poured into the bottom of the Petri dish. ***Agar*** is a derivative of a red algae called agar-agar. It is a solidifying agent. The true nourishment is added to the agar and consists of partially digested protein, carbohydrates, minerals, and other nutrients.

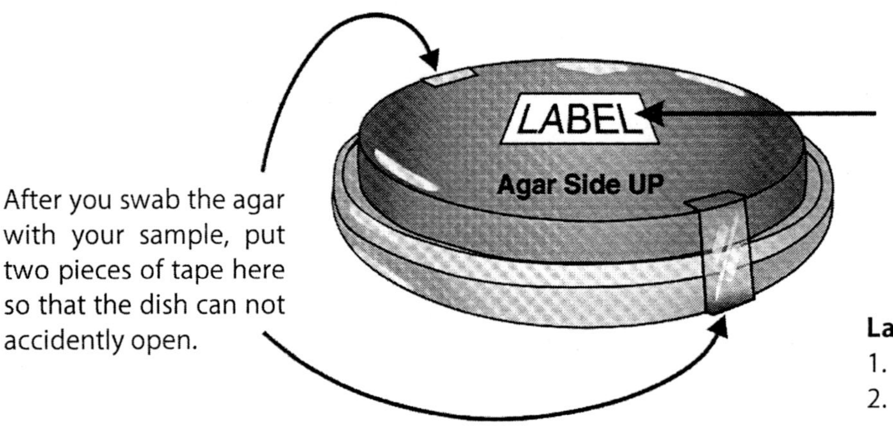

Figure 13.1. Proper labeling of a Petri Dish. Remember to tape and label the petri dish.

It is important that the label be placed on the bottom half of the Petri dish (the one containing the nutrient agar). When the label is placed on the agar side, it can't be accidentally separated from the bacteria. The Petri dishes will be incubated in an inverted (bottoms up) position. This prevent drops of moisture from dropping onto the agar.

Sampling for Bacteria

Do not obtain sampling specimens from your mouth, throat, groin, nose, or any area of your body except for healthy skin. Be very careful when carrying out this experiment. The environment contains a large variety of different microbes (bacteria and fungi). Some of these microbes may be ***pathogenic*** (disease causing).

Procedure

- You have labeled your dishes before sampling. Check your label for complete information. Print the Sample Description and your Class Name.
- A sterile moistened swab will be used to sample the environment and then to inoculate the sterile agar in the Petri dish.
- Swab the plate in this fashion:
  Gently roll the swab across the surface of the agar. Do not dig! Skimming the surface as you roll the swab back and forth will do the trick.
- Tape the two sides of the Petri dish together.
- Put a piece of tape around the entire class's agar plates so they don't get mixed up with those from other lab classes.
- Place the inverted plates in the tub labeled "To Be Incubated." Be sure that your plates are all inverted in the same direction (agar side up). The Petri dishes will be incubated during the next week.
- After you obtain your sample and inoculate your agar plate, discard the used swabs in the test tube or envelope provided. Then place the test tube or envelope in the laboratory area marked: "Danger—Contaminated Materials to be Autoclaved".

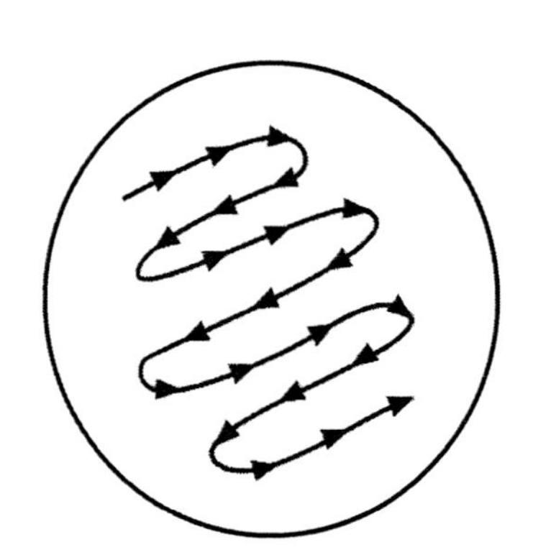

Swab the agar plate like this.

A special machine called an ***autoclave*** will be used to decontaminate these materials by using a process that involves steam sterilization under pressure.

Exercise #2
Some Basic Characteristics of Organisms

By the end of this lab you will know three very basic characteristics for the majority of organisms in each of two domains — ***Eubacteria*** and ***Eukarya***. Examples and traits of the Domain ***Archeobacteria*** are covered in a more advanced biology class.

First Characteristic: All cells can be divided into one of two types.

Prokaryotic
or
Eukaryotic

How do you tell?

- A prokaryotic cell is very small compared to a eukaryotic cell. A prokaryotic cell does not have a nucleus.
- A prokaryotic cell does not have any chloroplasts. If the cell does photosynthesis, then the light-catching pigments are spread throughout the cell water and not in plastids.
- A eukaryotic cell has many organelles (including a nucleus, chloroplasts, mitochondria, and others).
- A eukaryotic cell is big compared to a prokaryotic cell.
- ***Prokaryotic*** means "before nucleus," and the word ***eukaryotic*** means "true nucleus."

? Question

1. Remember the onion cells that you saw in an earlier lab? What kind of cells were they—prokaryotic or eukaryotic?

2. How do you know?

Second Characteristic: All living organisms are composed of one or more cells.

Unicellular
or
Multicellular

How do you tell?

- ***Unicellular*** means that the organism is a single cell, or it is a small chain or cluster of identical cells.

- ***Multicellular*** means that the organism is made of more than one kind of cell type.

? Question

1. The plant leaves that you saw in an earlier lab are part of what type of living organism? Unicellular or multicellular
2. How do you know?

Third Characteristic: All living organisms get energy for their life processes from a specific source.

Phototrophic
or
Heterotrophic

How do you tell?

- ***Phototrophic*** means "light-eater," and ***heterotrophic*** means "other-eater."
- A phototrophic organism will have chloroplasts or else the chlorophyll will be spread throughout the cytoplasm of the cell. These organisms require light for their survival.
- A heterotrophic organism does not have photosynthetic pigments or chloroplasts. It cannot do photosynthesis. This type of organism must eat some other organism or the by-products of that organism to get its energy. Sometimes, this category of organism is called a ***chemotroph*** because it gets its energy from chemicals.
- If you go on to advanced biology classes, you will learn a more detailed distinction in energy categories. For this class, we will combine all non-phototrophs under the one word heterotroph.

Exercise #3
Bacteria

Bacteria are the most numerous organisms on this planet. They exhibit more variety in the way they get their energy than any other group. They are an essential organism for the recycling of dead plants and animals. Last week you sampled for bacteria in various parts of the environment, and Petri dishes were inoculated.

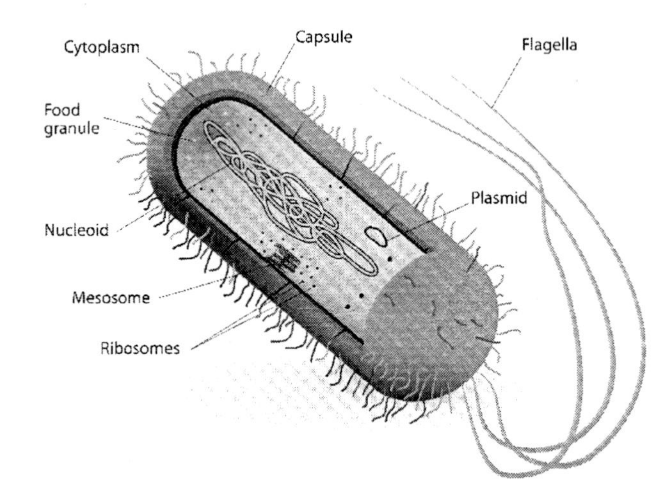

Figure 13.2. Structure of a bacterial cell. Even though these are prokaryotic cells, there are quite a few identifiable features.

Procedure

- Go get the agar plates from last week's sampling experiment.
- **Do not open the Petri dishes under any circumstances!** The bacteria and molds inside could be very dangerous to you in these high concentrations.
- These agar plates were kept in the dark ever since last week. This should help you decide whether bacteria are heterotrophic or phototrophic.
- Determine the answers to your experimental questions.
- You should be able to tell the difference between bacterial colonies and fungus colonies. (Fungus looks "fuzzy" and bacteria usually don't.)
- Describe the growth (color and shape) of the colonies.
- When you are finished, put all the agar plates in the container labeled "To Be Sterilized."
- Wash your hands!

Materials

- ✓ a compound microscope
- ✓ a prepared microscope slide of bacterial shapes

Procedure

- Examine the three different samples on this slide and determine the three basic shapes of bacteria. The bacteria may be clumped or hooked together, so look at the shape of individual cells.
- The three shapes of bacteria are:

  ____

  ____

  ____

- Return the slide when you are finished.

? Question

Now you should have enough information to determine the three basic characteristics of bacteria. (circle your choices)

Prokaryotic	Unicellular	Phototrophic
or	or	or
Eukaryotic	Multicellular	Heterotrophic

Exercise #4
Cyanobacteria and the True Algae

C***yanobacteria*** and the ***true algae*** are very different kinds of organisms, but both of them do photosynthesis. The objective of this Exercise is to determine the difference between the algae and the cyanobacteria. You will need to work with another microscope group in order to make comparisons of the two groups.

Materials

- ✓ One lab group is to get a prepared microscope slide of *Spirogyra*.
- ✓ The other group is to get a prepared microscope slide of *Nostoc*.

Procedure

- Work as a team with another group at your table. Each microscope group is to find their organism, and put it under the same high magnification.
- Both of these two organisms grow in chains of identical cells. *Nostoc* is a cyanobacteria, and *Spirogyra* is a true algae.
- Look back and forth between the microscope views until the most basic differences become obvious. Ignore the color of the stain.
- Remember: You are going to have to answer the three basic characteristics questions about both of these organisms. Make some quick notes and a sketch of both organisms.
- Ask your instructor about the ribbon-shaped structure inside the *Spirogyra* cells.
- Return the slides to where you got them.
- Talk it over with your group, and decide which features represent each group.

Nostoc

Spirogyra

? Question

1. Are Cyanobacteria

Prokaryotic or Eukaryotic	Unicellular or Multicellular	Phototrophic or Heterotrophic

2. Are True Algae

Prokaryotic or Eukaryotic	Unicellular or Multicellular	Phototrophic or Heterotrophic

Note: You should have concluded that true algae are unicellular, but we have examined only the unicellular algae. There are also multicellular forms, but they were not the first algae, and their features are studied in advanced botany courses.

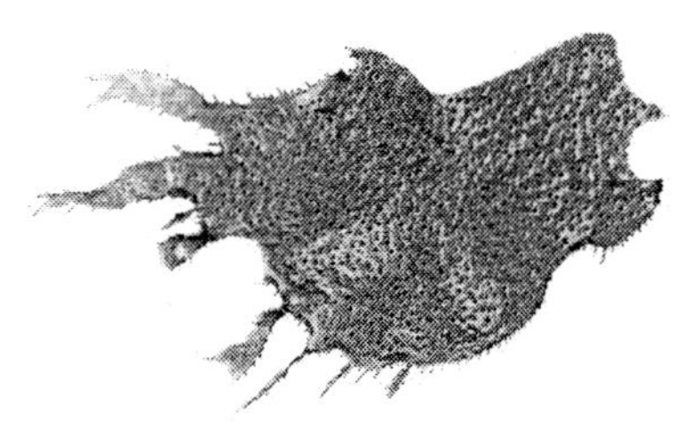

Red Algae
(one example of a multicellular algae)

Materials

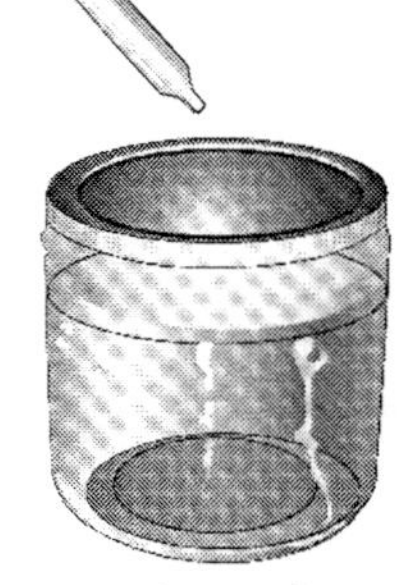

Use a drop from the "Live Mixed Algae jar.

- ✓ Use your compound microscope.
- ✓ Make a slide using a sample from the "Live Mixed Algae" jar. One drop of the green stuff will do it.

Procedure

- Find two different cyanobacteria in the sample. Show them to your instructor. What characteristics tell you that they are cyanobacteria?

- Find two different true algae in the sample. Show them to your instructor. What characteristics tell you that they are true algae?

Exercise #5
Protozoans

You saw Protozoans previously in the Cell lab. Now, it's time for you to determine their basic characteristics.

Materials

✓ Make a slide using a sample from the "Live Mixed Protozoa" jar. One drop from the bottom of the jar will do it.

Procedure

- Find at least two different kinds of Protozoans. Show them to your instructor.
- You may need to use Protoslo® to slow the swimming organisms enough to follow them with your microscope. Ask your instructor how to use Protoslo®.
- You may find some quite amazing attached Protozoans in the sample. We have a booklet that will give you the names of some Protozoans you are likely to see. Try to identify your Protozoans.

? Question

Talk it over with your group, and decide which features represent Protozoans.

Prokaryotic	Unicellular	Phototrophic
or	or	or
Eukaryotic	Multicellular	Heterotrophic

Exercise #6
Bread Mold and Mushrooms

Bread mold and mushrooms are strange beasts indeed. During this Exercise you will continue to determine basic characteristics of these groups, and you will discover some interesting structural features and a strange method of reproduction called ***spore reproduction***. This special type of reproduction helps the organism to get through tough times in their environment. They produce ***spores*** which are small cells with a very thick protective coating. Spores look like little balls. When the environment changes, these spores grow into the next stage in the life cycle of these organisms.

Bread Mold

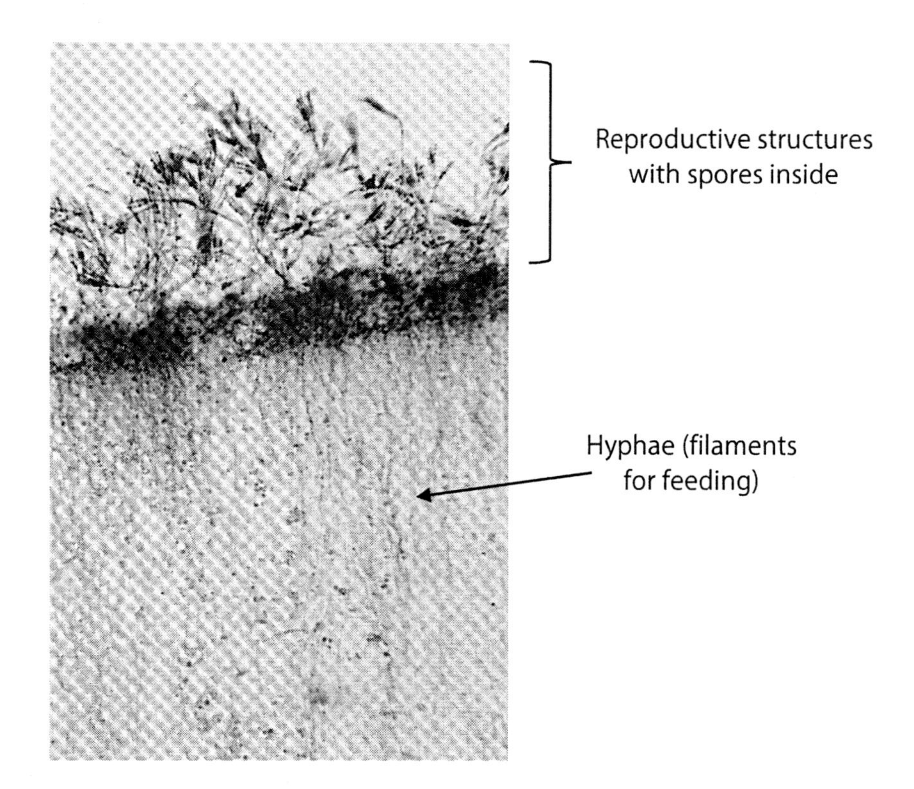

Figure 13.3. The basic structure of fungi. Most fungi have a mass of filaments called ***hyphae*** that grow through whatever food source the fungi is feeding on. They also have some kind of reproductive structure that produces ***spores***.

Materials

✓ Go the demonstration table with bread mold cultures growing on agar plates and on bread. Cut out a small part (¼" x ¼") of the mold, and put it on a slide (no coverslip).

Procedure

- Look at the bread mold samples growing on the bread. Make observations about how you think it is "making a living" and talk this over with your group. What are your conclusions?
- Examine this organism with the dissecting microscope and make observations about its basic characteristics.
- Draw a simple sketch of the structure of the bread mold.
- Find those "little balls" called spores. Make a wet mount slide of the balls and a few of the filaments. Use a compound microscope, and examine under higher magnification.
- Make a simple sketch of what you see. The filaments are used for feeding, and the little balls are spores.

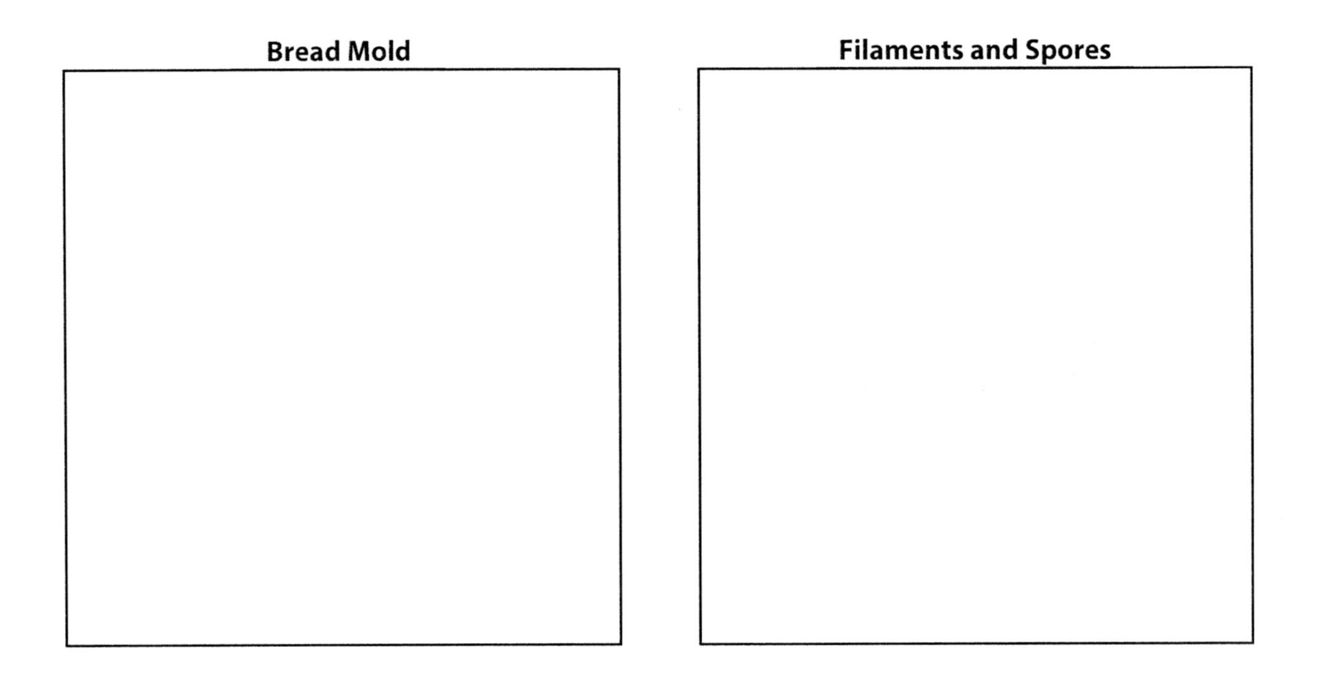

Mushrooms

Materials

✓ Remove a small piece of the brown-colored tissue from under the mushroom cap.

This picture shows the extensive hyphae network and the many mushrooms.

Procedures

- Make a slide of the mushroom tissue, and determine what that brown-colored tissue is.
- Make a simple sketch of what you see.

Mushroom

? Question

1. How do you think that this mushroom organism gets its energy for life? Talk it over with your group. What are your conclusions?

2. Do you think that this organism does photosynthesis?
   Why or why not?

3. Talk with your group and decide what features represent the bread mold and mushroom organisms.

Prokaryotic or Eukaryotic	Unicellular or Multicellular	Phototrophic or Heterotrophic

Exercise #7
Comparisons of Eubacteria and Eukarya

This lab has presented examples that define the basic characteristics of each of two domains — Eubacteria and Eukarya. You have answered the three characteristics questions about each of the groups presented.

? Question

1. Which cell type do you think is the most complex?
   Prokaryotic or Eukaryotic

2. Which cell arrangement do you think is the most complex?
   Unicellular or Multicellular

Procedure

- Using your answers to the above questions, list the four possible combinations of cell type and cell arrangement, placing them in order from simple to more complex. Fill in the domain names after reading the information below.

	Cell Type	Cell Arrangement	Domain
Simple ↓	____________	____________	____________
	____________	____________	____________
	____________	____________	____________
More Complex	____________	____________	____________

- Refer to the basic characteristics for each of the subgroups examined in this lab. You should see that one of the four possible combinations in the above list was not represented. Cross out that combination of traits. That is because the combination of prokaryotic and multicellular does not exist.
- You are left with three combinations. The first combination is Domain Eubacteria. The second two combinations are Domain Eukarya. Put those domain names on the correct lines in the list.
- Place each of the following five subgroups into the correct domain in the table on the next page, and list whether that subgroup is a heterotroph or a phototroph.
  Bacteria
  Cyanobacteria
  True Algae
  Protozoans
  Bread Mold and Mushrooms

Table 13.1. Domain Category and Source of Energy for Each Example Examined in Today's Lab.

Domain	Subgroup	Method of Energy
Archaebacteria (not covered)	Ocean Thermal Vent and Hot Spring Bacteria	Oxidation of Various Minerals
Eubacteria		
Eukarya		

You should be able to name the subgroups in each domain that you studied in this lab, and list the three basic characteristics of each. Your textbook or lecture class may define the basic characteristics of the domains in more detail than this lab did. Also, your textbook or lecture class may point out some of the exceptions to the generalizations presented in this lab.

Exercise #8
Lichens

Figure 13.5. Lichens grow in many forms and on many surfaces. Instead of being one organism, lichens are a complex association between two unrelated species.

A little more than 100 years ago botanists discovered that ***lichens***, strange plant-like organisms that are actually a cooperative relationship between two very different organisms. One of the organisms provides the basic structural framework of the lichen, and the other organism does the energy collecting (photosynthesis). Lichens are often found growing on rocks, tree trunks, or fallen branches on the forest floor.

The metabolic lives of the two very distantly related species in a lichen are intimately intertwined in many different ways. Lichens exhibit a high resistance to unfavorable environments including the cold regions, barren mountain rocks far above the timberline, and fully exposed rocks in the hottest of desert areas. There are several different forms of lichens, and we have some in the lab today: a crust-like form growing very close to the rock's surface, a leaf- like form, and a shrub-like or hair-like form.

Materials

- ✓ a probe
- ✓ a small sample of lichen from the station marked "Lichen Sample"

Procedure

- Crush a small amount of lichen with a probe in a drop of water on the slide. Cover and examine it with your compound microscope.
- Search the slide for evidence of two very different kinds of organisms.
- Show your instructor the two organisms, and answer the next six questions based on your observations.

? Question

1. Which organism does the photosynthesis?

2. Draw a quick sketch of that organism.

Photosynthetic Part of a Lichen

3. What is your evidence that this organism does the photosynthesis?

4. Which organism makes up most of the body structure of this lichen?

5. Draw a quick sketch of this organism.

Structural Part of a Lichen

6. What evidence do you have that this organism does not do photosynthesis?

We hope that you've enjoyed discovering some of the many organisms that inhabit the domains of Eubacteria and simple Eukarya. There are many more examples of strange microbes living around us. And there are many more of them than us. The study of these organisms is what ***microbiology*** is all about.

LAB

14

Mosses and Ferns

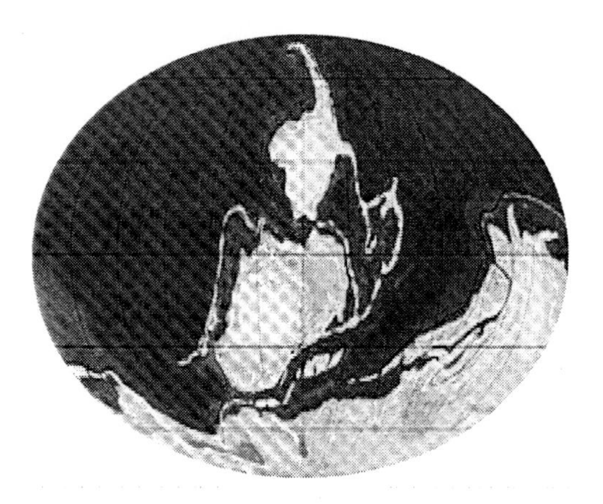

It was the time of Rhyniophytes and whisk ferns, club mosses, and horsetails. And the world was a swampy place. Ah, those were the days.

There was a lot of swampy land about 400 million years ago, and natural selection picked traits that allowed aquatic plants to move into that environment. Land would have been an excellent new opportunity for any plant. It provided "unlimited access" to sunlight for photosynthesis. Two basic questions for understanding this evolutionary event are:

- What are the traits that would be needed by land plants?
- What traits did the algae already have before the land invasion?

Last week we looked at algae and saw that they started as one-celled organisms and developed into more complex algae. The advanced types of algae were the most probable ancestors of land plants. We will see the traits needed on land as we look at mosses and ferns during this lab. We begin with a basic understanding of the algae life cycle on which all land plant reproduction is based.

Exercise #1 **Plant Life Cycles**
Exercise #2 **Which Works Better on Land – Haploid or Diploid?**
Exercise #3 **The Moss Plant**
Exercise #4 **The Fern Plant**

Exercise #1
Plant Life Cycles

Before plants were capable of invading land, the algae had already evolved a unique life cycle called ***alternation of generations***. During this cycle the algae alternated between a stage called ***gametophyte*** (which means "gamete plant") and a stage called ***sporophyte*** (which means "spore plant"). These two stages repeated one after the other, generation after generation.

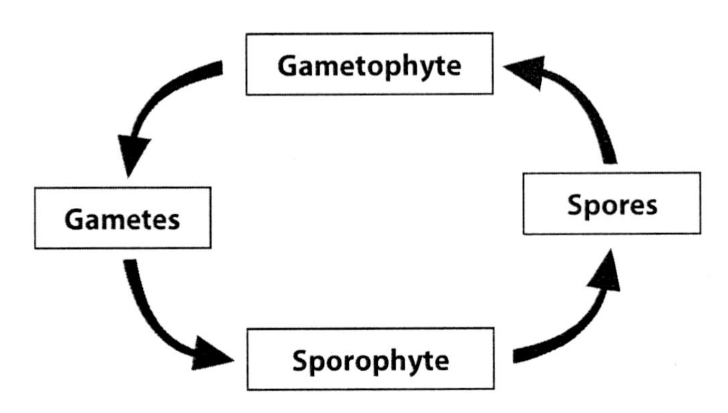

Figure 14.1. Basic Algae Life Cycle showing alternation of generations. The life cycle of algae is the basis of all land plant life cycles. There is alternation between gametophyte and sporophyte generations.

The gametophyte produces gametes which fuse (fertilization) to start the next generation called the sporophyte. The sporophyte produces spores which grow into new gametophytes. In early land plants like the mosses and ferns the gametophyte is a small plant that produces ***gametes*** (eggs and sperm), and the gametophyte is already haploid when it does so. The diploid sporophyte grows from the union of an egg and sperm. In plants the sporophyte performs meiosis to make ***spores***. This is in contrast to animals which produce eggs and sperm directly by meiosis.

- Alternation of generations allows plants to reproduce by using two different methods—spores and gametes.
- The plant life cycle seems a little strange to us, but if you understand its basics, all the rest about land plants makes sense.

- As algae gave rise to land plants, evolution changed the physical appearance of the gametophyte and sporophyte generations, and those changes led to success on land.
- The gametophyte is ***haploid***. This means that the cells have only one set of chromosomes.
- The sporophyte is ***diploid***. This means that the cells have two sets of chromosomes.

? Question

1. Put the names of the correct stage inside the Life Cycle boxes below. (The lines below the boxes are answered in question #7.)

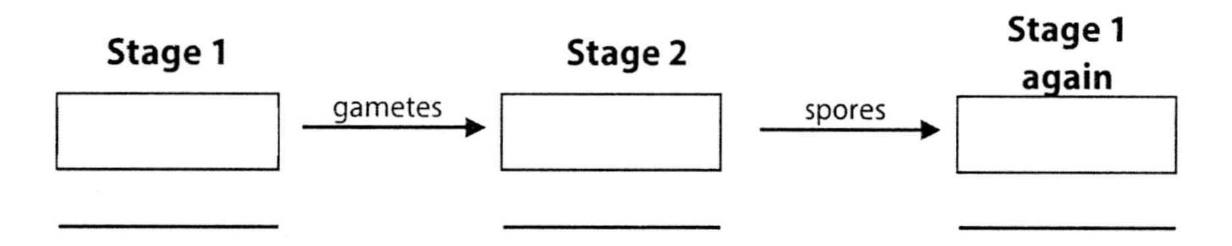

2. Which stage in the life cycle of algae produces eggs and sperm?
3. How many sets of chromosomes does an egg have?
4. How many sets of chromosomes does a sperm have?
5. When the egg and sperm unite, we are back to . . .
   one set of chromosomes or two sets of chromosomes
6. The fertilized egg will grow into a . . .
   gametophyte plant or sporophyte plant
7. Go back to your Life Cycle diagram in question #1, and write the words "diploid" or "haploid" on the line under the appropriate stages.

We have seen that the algae have two stages in their life cycle—the gametophyte stage and the sporophyte stage. One of these stages is haploid and the other stage is diploid. The algae used these two stages for different environmental and reproductive purposes. Now, the question is: Which of these two stages would prove to be best able to adapt to a dry land environment? Any thoughts? Let's proceed to Exercise #2.

Exercise #2
Which Works Better on Land— Haploid or Diploid?

Genes and Mutations

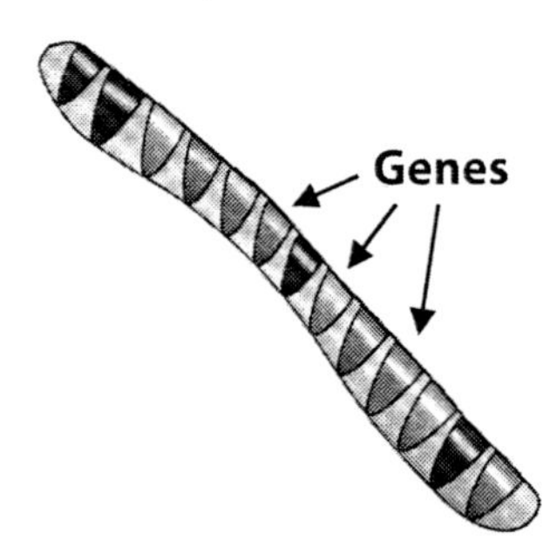

Each gene is carried on a chromosome and is responsible for a plant trait.

Plants have many genes. These genes are carried on the chromosomes. Each gene is involved with the making of a specific protein. Each protein is used in one of two ways:

- A protein might be part of a plant structure.
- A protein might be an enzyme.

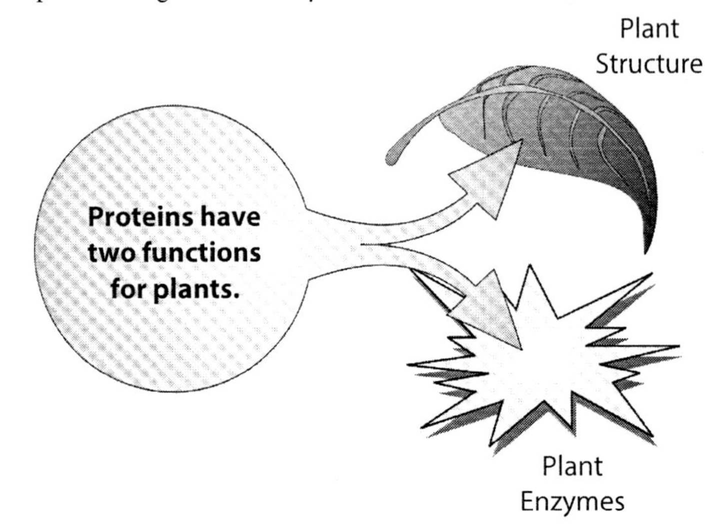

A plant has a certain number of chromosomes (the number depends on the species), and each chromosome has many genes. Each gene produces a plant trait. The more genes that a plant has, the more traits it has. Presumably, more traits give the plant a better chance to adapt to dry land conditions.

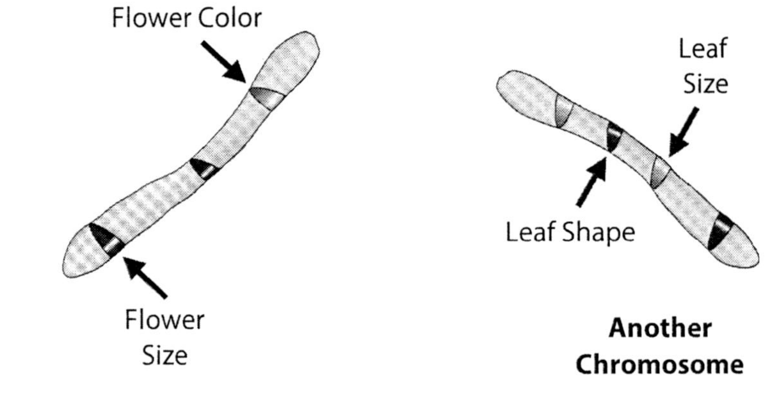

Another Chromosome

One of the Chromosomes

A mutation usually results in a destroyed gene. A gene can be destroyed by radiation, by chemicals in the environment, or by other spontaneous events involving the DNA of a cell.

A mutation is a change in the sequence of nucleotides in the DNA.

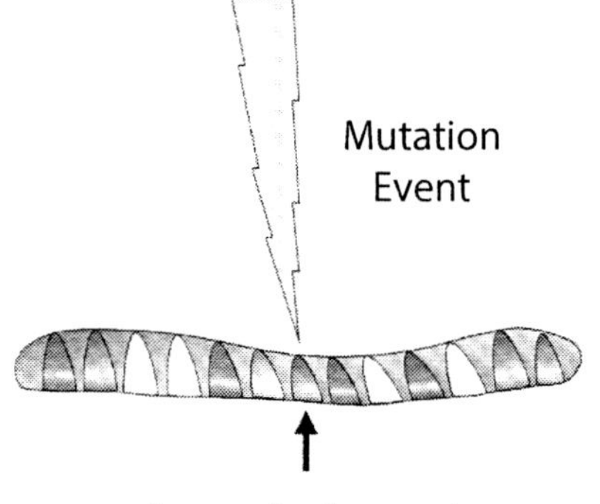

? Question

1. What will be lost if a gene is destroyed?

2. Are mutations almost always a good thing, or a bad thing? Explain your answer.

Limitations of the Haploid

You are ready for the next two puzzles comparing the limitations and advantages of haploid and diploid plants.

Procedure

- Imagine that you are a plant with a single set of chromosomes. That is, you are haploid. Remember: Haploid means that you have only one copy of each different chromosome in your species.
- Assume that mutation events occur at the rate of one mutation for every 1000 genes. The probability of a mutation would be $\frac{1}{1000}$. (The actual probability of mutations is dependent on many factors, and the value of $\frac{1}{1000}$ was picked so that you can make comparison calculations during this exercise.)
- Assume that every gene (every trait) is necessary for survival as a plant.
- Assume it takes more than 10,000 genes to be a complex plant with lots of structures and enzymes.

? Question

1. Assuming the previous information, what would be the safest number of genes for you if you lived as a haploid plant?
    a. 1000
    b. more than 1000
    c. less than 1000

   Check your answer to question #1 with your instructor before going on to the next questions.

2. Describe your basic structure as a haploid plant using the conclusion in question #1.
    a. You are a complex plant (lots of structures).
    b. You are a simple plant (fewer structures).

3. If you were an algae plant that was starting to move onto land through the process of evolution, then what would be the most likely requirement to ensure your survival in a land environment?
    a. Don't change any genes.
    b. Add more genes (new traits).

4. You should have decided that more genes enable you to develop more traits to help you survive on land. But, you are haploid! What risk would you be taking if you were to add many more genes?

5. Some algae evolved onto land by developing an emphasis on the haploid stage. These plants could not add very many more genes without risking extinction. Therefore, the algae plants that used this strategy were restricted to being . . .
    a. simple plants (few structures).
    b. complex plants (many structures).

Advantages of the Diploid

Some plants did move onto land with an emphasis on their haploid gametophyte stage. From what you have discovered, you should be able to deduce that their structure would be very simple. Now you know why we called the previous example, "Limitations of the Haploid."

All of the assumptions of the haploid example are also true for this next case except that now you are a plant with a double set of chromosomes. That is, you are diploid. Diploid means you have two copies of each different chromosome in your species. You have two of every gene.

? Question

1. Consider what would happen to a diploid plant when genes are destroyed by mutation. How many genes have to be destroyed in a diploid plant for one trait to be lost?

2. We assumed that the probability of a gene mutation was one chance for every 1000 genes. A general rule in statistics states that: "If you want to know the probability of two events happening together, then multiply their separate probabilities." Based on your answer to question #1, as a diploid plant, what is the probability of your losing both genes for the same trait because of destruction through mutation?

3. Consider your answer to question #2 and that mutation does destroy genes. How many genes can a diploid plant "safely" have?

   a. about the same number as a haploid plant
   b. fewer genes than a haploid plant
   c. many more genes than a haploid plant

4. Based on your conclusions in the previous questions, describe yourself as a diploid plant.

   a. You can be a complex plant (lots of structures).
   b. You are stuck being a simple plant (fewer structures).

5. How does your description above relate to your chances for doing well in a land environment? Explain.

Some plants did move onto land with an emphasis on their sporophyte stage. You should be able to deduce whether their structure would be simple or complex. Now you know why we call this puzzle, "Advantages of the Diploid." The time has come to test your detective skills with the moss plant and the fern plant.

Exercise #3
The Moss Plant

During this Exercise you will figure out the life cycle of the moss plant and identify some distinguishing adaptations and limitations of the moss plant. When you see green moss growing on a river bank or on a fallen tree, you are actually seeing only one of the two alternating generations in its life cycle. You must look closely at this generation to determine which it is — ***gametophyte*** or ***sporophyte***.

Materials

- ✓ a dissecting microscope
- ✓ a moss plant

Procedure

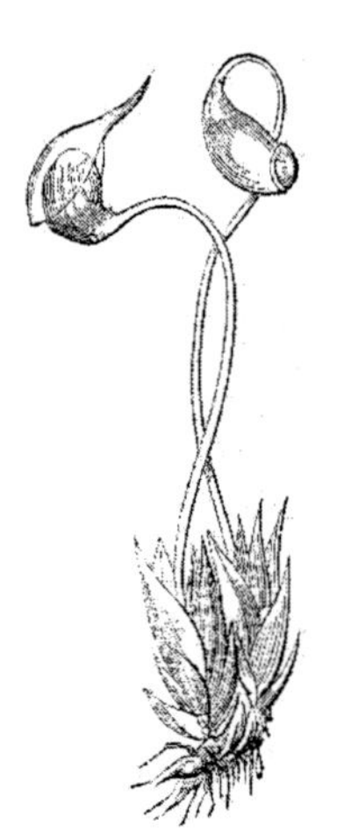

Moss Plant

- Look at the green parts of the moss plant under the dissecting microscope. Moss leaves don't have veins for transporting nutrients or stomata for regulating air flow like typical plants. Does this suggest that moss leaves are less complex than tree leaves? _____

- Leaf veins and stomata are land traits and require many genes to create them. How many "leaf genes" would the moss plant have compared to a typical tree (many or few)? __________

- Does the moss plant have much root? _____

  Assuming that large complex roots require more genes than small simple roots, how many "root genes" would the moss have (many or few)? __________

- In general, being bigger requires more genes than staying small. How many "size related genes" would the moss plant have? (many or few) __________

- Looking at the general features of the moss plant, you should be able to conclude that it has (many or few) ________ genes for land traits.

In Exercise #2 you concluded that being haploid restricts the plant to being simple, whereas being diploid allows for more complex structures. Now you are ready to identify the generations in the moss life cycle.

? Question

1. This stage of your moss plant is . . .
   a complex plant or a simple plant

2. Therefore, this stage of the moss plant is probably . . .
   haploid or diploid

3. This means that you are looking at the . . .
   sporophyte generation or gametophyte generation

4. Which means that this stage of your moss plant produces . . .
   gametes (eggs and sperm) or spores

5. Draw a quick sketch of your moss plant.

Moss Plant

Save your moss plant. You'll be needing it again.

Moss Sex Organs

Moss sex organs can be found at the tips of the branches. It is not possible to see the details of the moss sex organs by looking at a living plant. You need prepared slides and a microscope.

Materials

✓ compound microscope.
✓ microscope slide marked "Moss Archegonia"
✓ microscope slide marked "Moss Antheridia"

Both Mosses and Ferns have swimming sperm, and they require water for fertilization.

Procedure

- Don't worry about the fancy names of these sex organs. You will discover which one is male and which one is female as you continue.
- Female organs usually have one large egg cell in the middle of a vase-like structure.
- Male organs are usually round or oval in shape and have a dozen or more small round cells inside called ***sperm***. After a rain the moss sperm must swim to the egg for fertilization. This means that moss reproduction depends on a wet environment.
- Work with a partner so that one of you has the "Antheridia" slide, and the other has the "Archegonia" slide. You are viewing moss stem tips which look something like an artichoke sliced in half. The leaves at the stem tips surround the sex organs.

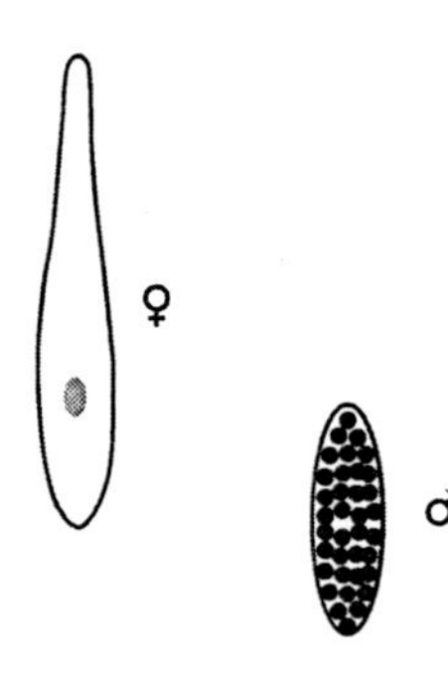

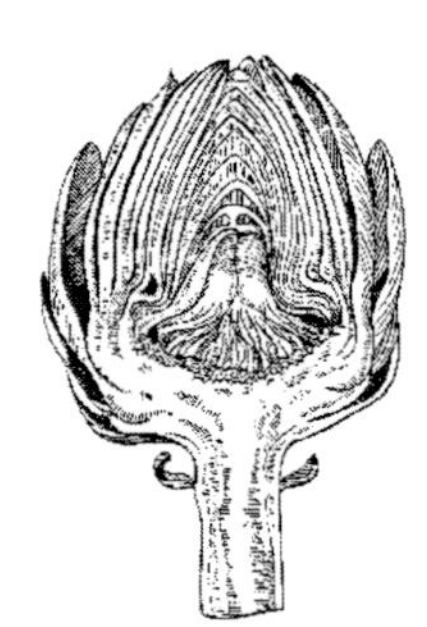

The Moss sex organs are at the tips of the branches and look a bit like a sliced artichoke.

- Finding the female organ is a bit of a challenge. The female sex organ looks like a vase or a bowling pin, depending on the species. If you are lucky, the thin section of the moss plant will cut exactly through the middle of the organ, but you may find odd-shaped sections. A large egg cell is a clue. Looking back and forth between both slides, as well as the slides of other students, may help. Find a female organ, and show it to your instructor. Draw a quick sketch of what you see.
- Search the other slide and find a male organ. Show your instructor. Draw a quick sketch of what you see.
- These sex organs are very simple in structure which shouldn't be a surprise to you. The sex organs are part of the gametophyte which is haploid. And, by now you realize that haploid forces a simple design (fewer genes). You can't make a fancy sex organ unless you use the diploid stage to do so.

Female Moss Sex Organs

Male Moss Sex Organs

What is that Funny-Looking Stalk

Materials

✓ a fine-pointed probe and small tweezers
✓ dissection microscope

Procedure

- Using your dissecting microscope, search for a funny-looking stalk coming out of the top of the moss plant you saved.

- Look at the top of the stalk. You should see a small capsule. The stalk with its capsule is actually the next generation stage in the life cycle of the moss. This next stage grows out of the female sex organ of the previous gametophyte stage. It grows from a fertilized egg. Draw a simple sketch of what you see.

Dissection Scope View of the Moss Stalk and Capsule

- If there is a cap on the capsule, pry it off and look at what is inside.
- Make a wet mount of the small balls, and look at them with your compound microscope.
- Draw a simple sketch of the balls.

Compound Scope View of the Little Round Balls

? Question

1. What are the little round balls inside the capsule? Hint: If you have trouble figuring out the answer, then look back at the Life Cycle in Exercise #1.

2. Now that you know what the little balls are, you also know that the "funny looking stalk" is actually the . . .
    a. gametophyte stage.
    b. sporophyte stage.
3. What will the spores grow into?

Moss Protonema

The spore grows into the next gametophyte generation. The spore is both a reproductive and a "survival" part of the moss life cycle. It is reproductive in function because many thousands of them can be made, and each can grow into a new gametophyte moss plant. Also, these spores are very tough and can easily survive through the dry part of the year. They will germinate when the environmental conditions get better.

Botanists tell us that the mosses developed from evolutionary changes in the growing spore of an algae. If this is true, then land traits were incrementally added to the algae gametophyte producing what we see today as a moss gametophyte. A logical implication of this theory is that the early growing moss spore should show some ancestral resemblance to the algae. The early stage of growth from the spore is called the ***protonema***.

Moss Protonema

Materials

✓ a microscope slide marked "Moss Protonema"

Procedure

- Using a compound microscope, search the slide for some branching filaments. The very first product of a growing spore is a branching chain of cells called the protonema. Normally, spores are produced by mosses at the end of the wet season. With the next rains these spores grow into the protonema which eventually matures into the new generation of gametophytes.
- Draw a quick sketch of the protonema.
- Compare the protonema structure to that of the filament algae (like *Spirogyra*) that you saw in the Microbes lab. Explain how this is evidence that mosses may have evolved from the algae?

- Diagram the Life Cycle of the moss plant below. Include: spores, gametes, sporophyte, and gametophyte. Draw a simple sketch of each generation as a visual reminder.

Exercise #4
The Fern Plant

During this Exercise you will figure out the life cycle of ferns. Also, you will determine what land adaptation traits distinguish them from the Mosses. And you should be able to conclude what their limitations are. These limitations are surpassed by the Conifers and Flowering Plants (next lab). When you look at a fern plant growing in a forest or in your yard, you are seeing only one of the generations in the life cycle of a fern. You must look closely at this generation to determine which stage of the cycle it is — gametophyte or sporophyte.

Materials

✓ a fern plant

Procedures

- Repeat the same logic steps we used to discover the moss life cycle in Exercise #3. Apply this method to discover the life cycle of the ferns. Look at the fern plant very carefully.

? Question

1. Decide: Is the fern plant simple or complex (compared to mosses)?
2. Therefore, is it haploid or diploid?
3. Now conclude: Which generation is the fern plant (gametophyte or sporophyte)?
4. Finally, what will it produce in this stage (spores or gametes)?

Materials

✓ a microscope slide of a cross-section of a "Fern Stem"

Procedure

- Find vascular tissue in the fern stem using your compound microscope. Vascular tissue is like the circulatory system in animals.

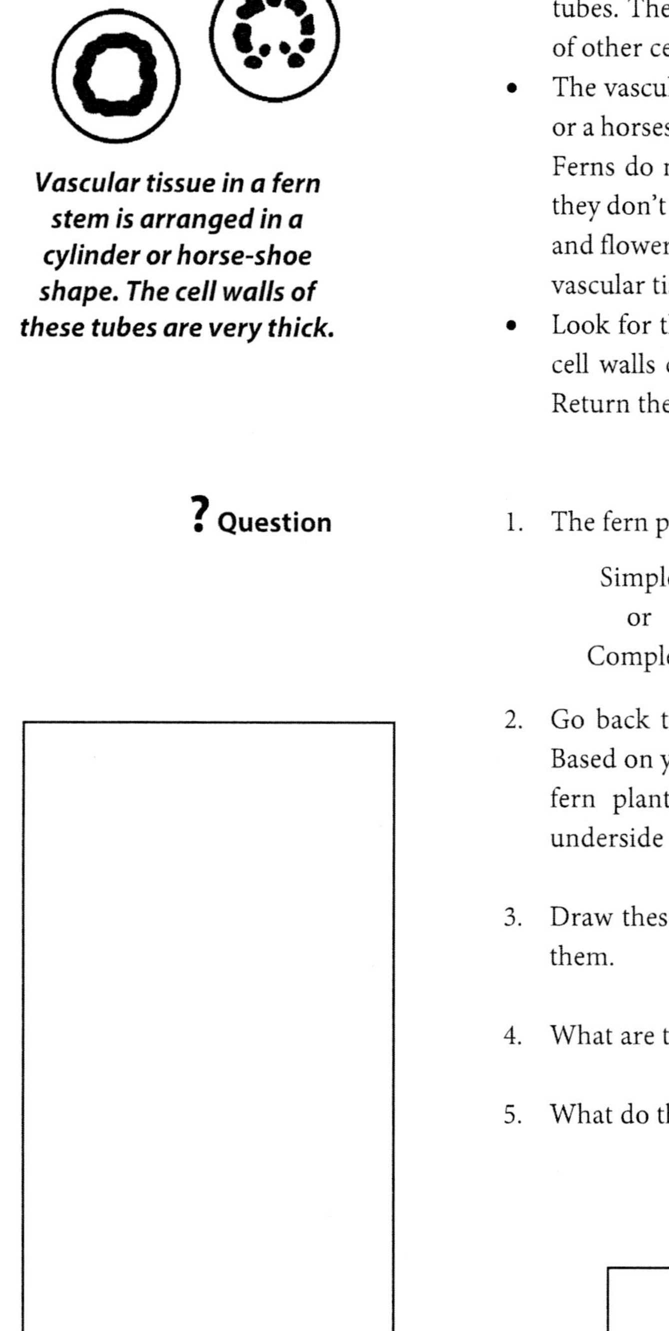

Vascular tissue in a fern stem is arranged in a cylinder or horse-shoe shape. The cell walls of these tubes are very thick.

It moves food and water up and down the plant through specialized tubes. The cells of these tubes have thicker cell walls than the walls of other cells in the stem.

- The vascular tubes may be grouped in clusters that show a cylinder or a horseshoe-shaped arrangement in the cross-section of the stem. Ferns do not have well developed vascular tissue in their roots, so they don't thrive as well in very dry environments compared to cone and flowering plants (see next lab). By contrast, moss plants have no vascular tissue at all. Ferns are better on land.
- Look for the vascular tissue cells in the stem and notice their thick cell walls compared to the other stem cells. Show your instructor. Return the slide to where you got it.

? Question

1. The fern plant that you normally see is . . . (circle your choices)

Simple	Haploid	Gametophyte
or	or	or
Complex	Diploid	Sporophyte

2. Go back to the demonstration table and examine the fern plant. Based on your conclusions, what does this stage in the life cycle of a fern plant produce? Find them! Hint: Look for them on the underside of leaves. Make a slide, and show them to your instructor.

3. Draw these structures and the part of the fern plant that produces them.

4. What are the small balls called? __________

5. What do these structures grow into? __________

What do they look like?

Where do you find them?

Fern Sex Organs

Spores are produced by the fern sporophyte. You found them on the underside of the fern leaves. These spores are similar in function to the spores of Mosses – for reproduction and for survival during harsh times of the year. Each spore grows into a new gametophyte when the wet season returns. That gametophyte is called the ***fern prothallium***. It is very small compared to the sporophyte and does not live long because its only purpose is to make eggs and sperm. Those eggs and sperm unite to produce the next sporophyte generation plant. We will depend on slides and preserved specimens to examine the prothallium and the fern sex organs.

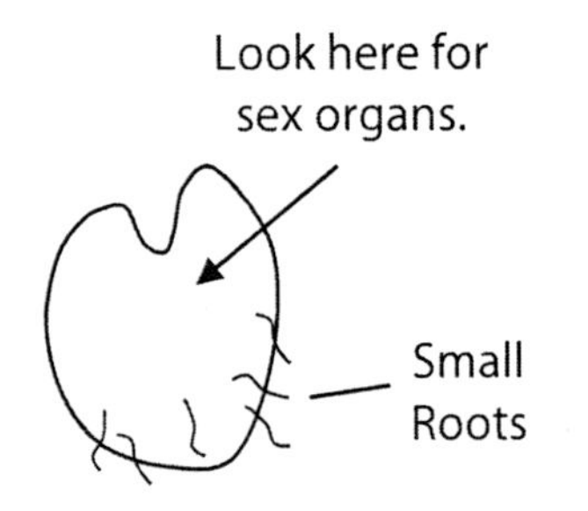

The prothallium looks something like this.

Materials

✓ a prepared microscope slide of "Fern Prothallium" (Antheridia and Archegonia)

Procedure

- Hold the slide up to the light and observe the overall size and shape of the prothallium.
- Use your compound microscope to examine the general structure of the fern prothallium. It grows from a spore that fell from the underside of a mature fern plant leaf. This very simple stage of the fern life cycle is normally found living on the forest floor, and it may be only 1 or 2 centimeters in size. You should be able to conclude which stage you are looking at. Is it the sporophyte or the gametophyte? ______________ As a reminder, write your answer next to the word "prothallium" above your drawing.
- How would you guess which is female and which is male? Find the male and female organs. Show your instructor. Draw the prothallium showing the male and female organs.

Male Sex Organ

Female Sex Organ

Fern Prothallium with Sex Organs (______________)

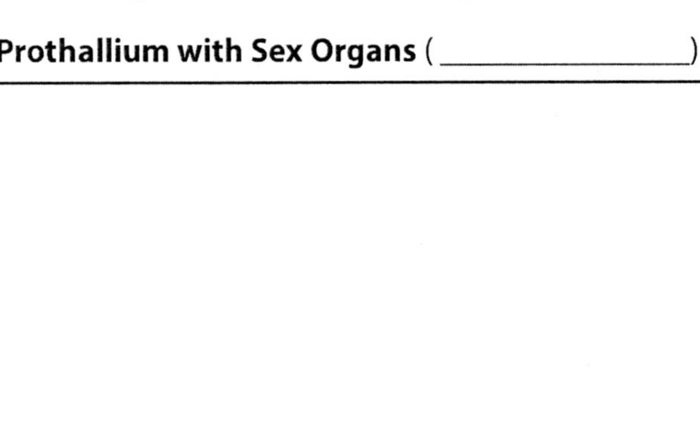

- The egg and sperm unite to form the beginning of the next stage in the life cycle. That stage is the fern plant you normally see.
  The ferns have swimming sperm just like the mosses. This means that both moss and fern reproduction are dependent on a wet environment. This limitation is overcome by the Conifers and Flowering Plants (next lab).
- Diagram the Life Cycle of the fern plant below. Include spores, gametes, sporophyte, and gametophyte. Draw a simple sketch of each generation as a visual reminder.

Life Cycle of the Fern Plant

LAB

15

Dry Land Plants

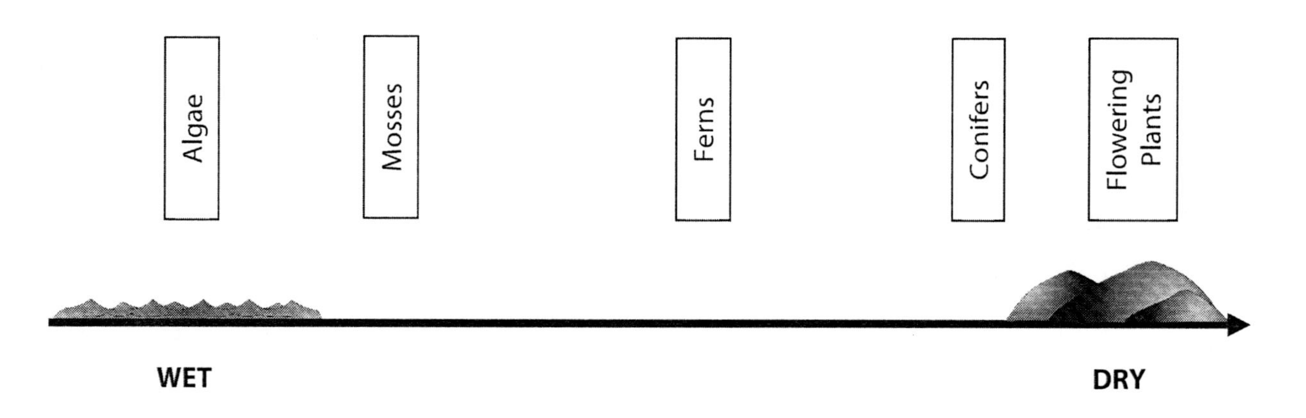

We finish the story of plants with two groups that are the most successful at living in dry environments—the ***cone plants*** and the ***flowering plants***. They are taken for granted because they are everywhere, but both came long after the mosses and ferns first invaded land. And the flowering plants are the newest with their success blossoming after 60 million years ago. Both groups developed special structural and reproductive mechanisms to master the challenges of living on land.

Exercise #1	**Review of Plant Groups**
Exercise #2	**Cones**
Exercise #3	**Flowers and Fruit**
Exercise #4	**Vascular Tissue in Stems and Roots**
Exercise #5	**Wood**
Exercise #6	**Leaves**

Exercise #1
Review of Plant Groups

Mosses and Ferns

? Question

1. Can you remember a structural and a reproductive feature of moss and fern plants that would limit their success in very dry environments?

   _______________ and _______________

2. The moss plant that we see growing on a log is haploid. How many sets of chromosomes does a haploid plant have?

3. A haploid plant is limited to a comparatively simple structure. How does the mutation of genes particularly hurt a haploid plant?

4. What is one example of a simple structure in moss plants?

5. What is simple about moss reproduction?

6. The fern plants that we see growing in the forest are diploid. How many sets of chromosomes does a diploid plant have?

7. A diploid plant can possess more complex structures. How does having a diploid set of genes contribute to a plant possessing more complex structures?

8. What is one example of a complex structure in fern plants that helps them to be better adapted to dry environments?

9. Remember that the ferns use a simple approach to reproduction just like the mosses. How does this limit fern plant success in dry environments?

Cone Plants and Flowering Plants

The gametophyte is the vulnerable stage in a dry land environment.

Both the cone plants and the flowering plants owe their reproductive success to an emphasis on the diploid stage of the life cycle. As you learned, an emphasis on the diploid stage allows the organism to have many more genes without taking a serious risk to the destruction of traits by mutation. (Refer to the "Moss and Ferns" lab for a more thorough review of mutation risks.)

Mosses and ferns drop their spores to the ground, and these spores grow into the gametophyte stage. The gametophyte, because it is haploid and simple, is the vulnerable stage in a dry land environment. And this stage requires water for fertilization in both the mosses and ferns. Conifers and flowering plants solve the reproductive challenges of dry land by not dropping spores to the ground. They keep the spores inside of cones or flower parts which allows the spores to grow into gametophytes in a safe environment. The gametophyte is protected by the sporophyte.

The diploid sporophyte is the dominant stage in the life cycle of Conifers and Flowering Plants.

Conifers and flowering plants also evolved a new reproductive method called ***pollination***. They don't have swimming sperm. The male sex cell (called ***pollen***) is carried to the female sex cell (called the ***egg***) by wind or insects or birds or mammals. This new method is a giant step forward for land plants.

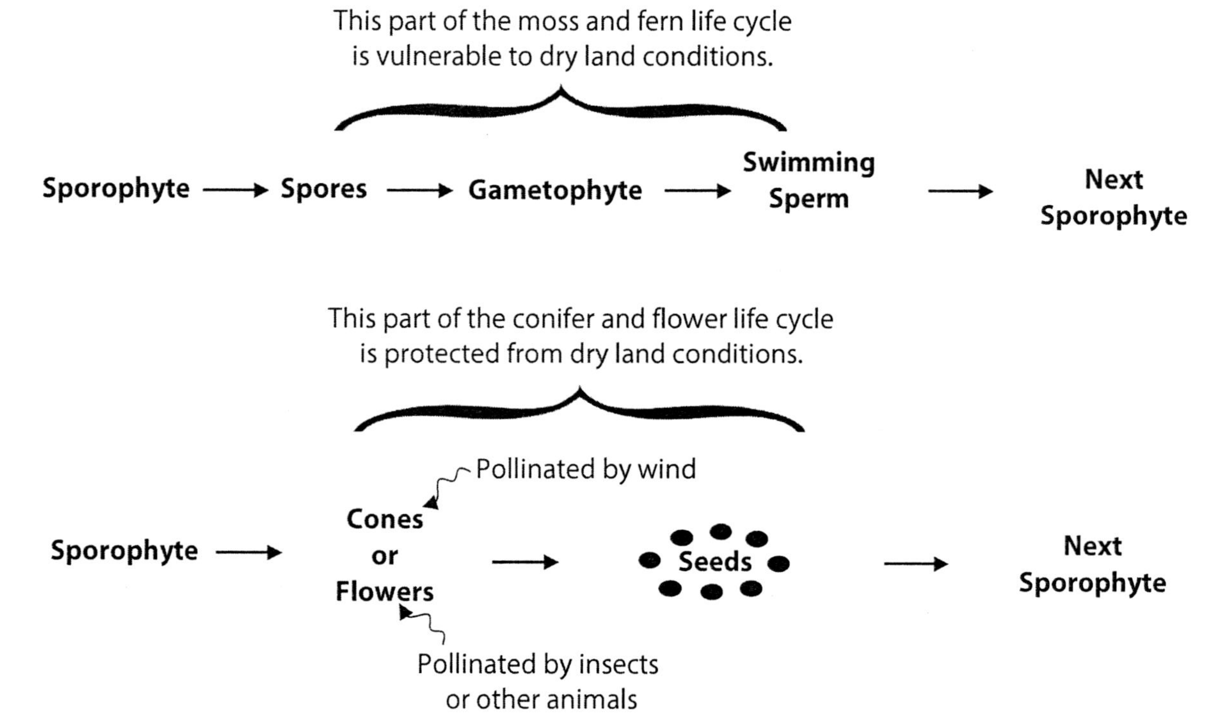

Figure 15.1. Comparison of Life Cycles. The life cycle of moss and ferns is compared to the strategies of the cone and flowering plants.

The conifers and flowering plants have seeds. A ***seed*** is an embryo sporophyte plant with a protective and nourishing covering. They are able to live a long time in the environment until it rains. Then the embryo plant grows into the adult plant. (Mosses and ferns do not have seeds. They only have a single-cell spore to weather the dry season.) Make sure you understand the differences in the general life cycles of mosses and ferns as compared to cone plants and flowering plants. There are structural and reproductive differences between the conifers and the flowering plants, and you will discover those differences in the following Exercises.

? Question

1. Which stage of the life cycle is emphasized in the cone plants and flowering plants?

   haploid or diploid

2. Which stage of the life cycle is most vulnerable in the dry land environment?

   gametophyte or sporophyte

3. Where do flowering plants and cone plants protect their spores and gametophyte stage?

4. Define pollination.

5. Do cone and flowering plants have swimming sperm like the mosses and ferns?

6. What is pollen?

7. Where is the egg in conifers and flowering plants?

8. What are seeds?

Exercise #2
Cones

In conifers, the vulnerable gametophyte stage develops inside the protective cones.

The word conifer means "cone-bearing" and refers to any plant that uses cones as its characteristic reproductive structures. Basically, the ***cone*** is where the gametophyte generation is produced and protected. In contrast, the moss and fern gametophyte has to live independently on the forest floor where it is in danger of drying out and dying.

There are both male and female cones (although the female cone is what we call a "pine cone"). The ***male cone*** produces ***pollen***, and the ***female cone*** produces ***eggs***. Pollen contains the male sex cell and is carried by the wind to the female cones. This is called ***pollination***. After pollination, the pollen produces tubes that grow towards the eggs. Fertilization happens when a pollen nucleus (haploid) is transported through the pollen tube and fuses with an egg nucleus (haploid). A diploid seed (pine nut) grows from that union.

What we think of as a pine cone is actually the female cone. There are male cones also.

Male Cones

Male cones are small (about 1 to 2 cm in length), and can be found in clusters at the ends of pine branches. They are designed for wind pollination.

Materials

- ✓ a compound microscope
- ✓ A small sample of pollen from the male cones on display. Look at the design of the male cones.

Procedure

- Draw a simple sketch of a cluster of male cones.
- Make a wet mount of the pollen and put it under high magnification.
- Draw a simple sketch of the pollen.

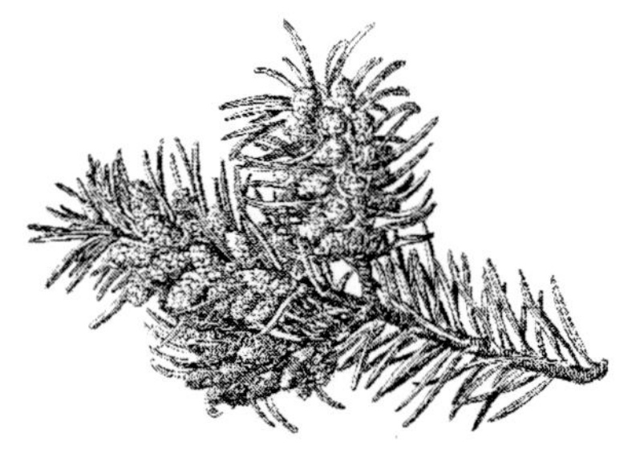

Male cones are small and in clusters at the tips of branches.

Male Cones | **Pollen**

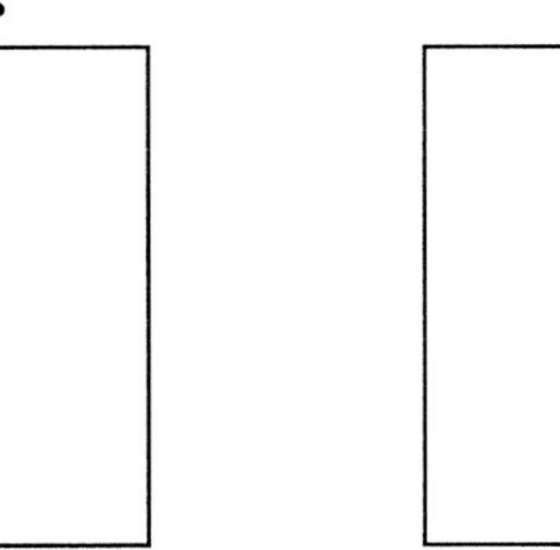

? Question

1. What does the word "conifer" mean?
2. Where is the gametophyte generation protected in the conifers?
3. Is what we call the "pine cone" male or female?
4. Why is so much pollen produced by each male cone?
5. How does the pollen get to the female cones?
6. What is the value of the two "ears" on the conifer pollen?
7. Which generation is the pollen? sporophyte or gametophyte
8. When the pollen nucleus unites with the egg nucleus what is produced?

The female cones are very small when they are young, so it is a very short distance for the pollen to grow to the eggs.

Growing Pine Pollen

Procedure

- Ask your instructor if the class will be growing pine pollen, or wait to grow flower pollen in the next Exercise.
- Several hours ago, some pollen from a campus pine tree was put into a weak sugar solution (food). A slide of the growing pollen cells is under the microscope on the Instructor's Table. Examine the slide, and draw what you see.

Growing Pollen Cells

? Question

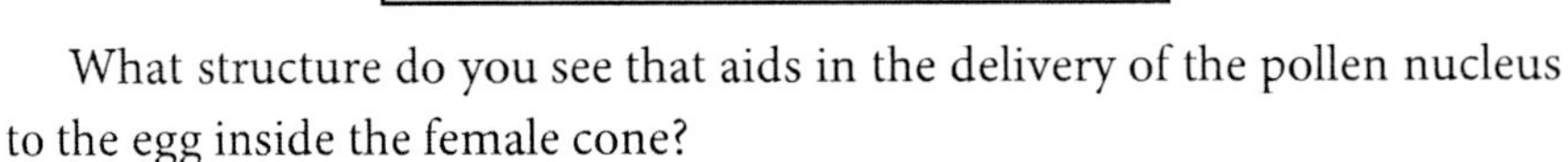

What structure do you see that aids in the delivery of the pollen nucleus to the egg inside the female cone?

Female Cones

Female cones are usually found near the tips of higher branches in the pine tree. They start out very small, but there may be more than 100 eggs in each of these cones. The pollen from the male cone blows onto these small female cones when the eggs are fairly close to the outside of the cone. Fertilization happens with the union of the male pollen and egg.

The fertilized egg becomes part of the seed (next sporophyte generation) and is protected by the female cone as it grows larger. Eventually the female cone is quite large (what you usually see on a pine tree), and the seeds are now ready to be released into the environment.

Look at the base of these projections for an ovule with an egg inside.

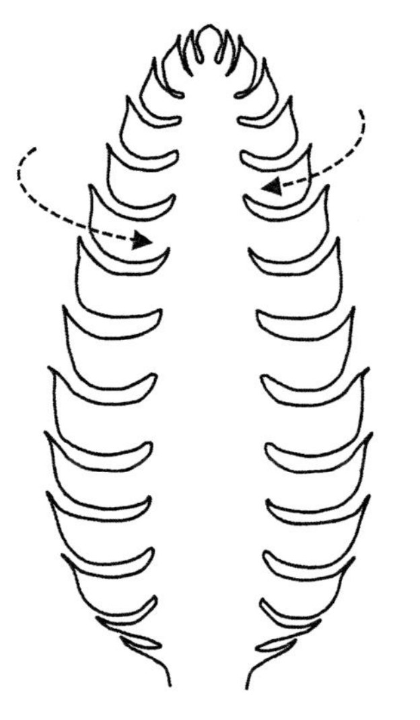

Young Female Cone

Materials

✓ a prepared microscope slide marked "Pine Megasporangia" (very young female cones)

Procedure

- Use low power magnification. Look for oval structures at the base of the projections along the outside of the female cone. (Refer to figure.) These are called ***ovules***, and each produces an egg. You may be lucky enough to find an ovule with a large egg cell inside. Some ovules will have developing seeds. Ask your instructor to help you see the progression from egg to seed.
- Draw a simple sketch of where egg is found.

Ovule (with egg inside)

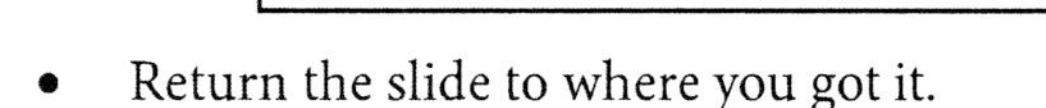

- Return the slide to where you got it.

? Question

1. Which generation (sporophyte or gametophyte) of the conifer life cycle actually produces the egg and is protected inside the female cone? (This is the same generation of the moss and fern life cycle that has to live unprotected and on its own.)

2. The seed is the next (sporophyte or gametophyte) generation?

Seeds

Seeds are a dry land feature of both conifers and flowering plants. There are no seeds in mosses and ferns. The seeds of conifers can be found inside the female cone about one year after pollination. Mature seeds are released into the environment and must survive on their own. Because conifers do not produce nutritive fruit surrounding their seeds, they are called ***gymnosperms*** (meaning "naked seeds"). The seeds of flowering plants are found inside the fruit, which gives those plants their name — ***angiosperms*** (meaning "covered seeds"). The fruit is discussed in more detail in the next Exercise.

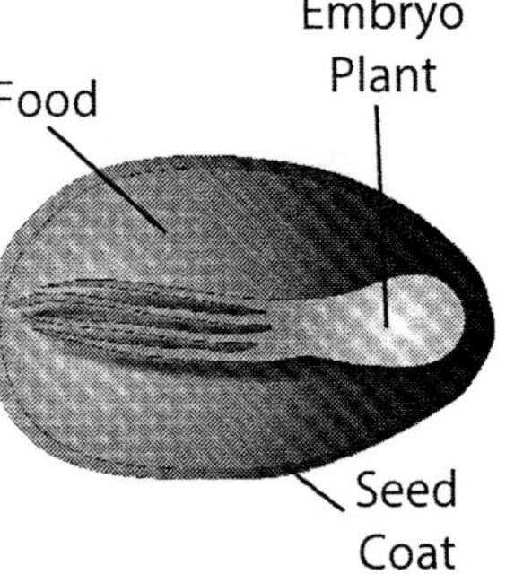

Structure of a Seed

Materials

- ✓ a dissecting microscope
- ✓ a pine seed (soaking in water)
- ✓ a dissecting probe

Procedure

- If your pine seeds haven't been shelled, then carefully crack the seed coat with pliers. Under the microscope, use your probe to carefully remove the seed tissue from the outer part of the pine seed until the ***plant embryo*** can be seen.
- Show your instructor.

? Question

1. Why use the name gymnosperms for the conifers?

2. Why use the name angiosperms for the flowering plants?

3. What function does the seed coat provide?

4. Why is the seed a better adaptation to a dry land environment?

5. The embryo inside the seed is which generation of the conifer life cycle?

   gametophyte or sporophyte

Exercise #3
Flowers and Fruit

In flowering plants, the gametophyte stage develops inside the flower.

Flowers are like "neon signs" to attract pollinators (usually insects, birds, or mammals). Color, fragrance, nectar, and shape are features that determine which pollinators will be attracted to a particular flower species. Examine each flower in the lab to see if you can determine what kind of pollinator might be involved with the reproduction of that plant. If a plant can involve another species to help with reproduction, then that plant will be more successful than a plant that must rely only on the wind for pollination (like conifers).

Materials

- ✓ a tree-tobacco flower
- ✓ Dissecting materials—a probe, a pair of tweezers, and a razor blade

Procedure

- The basic theme of most flowers is: male organs surround the female organ. Male organs are called the ***stamens***, and they contain the pollen. The female organ is called the ***pistil***, and it contains the eggs.
- Use the dissecting microscope to look at the open end of the tree-tobacco flower. Does any reproductive organ stick up higher than the others?
  What color is the tip of that organ?

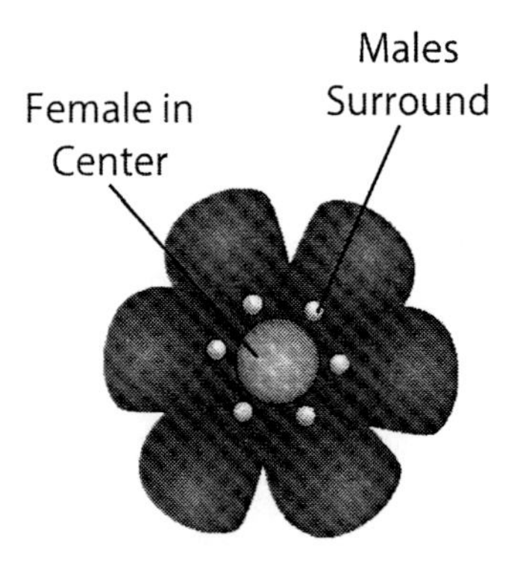

Hypothetical Simple Flower

- Carefully tear open the top half of the flower so that you can better see the reproductive organs. If you see several identical filaments, then you have found the males (stamens). Remember: There will only be one female (pistil). Can you now determine which organ sticks up highest in the flower?

- What would be the value of the female organ being higher than the stamens?

- Continue to tear the yellow petal until it is removed from the green base of the flower. Draw a simple sketch of the arrangement of pistil and stamens in the box provided.
- Using the compound microscope, make a wet mount slide of some pollen from the stamen.

Flower Pollen

- Draw a simple sketch of what you see. Does the flower pollen have "ears" like the conifer pollen?

- What does this tell you about the mechanism of pollination in this species?

Sketch of Your Flower

Pistil

- Optional: Your instructor may have you grow some flower pollen grains in sugar water on a microscope slide. (It takes 1–2 hours, and you must not let the pollen dry out during that time.)

- Cut open the green base of the flower with a razor blade, and find the eggs using the dissecting microscope. This part of the pistil is called the ***ovary***. You may see young seeds if your flower is older.
- Draw a simple sketch of the pistil with its eggs or seeds.

Materials

✓ a citrus flower without the fruit. (You can use the Tree-Tobacco fruits if citrus flowers and fruits are unavailable.)

✓ another citrus flower with the fruit

Procedure

- Don't dissect these citrus flowers. We need to save them for use in other classes.
- Look at the two flowers and determine what part of the flower grows into the fruit. Different species produce fruit from different parts of the flower, and the fruit usually attracts an animal to eat it.
- Return the flowers for other students to use.

? Question

1. The male reproductive organ in a flower is called a __________.
2. The female reproductive organ in a flower is called a __________.
3. What specific part of the flower becomes the fruit?
4. What kind of pollinators might this flower attract?
5. Explain how an animal can disperse seeds by eating a fruit.
6. Flowers typically grow in small clumps scattered in their habitat, whereas conifers typically grow in large groves of trees. How does this difference in distribution relate to your answer in question #5?

Exercise #4
Vascular Tissue in Stems and Roots

The vascular system in plants is analogous to the circulatory system in animals. It is a network of vertical tubes that extend from the leaves to the roots. Vascular tissue first appears in the fossils of ferns, but that conducting tissue is only moderately developed compared to conifers and flowering plants. Fern plants can't get water from deep in the ground. Conifers and flowering plants can, and their success in dry land environments is obvious.

As a general rule, the cell walls forming the vascular tubes are very thick when compared to other cells found in the plant. This is an important feature as you look at prepared cross-sections of plant parts. These cross-sections will make sense if you imagine the three-dimensional vertical tube structure that makes up the vascular tissue.

Materials

- ✓ a prepared microscope slide of a "Root Cross-Section"
- ✓ a prepared microscope slide of a "Herbaceous Stem Cross-Section"

Procedure

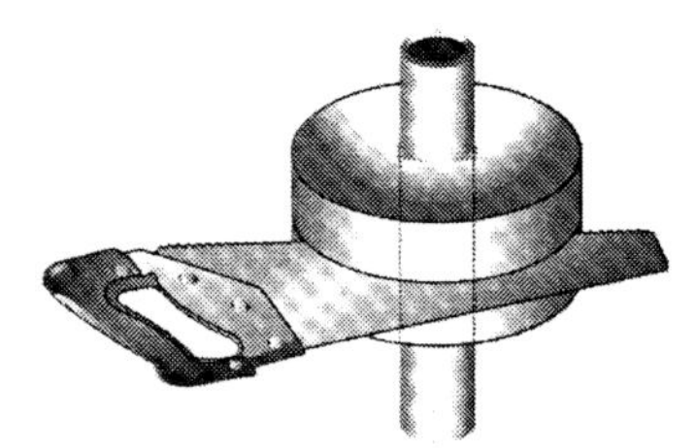

Non-Vascular cells are thin-walled.

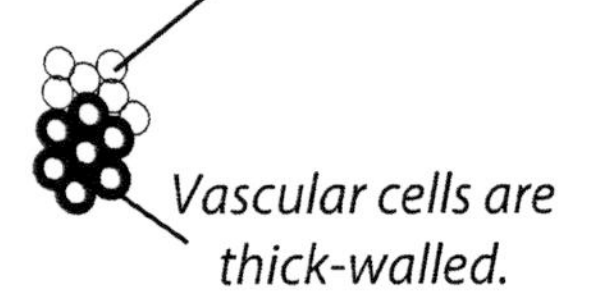

When looking at cross-sections of a stem or root, the vascular tubes appear as thick-walled cells. These are actually very long cells joining end to end to make a long tube.

- Examine the root cross-section slide under low magnification, and determine where most of the vascular tissue is located. The vascular tubes are thick-walled cells. (circle your choice)

  Center of Root or Toward the Outside of the Root

- Draw a simple sketch of the root cross-section showing the location of the vascular tissue.

Root

- The nutrients and water in the soil must diffuse into the vascular tubes. Now that you know where the vascular tubes actually are, what does this tell you about the size of the roots that do the absorption for a plant? (Is water absorbed by big roots or by the small roots?)

  Explain your answer.

- Examine the herbaceous stem cross-section under low magnification, and determine where the vascular tissue is located. Hint: Vascular tubes are arranged in "bundles". (circle your choice)

  Center of Stem or More towards the Outside of Stem

- Draw a simple sketch of the distribution of vascular tissue in the stem. This arrangement of vascular bundles is common in plants.

Stem

? Question

An unthinking person made a cut with a pocket knife all the way around the trunk of a tree. The tree died soon after. Explain why the tree died without being completely cut down.

Exercise #5
Wood

The wood that we use to build our houses comes from the stiff cell walls that are a part of the plant's vascular system. This part is called the ***xylem***, and it conducts water from the roots to the leaves. Xylem continues to be produced on the inside of an ever-expanding growth ring called the ***cambium***. Cambium is just inside the bark of a living trunk or branch. It is shown as a "dashed" circle in Figure 15.2. Xylem is continually produced on the inside of the cambium, and phloem is produced external to the cambium.

Xylem continues to conduct water even after it dies. It is somewhat like thousands of tiny straws that transport water and provide support for the tree at the same time. Water moves up the xylem by a process called ***capillary action***. Your textbook explains the details of capillarity and the osmotic "pushing" forces in the roots. In larger trees the middle section fills with resin and no longer transports water but functions only as a supporting structure. The thin zone of living vascular tubes produced on the outside of the cambium is called the ***phloem***. These tubes carry sugars from the leaves to the roots for storage. The outermost part of the bark is the remains of dead phloem that has been pushed out by the ever-expanding cambium.

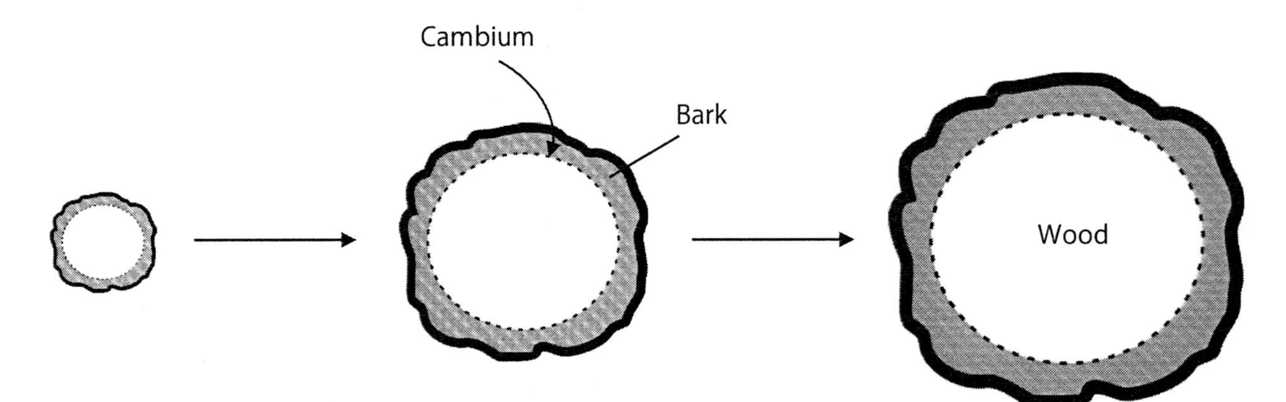

Figure 15.2. Growing tree showing the way that cambium adds to the thickness of the trunk and branches.

? Question

1. In a forest fire, sometimes the inside of a tree can be badly burned, forming a hollow in the tree trunk, yet the tree doesn't die. How do you explain this?

2. What specific part of the vascular tissue is the wood of the tree?

3. What specific cellular feature of vascular tissue makes it ideal as a building material for our houses?

4. What specific part of the vascular tissue is produced on the outside of the cambium?

5. What specific part of the vascular tissue is produced on the inside of the cambium?

Materials

✓ a prepared microscope slide marked "Woody Stem"
✓ a dissecting microscope and a compound microscope

Procedure

- Look at the slide under the dissecting microscope, and find the annual rings of tissue. These are called ***tree rings***.
- Find these same rings with your compound microscope and look carefully at the individual cells of the rings.
- Draw the tree rings in the diagram below, and show how the cell size changes within these rings. By the way, you should be able to see the vascular bundles that are towards the outside of the stem.

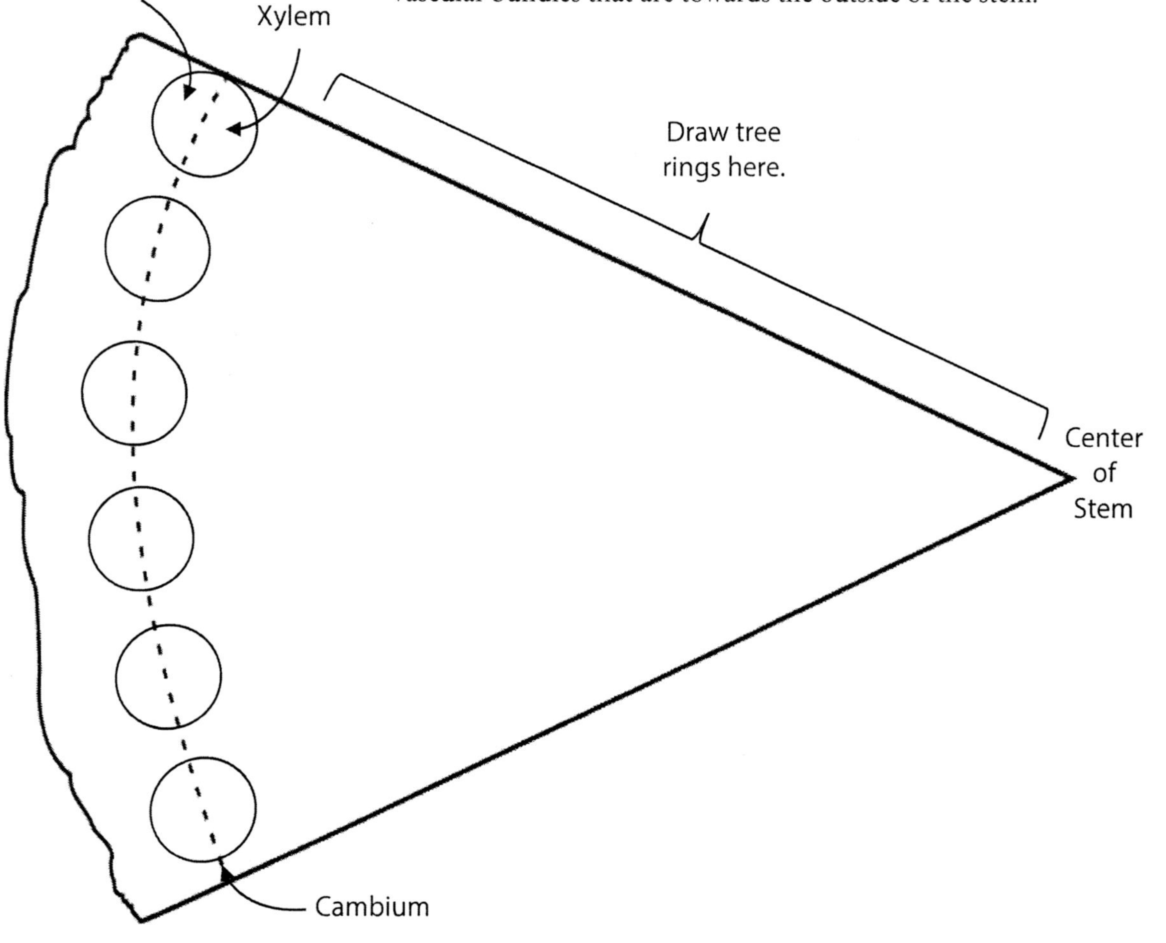

? Question

1. What is the direction of time as you move outward from the center of the tree stem?

2. Where are the youngest tree rings?

3. What would cause these annual rings to show a cycle of cell size change?

4. When would you see big cells?

5. When would you see small cells?

Exercise #6
Leaves

The most basic purpose of a leaf is to expose chlorophyll to light so that the plant can perform photosynthesis. There are a number of problems that a leaf has to overcome to be successful. Some of those challenges include:

- A leaf has to get water from the roots as one of the raw materials for photosynthesis, and it has to send sugars made during photosynthesis to the rest of the plant.
- A leaf has to get CO_2 as a raw material for photosynthesis, and the carbon dioxide has to be delivered to all of the cells in the leaf.
- A leaf has to catch light energy for photosynthesis, but it must not overheat from absorbing too much sunlight.
- A leaf cannot lose too much water from evaporation or the plant will dry up.

The next activities will demonstrate how the plant leaf solves the above four challenges.

Leaf Veins

Leaf veins are vascular bundles leading to various parts of the leaf. The pattern of veins depends on the type of plant and the shape of the leaf. However, in all cases these veins are used for carrying water to the leaf cells and for carrying photosynthetic sugars away from the leaf cells.

Materials

- ✓ A broad leaf from the mallow tree. (A leaf from this tree or another tree similar to it will be in the lab.)

Procedure

- Look at the leaf under the dissecting microscope, and determine the pattern of leaf veins. You already have some information about these structures from Exercise #4.
- Draw a simple sketch of the pattern of veins in the mallow leaf.

Leaf Veins

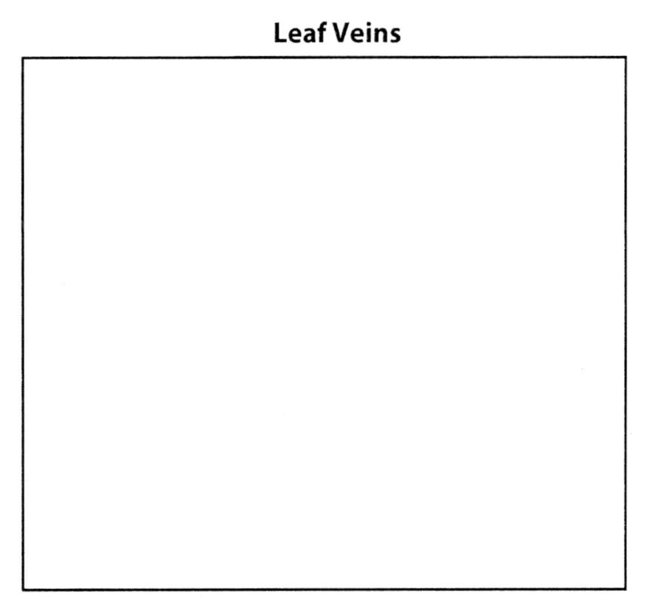

? Question

What is transported through the mallow leaf veins? (Be specific about the content, direction, and purpose.)

Leaf Stomata

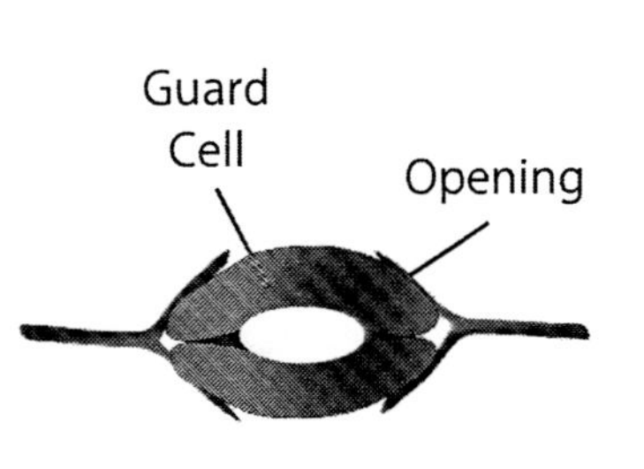

Stomata

Leaf ***stomata*** are openings on the underside of the leaves. We looked at stomata in the *Zebrina* plant during the Cell Lab. Even if you got a chance to see these structures before, make a leaf peel of the Mallow leaf and search for them again. (They don't look exactly same in all leaves.)

Stomata have the capacity to open and close regulating whether air flows freely into the leaf interior. The inside layers of leaf cells require this air to supply CO_2 for photosynthesis. The stomata are open as long as the plant has plenty of water. If the leaf begins to lose water (i.e. the soil is dry, or during the hot dry part of the day), the ***guard cells*** of the stomata shrink and close the opening. This conserves water in the leaves.

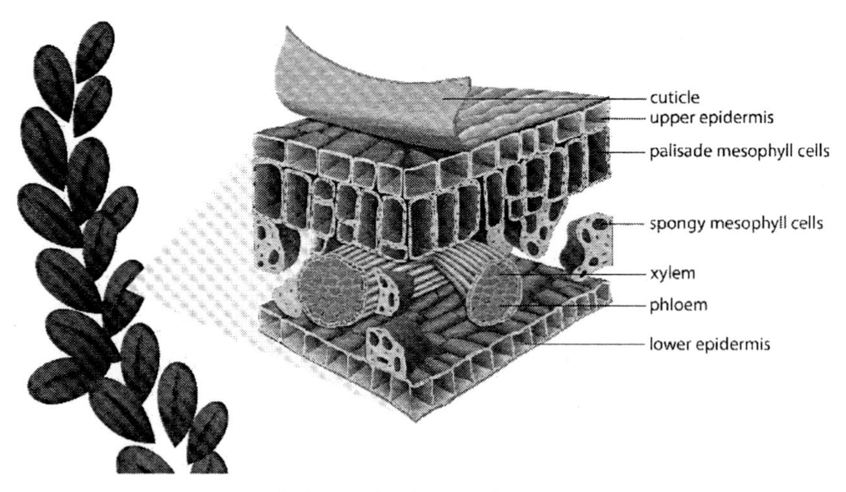

Figure 15.3. Anatomy of the leaf. The epidermis is a protective layer. The bottom epidermis has scattered stomata for controlling air flow into the leaf. Mesophyll cells do the photosynthesis.

A cross-section of the leaf is made like this.

Materials

✓ a prepared microscope slide of a cross-section of a leaf

Procedure

- First, locate the vein. It runs along the center of the leaf.
- Look for the stomata on the underside of the leaf. The guard cells are round in cross-section instead of being like "lips" in the leaf peel. Some of the guard cells are exactly sectioned through the middle so that you can see the stomata opening.
- Draw a simple sketch of the structural relationship between the stomata and the clear spaces inside the leaf. You will have to look around the slide to visualize the 3-D structure. Label the parts of the airway passages throughout the interior of the leaf.

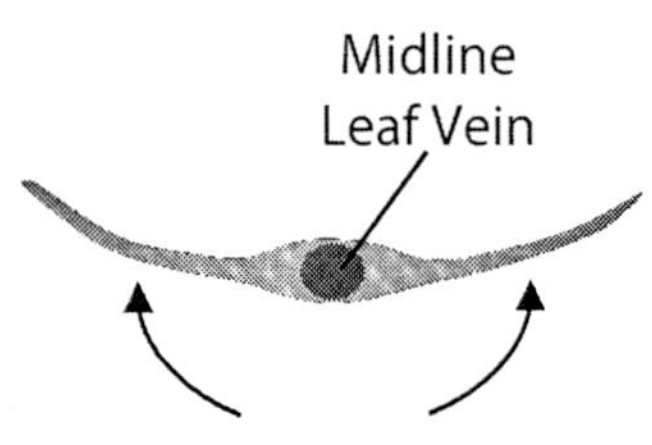

The cross-section will look something like this under the microscope.

Leaf Cross-Section

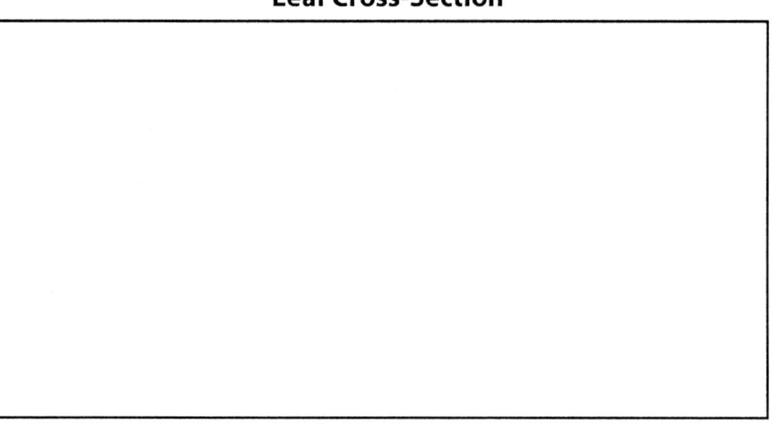

Comparisons of Leaf Types

The lab display has four leaf types to demonstrate different leaf adaptations. Look at each leaf, and make simple sketches of them. Your lab may have local plants to represent the four leaf types.

Mallow (broad-flat)

Pine (needle-like)

Jojoba (upward pointing)

Buckwheat (very small)

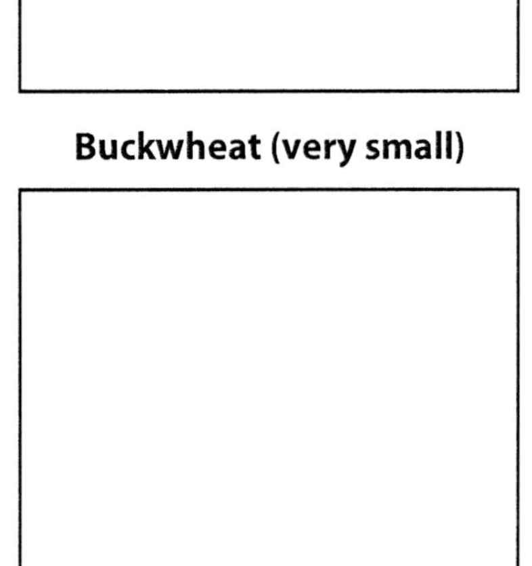

? Question

1. Which leaf design is best for catching the most light?
2. Which leaf design would absorb the most heat?
3. Which leaf design loses water the fastest and must live in a wetter environment than the rest?
4. The jojoba leaf does much better than the pine leaf in very dry environments. What features of the jojoba plant help their leaves lose less water than the pine leaves?
5. Notice the orientation of the jojoba and buckwheat leaves. What time of the day do these two leaf types absorb more light?

   What part of the day do they absorb less light and heat?

   What advantage is there to this kind of leaf orientation to the sun?

LAB

16

Survey of Animals

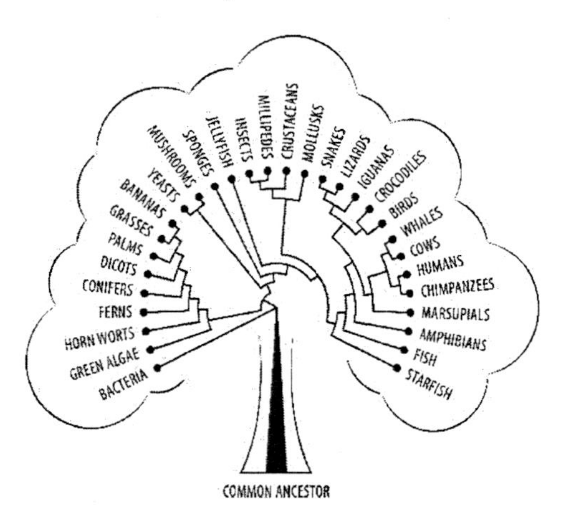

The evolutionary story presented in textbooks gives the impression that the most complex forms of life were created sequentially and that only the best and most complex survived until today. This portrayal is more a projection of progressive philosophy than true to the facts. Evolution doesn't seem to "climb the ladder" toward perfection. In fact, there is evidence to the contrary. For example, the reptiles as a group have enjoyed a much longer and more diverse history, yet the mammals dominate them today. And the mammals evolved from an early reptile form as did the birds. Evolutionarily speaking, which is the more advanced group?

Before placing yourself at the top or bottom of the evolutionary heap, consider that one animal group far out-numbers all of the other animal groups combined. That group is the ***arthropods***. By number and diversity, they are the dominant animal in both the aquatic and land environments. Today's lab is a survey of the Animals—organisms that are ***eukaryotic***, ***multicellular***, and ***heterotrophic***. You will investigate some of the defining characteristics and evolutionary relationships that divide these animals into subgroups called ***phylum*** and ***class***. You will explore both the ***invertebrates*** (animals without a backbone) and the ***vertebrates*** (animals with a backbone).

Exercise #1	**Taxonomy, Classification, and the History of Life**
Exercise #2	**Basic Body Plan**
Exercise #3	**Types of Digestive System**
Exercise #4	**Segmentation**
Exercise #5	**Skeletons**
Exercise #6	**Vertebrate Land Adaptations**

Exercise #1
Taxonomy, Classification, and the History of Life

Origin of Animals

Modern biology tells the history of life as best as we can with the scientific evidence we have. There are a few accepted generalizations.

- The fossils of simple one-celled organisms, called ***prokaryotes*** (before the nucleus), date back to almost 4 billion years ago.
- The oldest fossils of more advanced cell-types, called ***eukaryotes*** (have a nucleus), date between 1–1½ billion years ago.
- Fossils of multicellular organisms appear between ¾ and 1 billion years ago.
- The DNA comparisons of today's organisms suggest that there could have been at least four different paths of multicellular life.

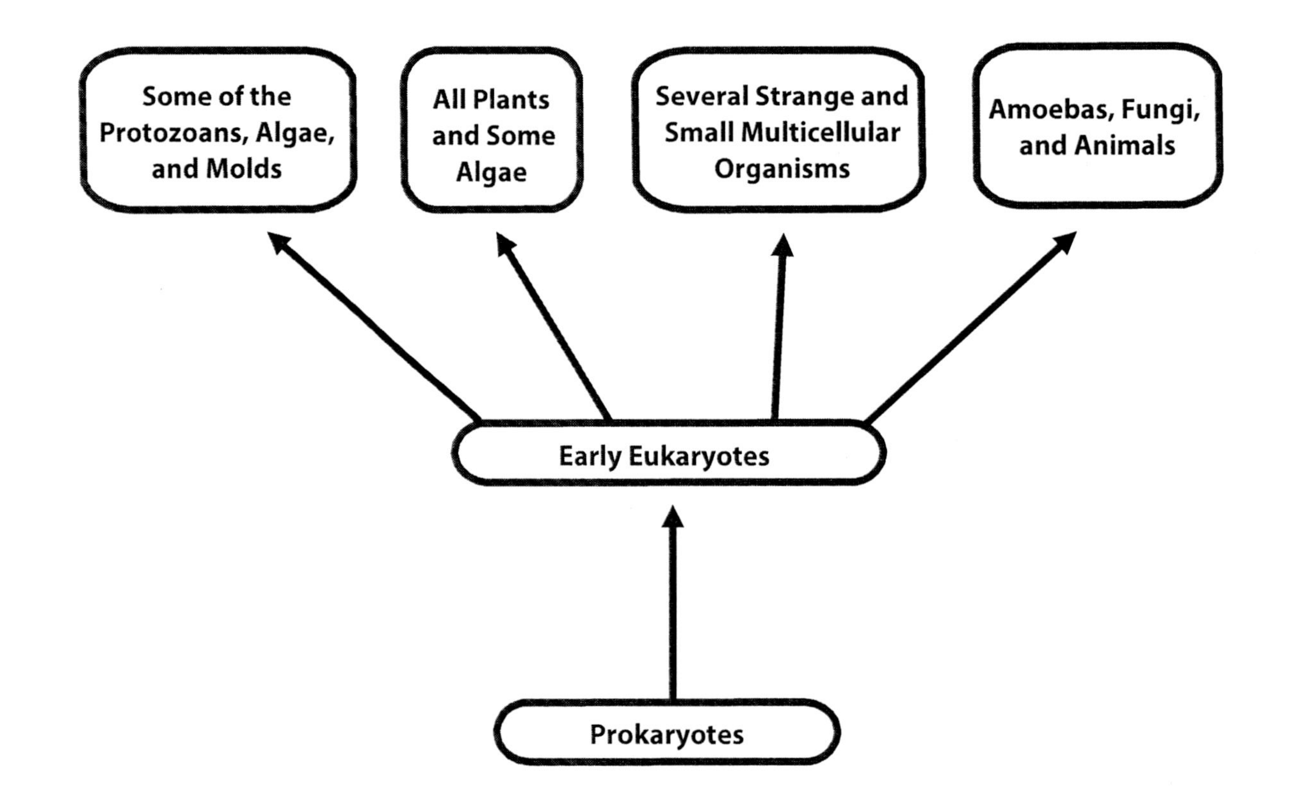

Figure 16.1. DNA comparisons of today's organisms suggest there are at least four different paths of multicellular life. One can only guess how many paths there were between prokaryotes and eukaryotes.

All of this complex branching, with extinct branches and new branching, etc. happened over and over so many times that we will never know the whole story of life. We will be working, reworking, and retelling this story for a long, long time to come. But we do have three kinds of evidence that will help us to solve parts of this impossible puzzle.

- Fossils
- Features of living animals
- DNA of living animals

The "tips" of the uppermost branches are all that survived and are alive now. Most paths were unsuccessful and became extinct.

? Question

1. An animal without a backbone is called an ________________ .
2. An animal with a backbone is called a ________________.
3. Which is the most successful animal group? (by number and diversity)
4. The oldest fossils of prokaryotic cells are about how old?
5. The oldest fossils of eukaryotic cells are about how old?
6. The oldest fossils of multicellular organisms are about how old?
7. List the three kinds of scientific evidence that help us to discover the history of animal life.

Taxonomy and Classification

Science is knowledge arranged and classified according to truth, facts, and the general laws of nature.

-Luther Burbank (1849-1926)
Horticultural scientist

Taxonomy is the process of identifying genetic traits (called ***characters***) with the goal being to name and identify all creatures. The general way of naming an organism is called the ***binomial system***, and it uses genus and species. ***Genus*** is a collection of different species that share many common features, but these different species cannot interbreed to produce fertile offspring. Humans, for example, are in genus *Homo* and species *sapiens*. Our taxonomic name is *Homo sapiens*. No other species but humans are *sapiens*. In fact, no other hominid still exists but *Homo sapiens*. In taxonomy there are a number of categories starting with the most general groups (Domains) and ending with the most specific groups (species).

The modern method of grouping organisms is by Phylogenetic Classification. ***Phylogeny*** is the description of the evolutionary history of a taxonomic group. In other words, it is the story of the steps in evolution that led to a particular group of organisms. We have recently discovered that mammals are evolved from an early reptile group, so the modern phylogenetic classification would have Mammals listed under Reptiles instead of alongside of them as in earlier taxonomic classification.

Table 16.1. Phylogenetic Description of *Homo sapiens*.

Category	Example
Domain	Eukaryotes
Kingdom	Animalia
Phylum	Chordata
Class	Reptilia
Sub Class	Mammalia
Order	Primate
Family	Hominid
Genus	*Homo*
Species	*sapiens*

A recent method for illustrating phylogenetic classification is called ***cladistics***. It uses diagrams of evolutionary history called ***cladograms*** (branch diagram). Each cladogram is a record of significant new characters (mutations) that caused separation of species along the "branches" of evolutionary history. Cladograms organize taxonomic groups not only by what they have in common, but also by how they differ from each other.

Each "bar" in the cladogram represents a defining evolutionary character or trait. The placement of any "bar" on the horizontal line of the cladogram means that the species on the right side of it (including vertical branches) have that character trait or some modification of it. Any "bar" on a vertical branch defines only the species above that bar.

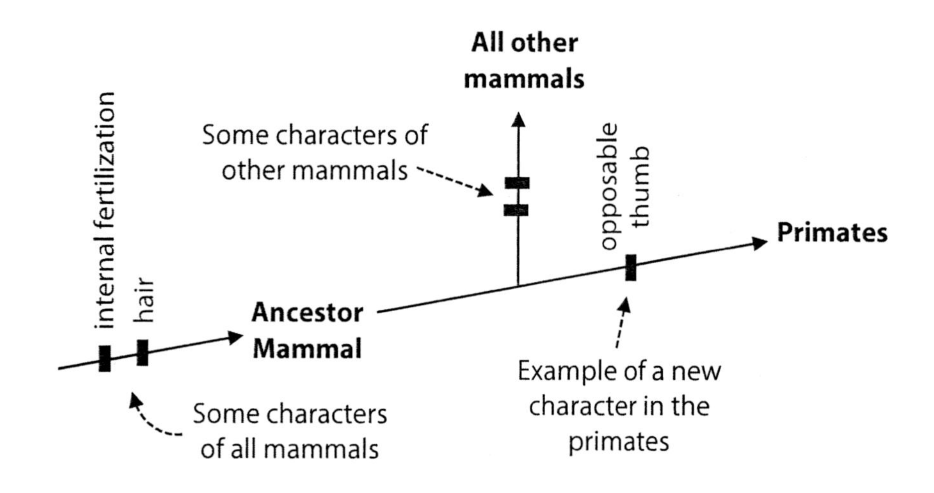

Figure 16.2. Example of a cladogram. Notice the placement of "bars" in this chart.

Cladograms present a detailed history of life, but every new evolutionary character has to be correctly placed in the chart. There are rules for this diagram method of phylogenetic classification. Evolutionary biologists must present direct evidence or a convincing theory when they decide to place a character on a specific branch of a cladogram.

Rule 1: A unique character happens only once. No organism had that particular feature before that point in evolution. All of the taxonomic groups on the branch beyond the new character trait either have that feature or some direct modification of it.

Rule 2: Some features (like wings, or a complete digestive tract, or segmentation) were so valuable for survival that they evolved independently more than once during the history of life. For example, insects, bats, and birds all have wings. However, those wings were created by modifications of different genes. They are not derived from the same ancestral gene, and they develop differently in the embryo. Therefore, each new kind of wing would be a different "bar" in the complete cladogram of animals.

Rule 3: Some evolutionary events are called ***reversals***. A group can lose a character it had before. For example, early land vertebrates had four legs, but one of the groups lost its limbs and became the ancestor of snakes. We would be mistaken to think that snakes are ancestors of legged animals, when actually they are the descendants.

Rule 4: The more similar the DNA of two groups, the closer they are to a common ancestor. Group placement in a cladogram must reflect these DNA comparisons or the cladogram is incorrect.

? Question

1. The process of describing or identifying animal characters with the goals of naming them is called ________________.

2. The binomial naming system includes ________________ and ________________.

3. Arranging organisms into groups and showing their evolutionary history and common ancestors is called ________________ classification.

4. Another way of telling the history of life is by using a diagram that illustrates evolutionary history with defining characters. That phylogenetic diagram is called a ________________.

History of Life

Fossil evidence for the appearance of various life forms is summarized in Table 16.2. Use this information to label the chart presented in Figure 16.3. Identify and label the group names for each of the "box-arrows) in the chart. Note: The length of the arrows indicates the time period that each group has existed, and the enlarged box of the arrow shows its time of greatest success.

Table 16.2. Summary of Fossil Evidence for Various Life Forms.

Group Name	Earliest Fossils	Time of Greatest Success
Anaerobic Life	4 billion years ago	3.9 to 3.3 billion years ago
Stromatolites (photosynthetic and aerobic bacteria)	4 billion years ago	3 billion to 700 million years ago
Eukaryotic Cells	4 billion years ago	1 billion years ago to today
Invertebrate Animals	4 billion years ago	500 million years ago to today
Early Chordates	600 million years ago	(shown as arrow only)
Jawless Fish	500 million years ago	480 to 370 million years ago
Cartilage Fish	400 million years ago	380 to 240 million years ago
Bony Fish	380 million years ago	66 million years ago to today
Lobe-Finned Fish	370 million years ago	(shown as arrow only)
Lung Fish	370 million years ago	(shown as arrow only)
Amphibians	350 million years ago	320 to 240 million years ago
Early Reptiles	300 million years ago	230 to 210 million years ago
Dinosaurs	210 million years ago	180 to 66 million years ago (extinct)
Mammals	200 million years ago	66 million years ago to today
Birds	180 million years ago	66 million years ago to today
Humans	1.5 million years ago	today

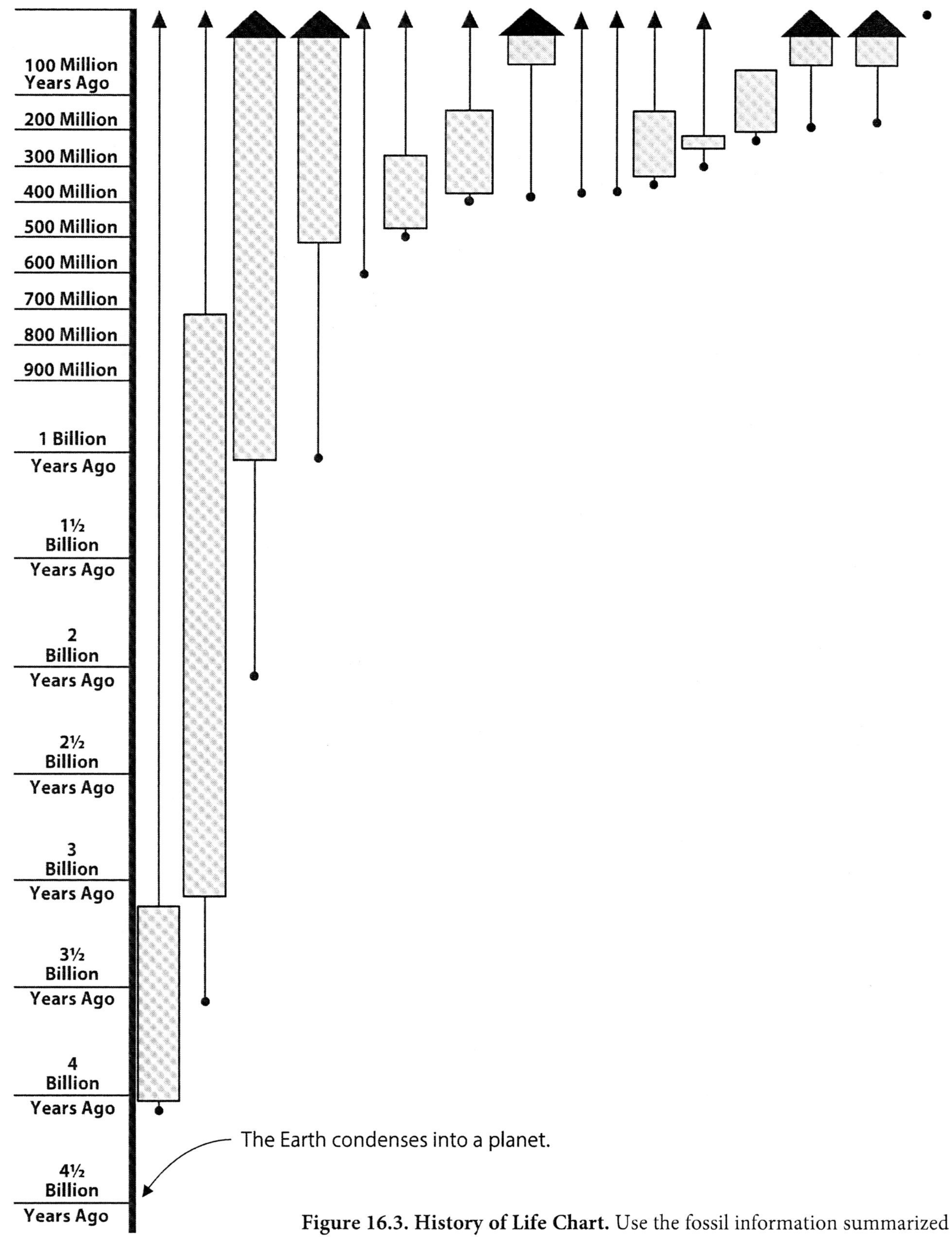

Figure 16.3. History of Life Chart. Use the fossil information summarized in Table 16.2 to label all of the "box-arrows" in this chart.

Kingdom Animalia

The classification systems we have been discussing are ways to understand and describe ***Kingdom Animalia***. The following list is a preview of some of the animals you will cover in the following Exercises.

Phylum Porifera .... sponges
Phylum Cnidaria .... jellyfish and hydra
Phylum Platyhelminthes .... flatworms
Phylum Nematoda .... roundworms
Phylum Mollusca .... mollusks
Phylum Annelida .... segmented worms
Phylum Arthropoda .... arthropods
Phylum Echinodermata .... echinoderms
Phylum Chordata .... *Amphioxus*

Subphylum Vertebrata
Class Agnatha .... jawless fishes
Class Chondrichthyes .... cartilage fishes
Class Osteichthyes .... bony fishes
Class Amphibia .... frogs
Class Reptilia .... lizards
SubClass Aves .... birds
SubClass Mammalia .... mammals

Exercise #2
Basic Body Plan

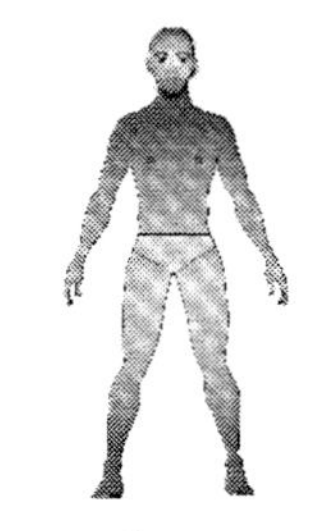

Bilateral Symmetry means "two sides".

There are three basic body plans found in the animal groups—no symmetry, radial symmetry, and bilateral symmetry. An animal with ***no symmetry*** grows in all directions, usually on the surface of something else, and typically has no particular defining form. These organisms do not move in response to stimuli in their environment because they don't have a nervous system. Each part of the organism can act independently of the other parts. We will look at only one group with this design: the sponges.

An animal with ***radial symmetry*** can be divided into two equal-looking halves by any cut through the longitudinal axis (like cutting a pie or a ball in half). These animals move or react to the environment equally in all directions. This design does not have a head end and a tail end and does not search for and locate its food as effectively as the next group does. Their sense organs are dispersed over the surface of the animal. To get food, they are usually stationary and capture food that passes by or they are carried by water currents. Jellyfish and sea anemones are examples of radial symmetry. Some animals, like the Starfish, look like they have radial symmetry, but they actually have a modified bilateral symmetry.

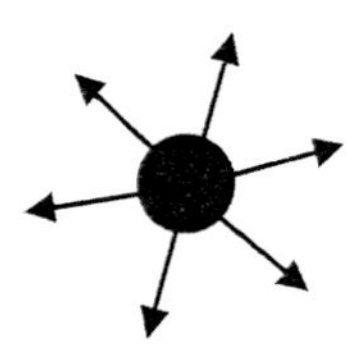

Radial symmetry is more like the design of an orange, or basketball, or apple pie.

Bilateral means "two sides", and there is only one way to cut this animal into two equal-looking halves. Bilateral design responds to the environment by using directional movement. Its head end is the location of the sense organs, and it orients to different kinds of environmental stimuli such as light and food. Flatworms are the first example of bilateral symmetry that we will consider.

Sponges (No symmetrical Body Plan)

The oldest animal fossils discovered so far are sponges from the late Precambrian Era (nearly 600 million years ago). But a biochemical residue of them has been found in geologic sediments more than a billion years old. Of all organisms, the sponges probably represent the best example of the first multicellular animals on our planet. They have no symmetry. The phylum name for sponges is ***Porifera*** which means "to have holes." Water flows in through some of the openings and flows back out of others. As the water percolates through the sponge, various small particles of food and bacteria are removed from the water and engulfed by special amoeba-like cells. The water canal system through a sponge is one of three types from very simple to quite complex. And the skeletal structure is provided by a meshwork of ***spicules*** made of either glass, calcium, or spongin (a protein fiber).

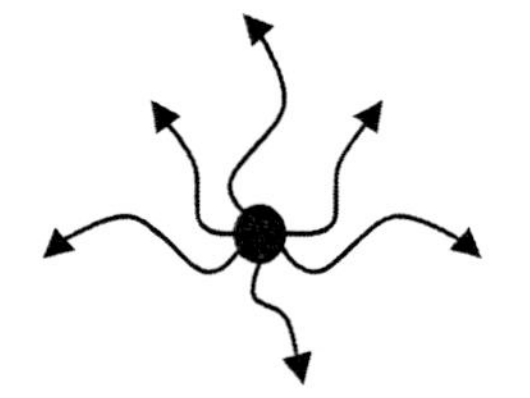

Organisms with "no symmetry" do not have a particular body plan like radial or bilateral creatures.

Please return all specimens to the specimen table so that other labs may also use them. Do not throw away any preserved animals unless you are instructed to do so.

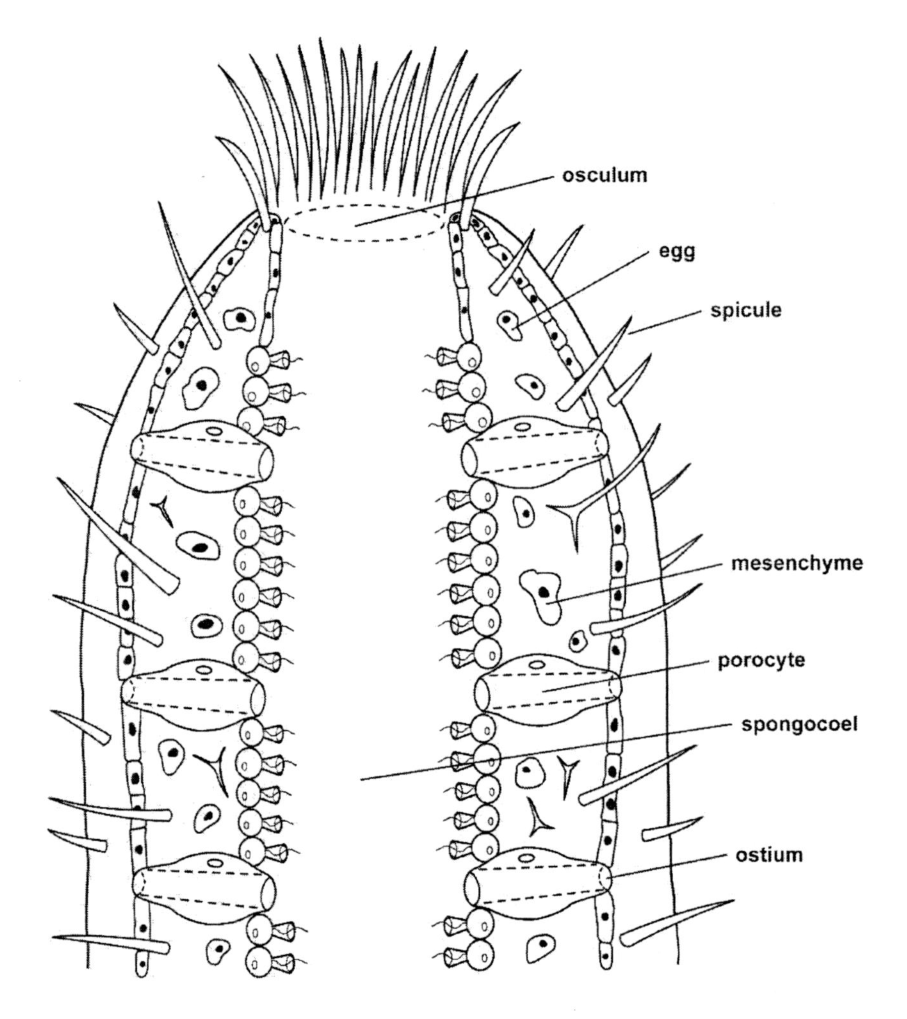

Figure 16.4. Detailed anatomy of a simple sponge.

Materials

- ✓ dissecting microscope
- ✓ small glass dish for the specimen
- ✓ small piece of a preserved sponge

Procedures

- Place a piece of the sponge onto the dish, and add enough fluid to cover the specimen.
- Use the dissecting microscope to observe the general symmetry and structure of the sponge. Notice the many holes.
- Draw a picture of the sponge and label its parts.

Sponge

- Return the sponge sample for other students to use.

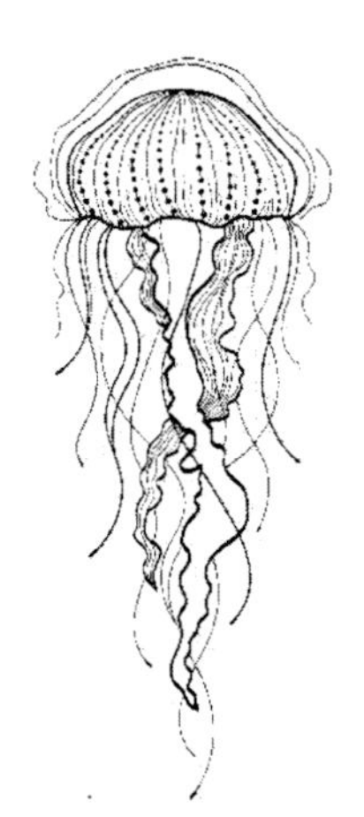

Phylum Cnidaria includes jellyfish, sea anemones, corals, and some other colonial forms.

Jellyfish Types (Body with Radial Symmetry)

Jellyfish and sea anemones belong in a phylum named ***Cnidaria*** which means "to sting." There are two unique evolutionary characters found in this group: ***stinging cells*** and ***radial symmetry***. Radial symmetry orients an animal to its environment in all directions at the same time. The jellyfish types and all groups following them also have tissues. ***Tissues*** are groups of identical cells that work together to perform a particular function. The Cnidarians have two primary tissue layers with various other cell types scattered through the organism.

These organisms feed by stinging their prey and pulling them into a ***gastrovascular cavity*** where they are digested. There are two body forms: ***polyp*** (the sea anemone) and ***medusa*** (the jellyfish form). Many species alternate between both forms as parts of their life cycle.

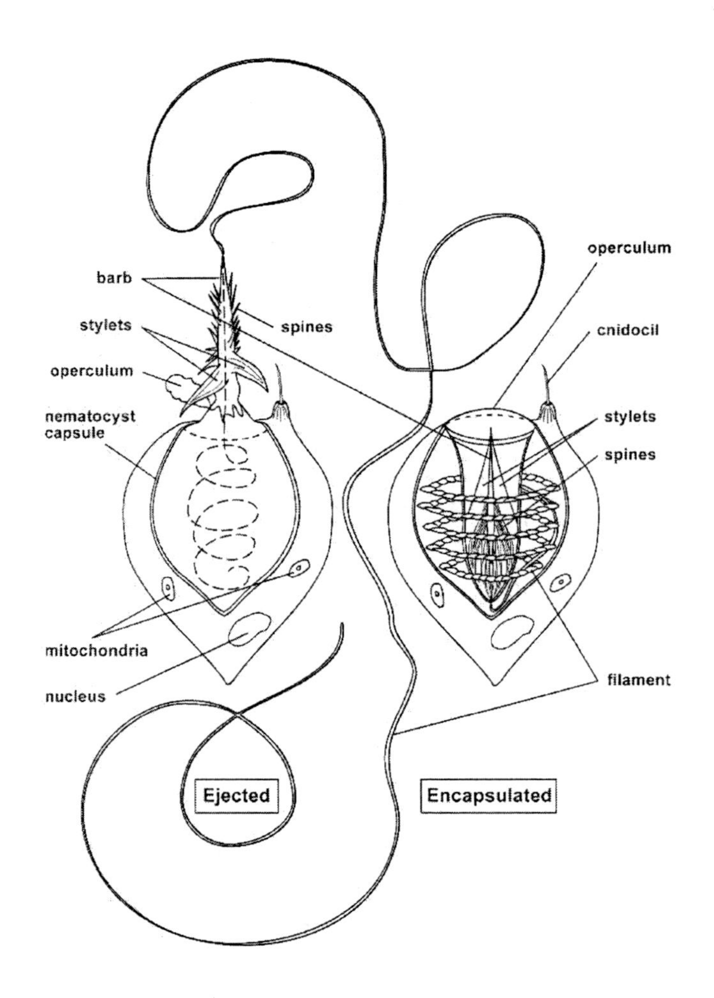

Figure 16.5. Structure of the Cnidarian stinging cell. All jellyfish, sea anemones, and other cnidarians have these cells for feeding and defense. It is a defining character of Phylum Cnidaria.

Materials

- ✓ dissecting microscope
- ✓ living hydra or preserved small jellyfish

Procedure

- Compare the two body forms at the demonstration table.
- If you are to examine a living hydra, follow the specific directions of your lab instructor. Otherwise, obtain a preserved specimen of a small jellyfish, and follow the same general procedure you used to examine the sponge.
- Draw a picture of your Cnidarian.

Hydra or Jellyfish

- Return your specimen for other students to use.

Flatworms (Body has Bilateral Symmetry)

Early animals like sponges grew with no particular symmetry. They were followed by the jellyfish types with radial symmetry. The next fossil groups appear with ***bilateral symmetry***. An early group illustrating that feature is the ***flatworm***. Flatworms are in the Phylum Platyhelminthes, which means "broad, flat stem". These worms are ***flat***, and we'll use that character as their unique trait for their cladogram later. Although the word ***bilateral*** means "two sides," the importance of this symmetry is that it results in a head end and a tail end. Bilateral animals are directional. As such, bilateral animals are better designed to focus on where they are going with a head end. They effectively search through their environment.

Flatworms have ***three primary tissue layers***, and they have ***organs***. The earlier groups did not have organs. An organ is made of a group of tissues working together for a united function. However, because flatworms are not complex organisms, they do not have organ systems like a circulatory system. This limitation means that flatworms depend on diffusion to move nutrients and waste products around their bodies. They are forced to be flat because they depend on diffusion.

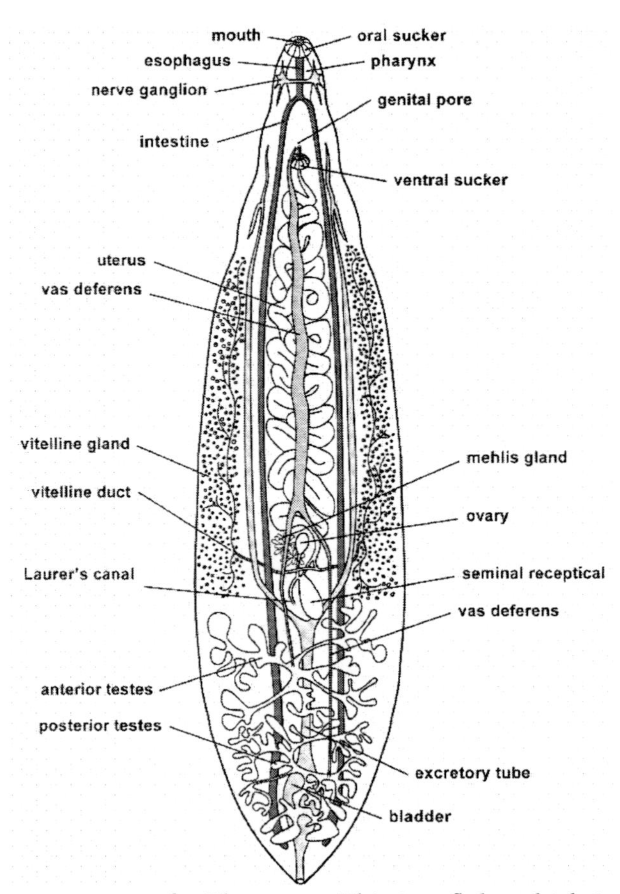

Figure 16.6. Anatomy of a Flatworm. This is a fluke which is a parasitic flatworm like the tapeworm. Another example of Phylum Platyhelminthes is the free-living group called planarians.

Materials

- ✓ dissecting microscope
- ✓ Plasti-mount of a flatworm
- ✓ living planarian flatworm

Procedure

- View the plasti-mount flatworm from the top and the side. You should see why they are called flatworms. Using the dissecting microscope, identify the organs or structures that are shown in the anatomical diagram in Figure 16.6.
- If there are living planarian flatworms, carefully remove one with a probe or dropper. Include a small amount of water from the container for the flatworm to move through. Examine with the dissecting microscope.
- Draw a picture of the flatworm.

Flatworm

- There are other parasitic flatworms (tapeworms and flukes) at the demonstration table. Examine them.
- When you have finished, please return all specimens to their appropriate containers for other students to use.

? Question

1. Which type of symmetry is displayed by the sponge?
2. The taxonomic name of the sponges is Phylum Porifera, which means "hole-bearing." Describe the holes in your sponge.
3. Which type of symmetry is displayed by the hydra and jellyfish?
4. Jellyfish and hydras are in the Phylum Cnidaria, which means "stinging cells." How are stinging cells used by these animals?
5. Compare the sponge anatomy in Figure 16.4 to the flatworm in Figure 16.6. What are the most general differences that you notice?

6. Which type of symmetry is displayed by the flatworm?

7. What are the two most obvious characteristics of flatworms?

8. Put the three animal subgroups (hydra, flatworm, and sponge) in order of increasing complexity or symmetry.

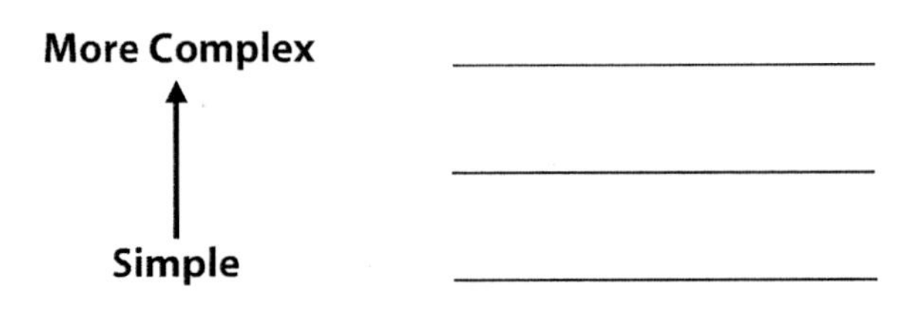

Exercise #3
Types of Digestive System

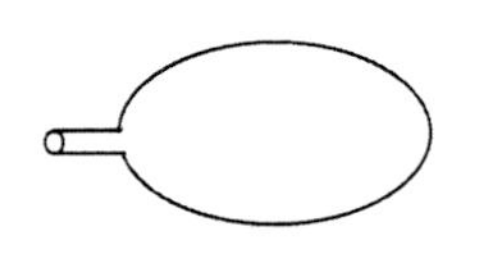

An incomplete gut has one opening into a digestive chamber.

There are two basic designs for the digestive system within the Animal Kingdom—***incomplete gut*** and ***complete gut***. The ***incomplete gut*** is a digestive system consisting of a tube with only one opening to the outside. The food is taken into the tube through the mouth and is digested. After digestion is completed, the waste products are ejected from the animal through the same opening. Animals with an incomplete gut are not advanced enough to have excretory and circulatory systems. Therefore, they are dependent on diffusion to obtain nutrients from digestion and to get rid of waste products. Example phyla are sponges, jellyfish, and flatworms.

Animals with a complete gut are better at digestion.

A ***complete gut*** digestive system consists of a tube extending through the organism. There are two openings — one for taking in food and another for eliminating the waste products of digestion. These animals are more advanced and do have excretory and circulatory systems to transport nutrients around the body and to remove wastes. The complete gut trait is much better at processing food, and it evolved independently at least two different times — once in the evolutionary branch leading to roundworms, mollusks, annelids, and arthropods (the Protostomes); and again in the branch leading to the echinoderms and chordates (the Deuterostomes). Those two evolutionary branches differ in the way their digestive system forms in the embryo. ***Protostome*** means "first mouth", and the first opening in the embryo becomes the mouth. ***Deuterostome*** means "second mouth" because the first opening becomes the anus and the second opening becomes the mouth.

Roundworms (Simple Complete Gut)

There are two types of complete digestive systems-protostome and deuterostome. The Protostomes include the roundworms, mollusks, annelids, and arthropods.

Roundworms are in Phylum Nematoda which means "thread-like." They are ***protostomes*** and are very small—just visible with a magnifying glass. Nematodes consume dead plants and animals and are important in the ecological recycling of organic matter in both soil and water. Some are parasitic, and the lab example roundworm is *Ascaris* (dog roundworm) because it is unusually large and easy to dissect.

Special evolutionary traits of roundworms include the ***complete digestive tract*** and the ***round body shape***. Roundworms and all the following phyla are not forced to have a body shape dependent on diffusion like earlier groups. A body shape other than thin and flat is allowed because roundworms have body systems. ***Systems*** are groups of organs that work together for a united function—like digestive, circulatory, and excretory. If you have body systems, then there are complex and effective methods for transporting materials around the body. Then body shape can be designed for something else. One of those possible designs is for burrowing which the roundworms can do very well.

The roundworms have a complete digestive system, a round body shape, and body systems.

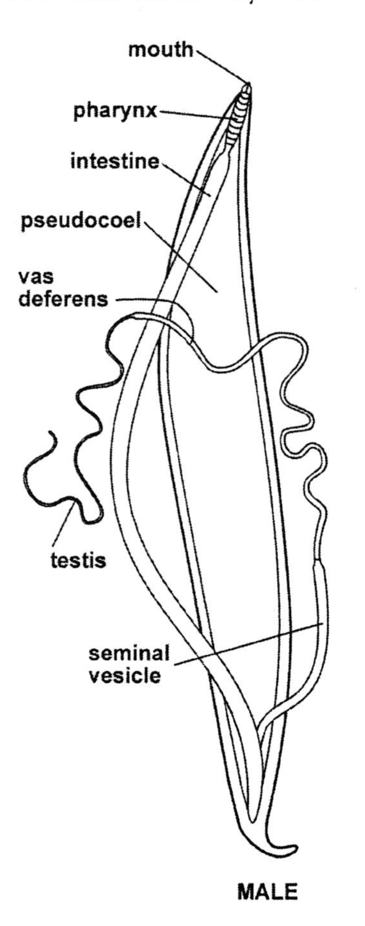

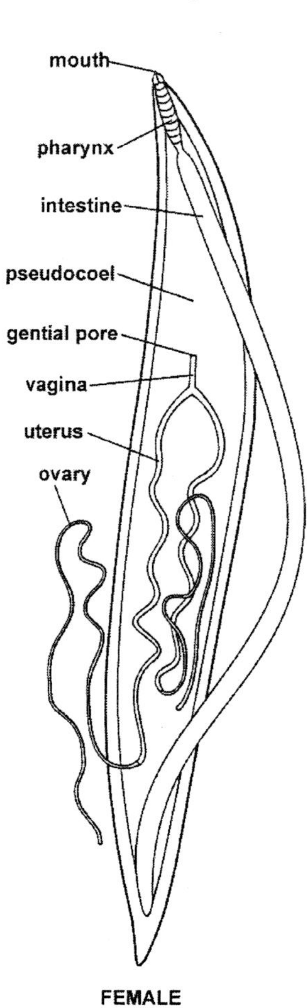

Figure 16.7. Anatomy of the roundworm *Ascaris*.

Materials

- ✓ dissecting microscope
- ✓ preserved *Ascaris* specimen
- ✓ dissecting equipment

Procedure

- The instructions at the demonstration table will indicate whether you are to do your own dissection of *Ascaris*, or if you will observe a previously dissected specimen instead. Now, observe the dissected *Ascaris* roundworm. Use Figure 16.7 as a reference.
- Notice the general appearance of the roundworm and compare it to the form of the flatworm. (Compare Figure 16.6 and Figure 16.7.) What obvious difference do you notice?
- Try to identify some of the internal organs as indicated on the anatomical diagram.
- Draw a simple sketch of the roundworm, and label some of its organs.

Roundworm

- If you dissected, please discard all remains appropriately.

? Question

1. Sponges are full of holes, jellyfish are big open sacs, and flatworms are flat. How are these features important in facilitating the process of diffusion?

2. Roundworms are round. Is a round shape a better or worse shape for diffusion compared to a flat shape? Explain.

3. If the basic body design of an animal like the roundworm is not good for diffusion, then how do they solve the problem of transporting nutrients and waste products around the body?

4. The shape of roundworms is best for

   Swimming or Burrowing

5. Which digestive system would you call an "eating machine"—one more efficient at getting nutrients?

   Incomplete gut or Complete gut

6. If an animal is more efficient at getting nutrients, then it will be

   More successful or Less successful

7. Draw a simple sketch of an incomplete digestive tract.

8. Which animal groups have an incomplete gut?

Example	Phylum

9. Draw a simple sketch of a complete digestive tract.

10. Which animal group (of the ones so far) has a complete gut?

Example	Phylum

11. Which digestive system type do you belong to?

Exercise #4
Segmentation

The next big jump in evolution is segmentation. This body design results from a radical change in the normal development of an animal. It is a unique evolutionary feature that allowed a "new" kind of animal to develop. In ***segmentation***, blocks of embryonic tissue are duplicated instead of remaining as one single block.

Original Animal

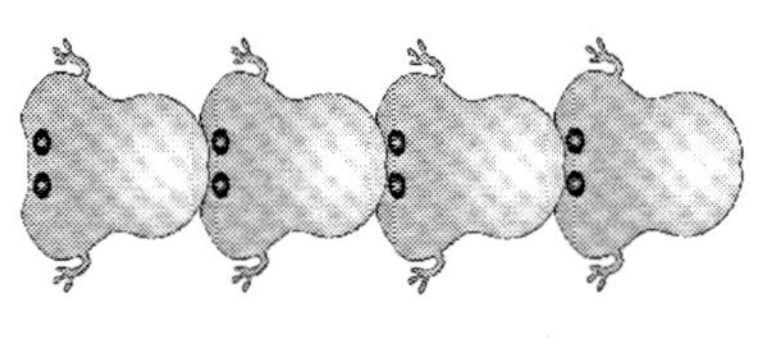

Segmented Animal

These repetitive blocks result in an organism that is a combination of ***segments***. This "new" animal is like making a chain of the original animal, linked together, one after the other. Apparently, segmentation is so valuable to animals that it evolved at least twice — once in the evolutionary branch leading to annelids and arthropods, and again in the branch leading to echinoderms and chordates. But these segmentation traits were from completely different mutations each time that segmentation evolved. This is another example of a trait, like flying, that evolved more than once in the history of life.

Segmentation appears to have been a profound error during embryonic development that led to a new line of animals.

One of the ways to understand the significance of segmentation is to watch it happen in a segmented animal. Figure 16.8 shows how segmented worms typically develop. What starts out as one kind of animal at stage one becomes something unbelievably different at stage six.

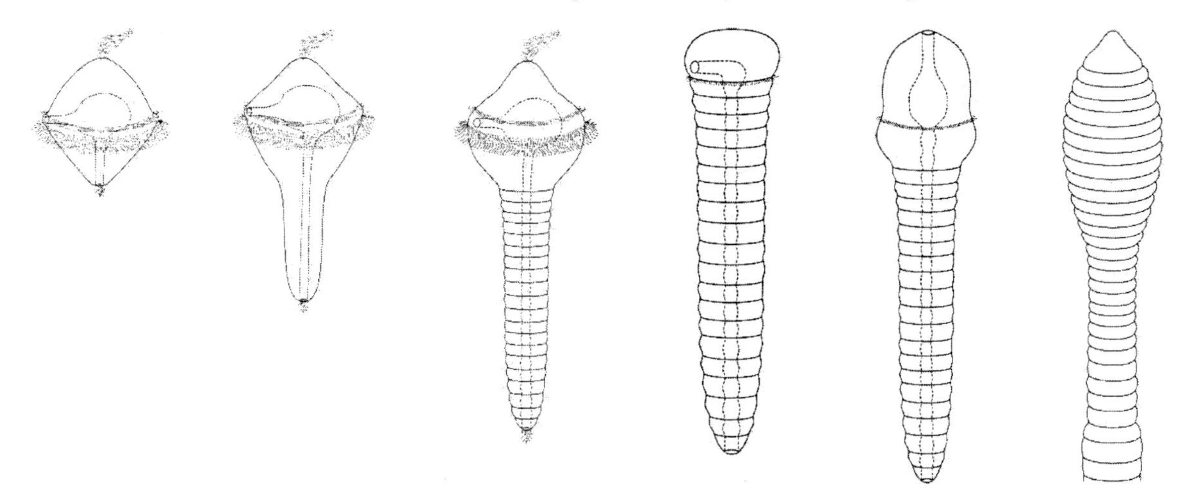

Figure 16.8. Development of the segmented worm from an unsegmented larval stage to a segmented body form.

Materials

✓ Go to the specimen table.

Procedure

- Put a roundworm, an earthworm, a millipede, and a crayfish next to each other in that order. You have just laid out an important evolutionary arrangement!

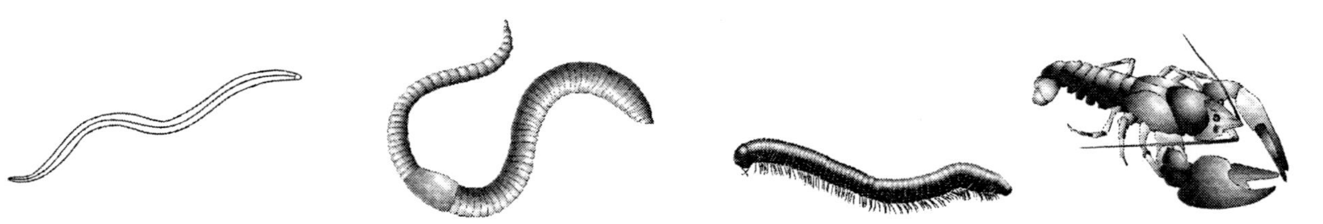

- The earthworm is called the ***segmented worm*** (Phylum Annelida, meaning "ringed"). The millipede and crayfish are the Phylum Arthropoda (meaning "jointed foot"), the most successful animal group today. Arthropoda includes many body designs based on segmentation.
- If your lab class is doing an earthworm dissection, refer to Figure 16.9 for assistance.

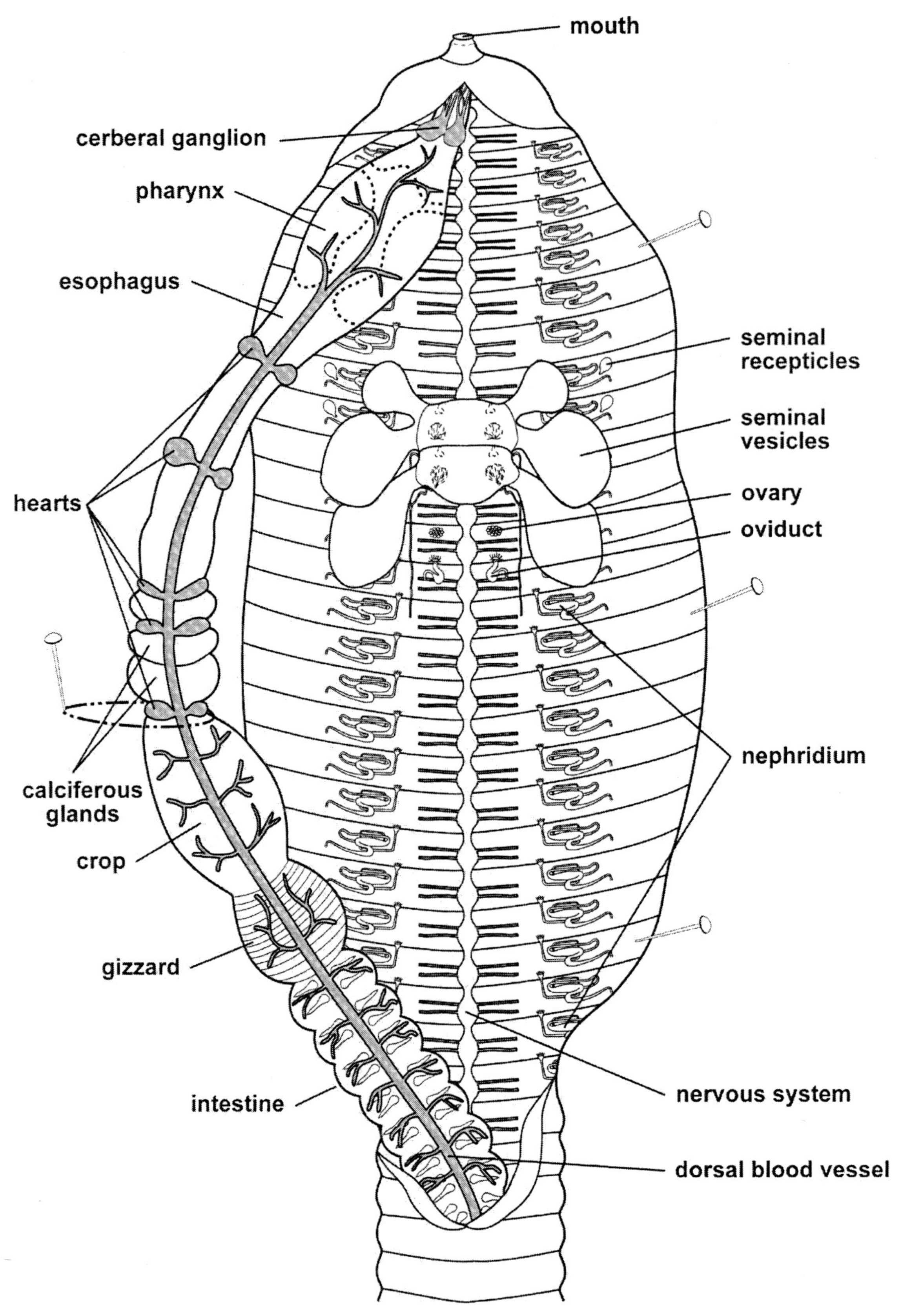

Figure 16.9. Anatomy of the Earthworm. Use this diagram to assist your earthworm dissection.

? Question

1. A segmented animal would be (smaller or larger) than the early "pre- segmented" animal form?

   Smaller or Larger

2. What would be the benefit to the animal that developed from the above change?

3. Based on your observations of the segmented worm, what evolutionary changes happened to the individual segments?

4. Use your knowledge of basic body plan, digestive systems, and segmentation to put the animal subgroups in order of increasing complexity (hydra, sponge, arthropod, roundworm, flatworm, and segmented worm).

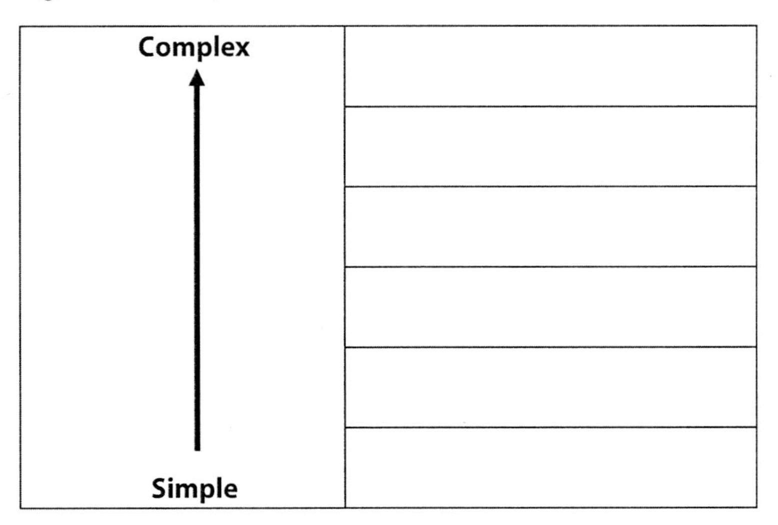

5. Do you think the evolutionary development of segmentation applies to human development? (Ask your instructor about the appearance of early embryonic development in humans.)

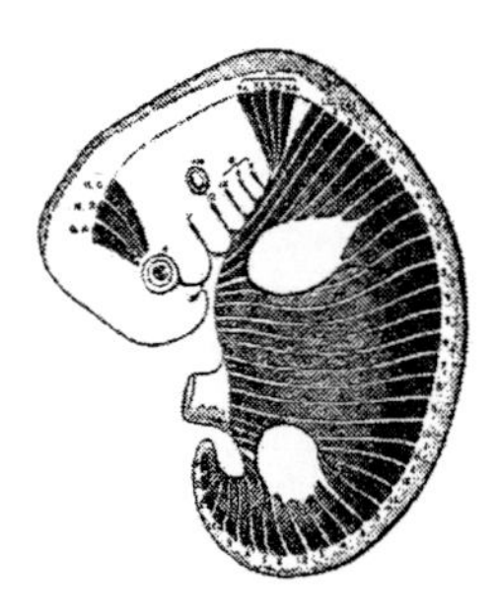

Human Embryo

Depending on your point of view, all organisms are either exceptions to the rule or examples of the rule.

Other Phyla

Evolution is not a "directional game," even though we find it easier to talk about the process that way. Evolution is more like an experiment. Some experiments succeed; others fail. And some go on and on because there are no big obstacles to prevent their continuation. There are dozens of animal phyla that are not covered in this lab. Many creatures resulting from evolution's experiments look like exceptions to the rule or special examples, but all organisms are exceptions depending on your point of view. Mollusks and echinoderms are two successful groups that should be mentioned in our brief story of animal life.

Mollusks

Phylum ***Mollusca***, meaning "the soft ones," includes animals that have a shell during some stage of their lives. They evolved at about the same time as segmentation was evolving, perhaps a bit before the segmented worms. Mollusks are very successful and have many species and body types.

Procedure

- Go to the demonstration table and look at the various body type subgroups in Phylum Mollusca. List five different body types.

____

____

____

____

____

Echinoderms

Members of the Phylum ***Echinodermata***, which means "spiny skin," are very complex animals that have returned to a modified radial symmetry. (Originally, they came from a bilateral marine group.) Echinoderms have an unusual arrangement of tube-feet that are moved by a ***hydraulic system*** within the body. The ***larval stage*** of echinoderms is more like the larva of simple chordates. And the complete gut seems to have developed from the same evolutionary mutation in both groups. These embryonic similarities suggest that echinoderms are one of the closest invertebrate relatives to Phylum ***Chordata*** (the animal group to which you belong).

Procedure

- Go to the demonstration and look at the various body type subgroups in Phylum Echinodermata. List five different body types.

____

____

____

____

____

Table 16.3. Practice Review of Invertebrate Evolutionary Traits.

Matching Invertebrate Traits Make the correct match between the two columns.	
Group Names	**Defining Features**
Sponge _____	**A.** This group has blocks of duplicated embryonic tissue that develop into different parts in the adult organism.
Jellyfish Types _____	**B.** This invertebrate group is more closely related to chordates than to most other invertebrates. Their larval stage is bilateral, but the adult body type has modified radial symmetry.
Flatworm _____	**C.** This group is by far the most successful animal group living on earth today. And they have an exoskeleton.
Roundworm _____	**D.** This group includes clams, squids, octopuses, slugs, and snails. There is a shell during some stage of the animal's development.
Mollusk _____	**E.** This animal has stinging cells, and it has radial symmetry in the adult form.
Segmented Worm _____	**F.** This animal is the first multicellular animal of all the groups covered in this lab.
Arthropod _____	**G.** This is the first group with a complete digestive tract – food in one end and waste out the other end.
Echinoderm _____	**H.** This is the first group with a head-end and a tail-end. They have bilateral symmetry.

Taxonomic Groups
Echinoderms
Sponges
Roundworms
Jellyfish Types
Mollusks
Flatworms
Segmented Worms
Arthropods

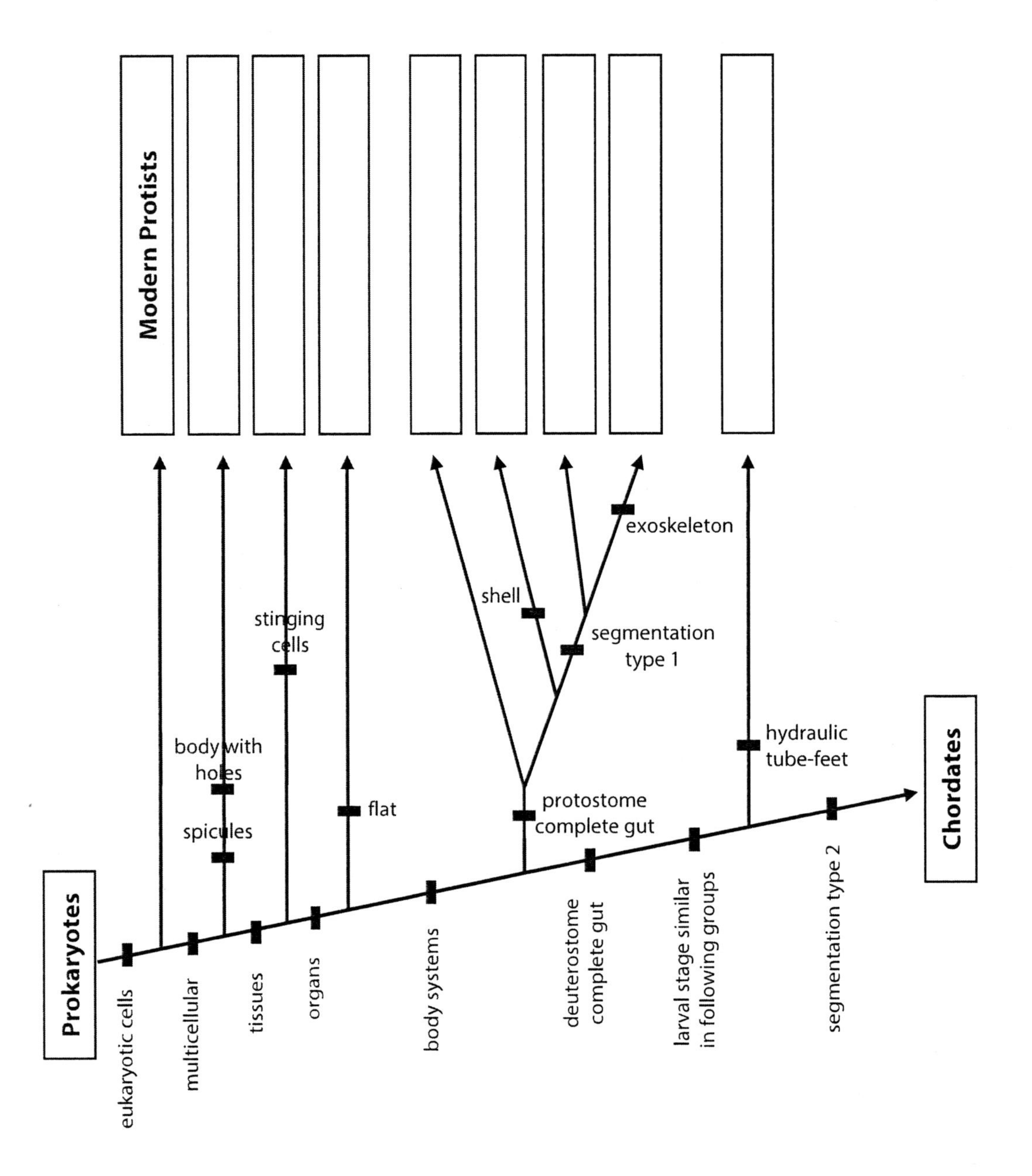

Figure 16.10. Cladogram for invertebrates covered in this lab. Use the clues from the evolutionary characters shown in this cladogram to fill in the correct group names in each of the boxes.

Exercise #5
Skeletons

Animals with an exoskeleton stay small; those with an endoskeleton can be big.

A ***skeleton*** is a supporting structure made of different pieces that can be individually moved by muscles. The evolutionary success of animals with skeletons follows their capacity for refined movement. The better you are at moving, the better off you are in terms of survival. There are two basic kinds of skeletons in the animal kingdom—the exoskeleton and the endoskeleton. Both types are successful. The main difference is in the size of the organism that can be supported by each.

Exoskeleton (Most Successful Invertebrates)

"Exo" means outside. An ***exoskeleton*** is on the outside surface of the animal. The arthropods are the animal group that most successfully utilized the exoskeleton. Name an example of an arthropod: ________________

Endoskeleton (Vertebrates)

"Endo" means inside. An ***endoskeleton*** is inside the animal and grows with it. There are three variations: notochord, cartilage skeleton, and bony skeleton.

Internal Flexible Rod (Notochord)

A notochord doesn't exactly fit our definition of skeleton because it is a one-piece structure, but it is the precursor of an internal skeleton. It has many muscle fibers attached along its length for controlling swimming movements. The first animal with a notochord would have been something like the current-day *Amphioxus*. This small fish-like chordate has ***gill slits***, a ***dorsal nerve cord***, and a ***notochord***. These evolutionary traits link them to all other chordates. The early chordates were followed by the Jawless Fish, which has a larger notochord and brain. Jawless fish dominated the seas 400 million years ago, and some of them were extremely large (10+ meters).

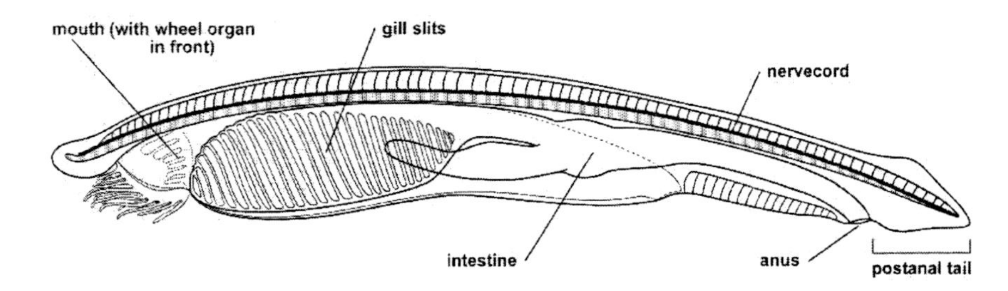

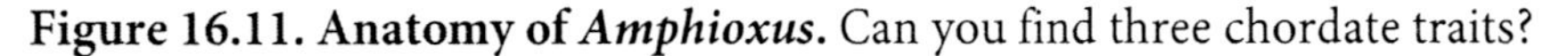

Figure 16.11. Anatomy of *Amphioxus*. Can you find three chordate traits?

Cartilage Skeleton

This type of skeleton is made of pieces of cartilage. Cartilage is a softer supporting material than bone. An example of cartilage is the flexible part of your nose and ear. The earliest cartilage skeleton occurred in the ***Cartilage Fish*** group (sharks and rays). Although the cartilage skeleton is less refined than the bony skeleton, it did allow for the evolution of a jaw and better fin movements compared to earlier chordates with only a notochord.

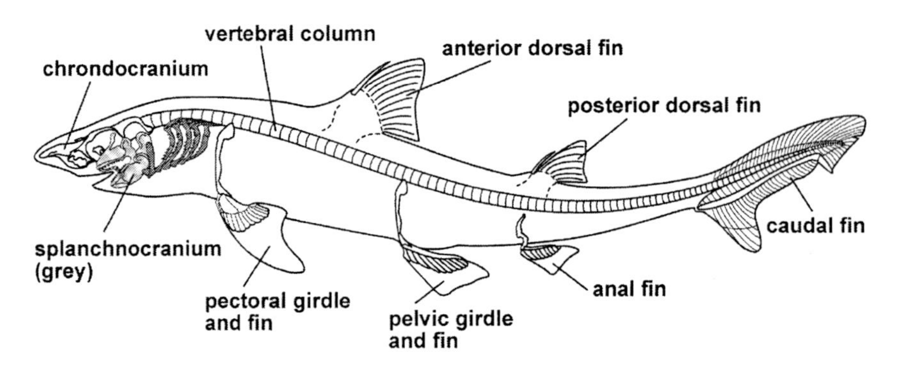

Figure 16.12. Structure of the cartilage skeleton of a shark. There is more specialization of the skeleton compared to the notochord of Amphioxus or jawless fish.

Bony Skeleton

This type of skeleton is made of many hard bony pieces that can be individually moved by separate muscles. Movement is much more refined than what a cartilage skeleton will allow. Bony Fish are the first evolutionary group with a bony skeleton, and they are the dominant fish found in today's seas. Theirs was the beginning skeleton for all other vertebrates that followed.

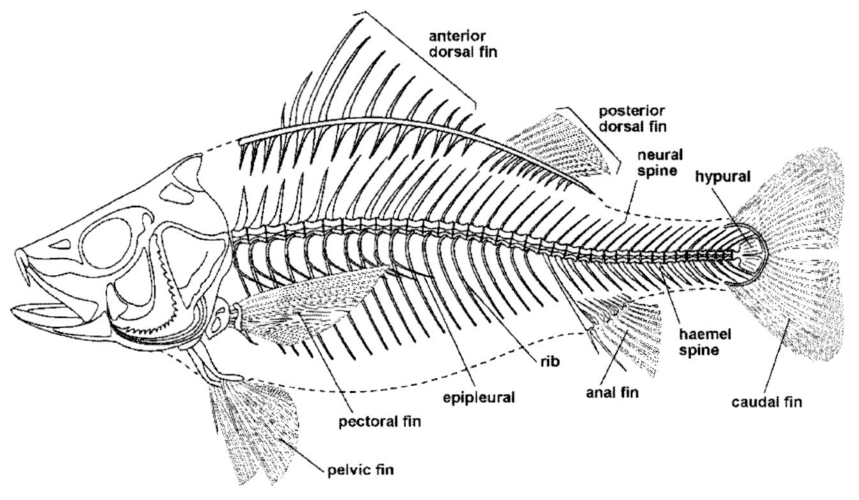

Figure 16.13. Skeleton of a Bony Fish. This is a more refined skeleton than the cartilage skeleton, because bone is a much better building material.

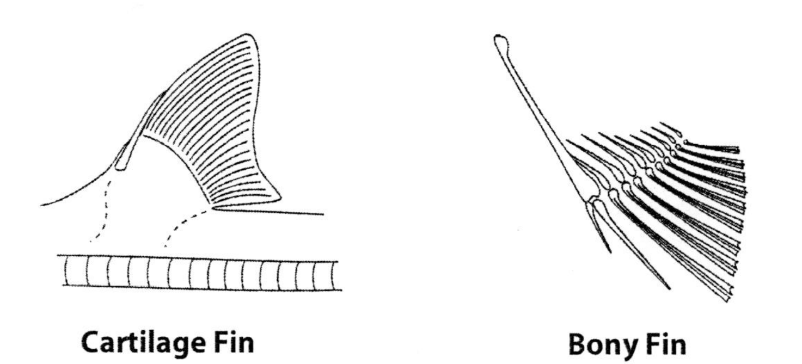

Figure 16.14. Comparison of cartilage fin and bony fin. Bony fish are much better at swimming than sharks, and they are much better at specialized eating for the same reason. Bones are better than cartilage, and all land vertebrates have this type of skeleton.

Procedure

- Go to the specimen table. Put an insect, an *Amphioxus*, a jawless fish, a cartilage fish, and a bony fish in that order.
- If Amphioxus slides are available, examine them with your compound microscope.
- Put the cartilage skeleton and the bony skeleton plastic mounts next to the appropriate fish. Study these animals carefully, observing their particular skeletal structures.

? Question

1. Compare the insect to the other four animals. What is the most obvious difference?

2. What kind of skeleton does an insect have?

3. Which type of skeleton might present a problem to a growing animal?

   exoskeleton or endoskeleton

4. Explain why you think it would be a problem.

5. Which type of skeleton is designed to grow along with the animal?
   exoskeleton or endoskeleton

6. Can a fish have a jaw if it doesn't have a multi-piece skeleton?

7. What does that tell you about an animal we call a jawless fish?

8. List two examples of animals with a one-piece internal flexible rod?

9. Which type of skeleton is capable of more refined swimming movements?
   cartilage skeleton or bony skeleton

10. Explain why you think so.

11. Do you think that Amphioxus can swim as well as an aquarium fish?

12. Explain why or why not.

Exercise #6
Vertebrate Land Adaptations

About 350 million years ago the vertebrate fish had already evolved, and there were large areas of the planet covered with shallow seas and lowlands. These conditions favored the evolution of any group with features suited to land. But several environmental challenges had to be overcome before any organism could survive in a land environment. Among those challenges were: movement on land, breathing air, preventing water loss, and reproduction.

Procedure

- Go to the specimen table.
- Put a bony fish, a frog, a lizard, a bird, and a mammal in that order. This order represents the series of land adaptations that were made by the vertebrate groups.

? Question

1. Observe and compare the bony fish to the frog. What is the most obvious structural difference?

2. Those structures found on the frogs evolved from what parts of the fish?

3. The frog is in the Phylum ***Chordata*** (meaning "notochord"). It is part of the subphylum ***Vertebrata*** (meaning "having a backbone") and is a member of the Class ***Amphibia*** (meaning "living both lives"). What process in the amphibian life cycle depends on water? (Recall the situation you studied with the lifecycle of moss and fern plants.)

4. The fact that a frog can live out of water and a fish cannot tells you that the frog has solved the problem of _______________.

5. What structure in the frog solves that problem?

6. Compare the frog to the lizard. What is the most obvious difference?

7. The lizard belongs to the vertebrate subgroup Class ***Reptilia***, which means "creeping." Modern reptiles have legs that are positioned out to the side of the animal which makes them appear to creep as they move. One of the early reptiles had legs positioned under the body, and that was the reptile group that gave rise to what are now mammals. The land success of reptiles resulted from two improvements over the amphibians—***waterproof skin*** and significant changes in reproduction. Reproductive improvements included ***internal fertilization*** and the ***amniotic egg***.

   What two features allowed the reptiles to be successful in dry environments?

8. How did the two special changes in reproduction improve land success for the reptiles? (Ask your instructor.)

9. Compare the lizard to the bird and the mammal. What are the most obvious external differences?

10. Discuss the advantages of having insulating features that allow for regulation of body temperature.

11. Birds and mammals are descended from early reptile groups and are classified as subclasses under Class Reptilia. The subclasses are ***Aves*** (which means "birds") and ***Mammalia*** (which means "breast"). Mammals are given their name because they are the only animals with mammary glands for feeding their young. And their embryos develop with the aid of a special structure called the ***placenta***. The external features of birds (feathers) and mammals (hair) are important evolutionary character traits that separate them from the other reptilian groups. Discuss how these external differences allow birds and mammals to live in places where reptiles can't survive.

12. Using your knowledge of skeletons and land adaptations, put the subgroups of chordate animals in their order of complexity (reptile, jawless fish, mammal, amphibian, bony fish, bird, and *Amphioxus*).

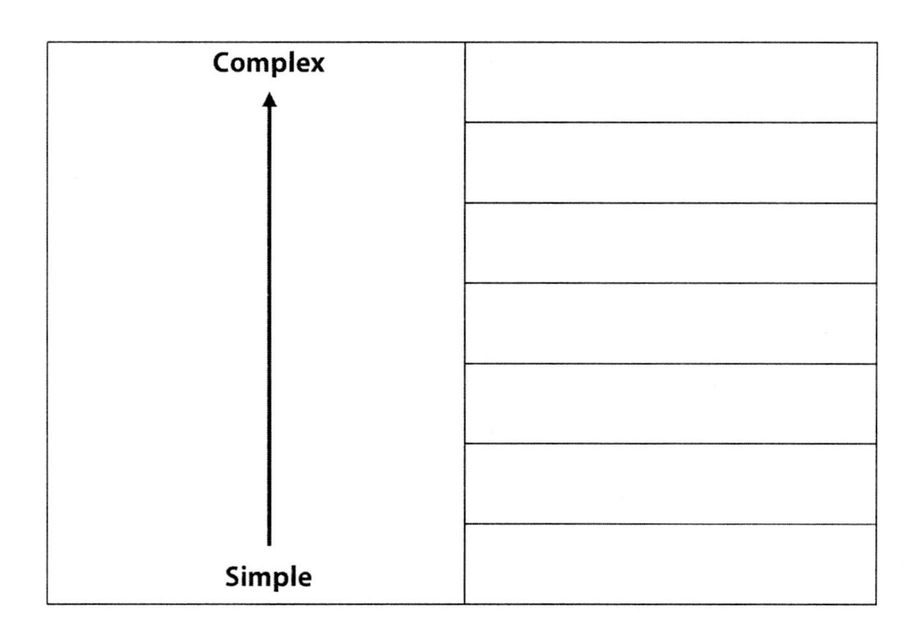

Table 16.4. Practice Review of Vertebrate Evolutionary Traits.

Matching Vertebrate Traits Make the correct match between the two columns.	
Group Names	**Defining Features**
Amphioxus _____	**A**. First animal with both legs and lungs.
Jawless Fish _____	**B**. One of the two groups that can regulate body temperature.
Cartilage Fish _____	**C**. Has hair. It is the only group with internal development of the embryo that is supported by a placenta.
Bony Fish _____	**D**. A small (5-7cm) fishy-looking creature (but not a fish) that has a notochord, gill slits, and a dorsal nerve cord.
Lung Fish _____	**E**. First animal with internal fertilization. First to lay an energy rich and protective egg to support the next generation.
Lobe-Finned Fish _____	**F**. First fish with a lung.
Amphibian _____	**G**. First fish with limbs.
Reptile _____	**H**. The most successful fish today.
Bird _____	**I.** The first fish with a skeleton and a jaw.
Mammal _____	**J.** A fish creature that is about a foot long or so today. This group has a vertebral column and brain but no skeleton beyond that. It dominated the seas 400 million years ago, and some were very large.

New Evolutionary Characters

A. Mammary glands
B. Internal fertilization and amniotic egg
C. Lung
D. Hair
E. Lobe-fins
F. Cartilage skeleton (including jaw)
G. Brain
H. Feathers (modified scales)
I. Notochord
J. Four limbs
K. Waterproof scales
L. Bony skeleton

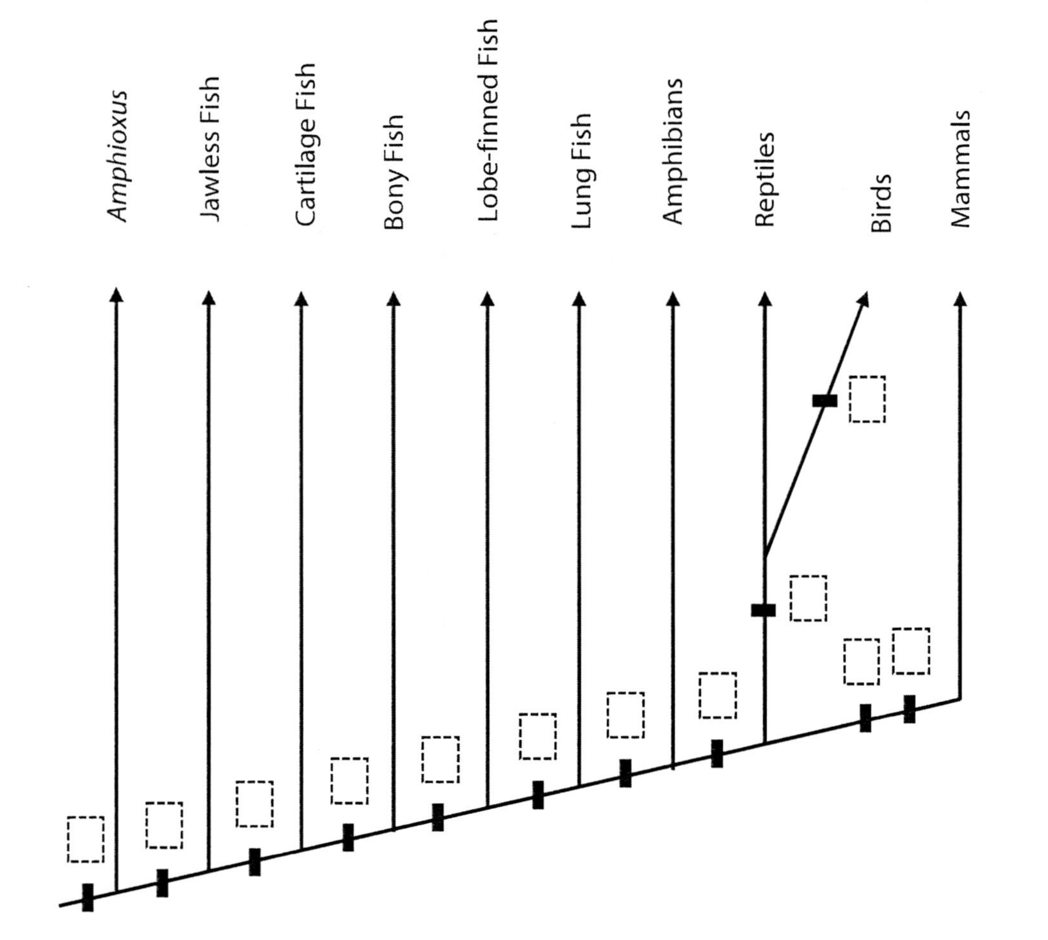

Figure 16.15. Cladogram for vertebrates covered in this lab. Taxonomic groups are completed for this cladogram. Label the evolutionary character at each point indicated in the diagram. (Put the correct letter in each box.)

LAB

17

Diet Analysis

Americans spend more money on high-Calorie fast foods and low-Calorie diet foods than anyone else in the world. More books have been written on body weight than on any other health issue. We enjoy the highest availability of foods, but processed foods far outsell fresh vegetables and fruits. What seems like the richest-fed nation is really a country with too many people suffering from diet-based diseases and excess overweight.

The curses of processed foods follow their blessings. They taste so good that we eat too much. They have extra salt (better taste and long shelf-life), and we surpass our daily need with one entrée. They are convenient, and we pass by the needed natural foods. Stripped of indigestible components (fiber, etc.), what we call "junk food" is really too high in Calorie value. Our body is designed for eating low Calorie food that has a lot of indigestible material in it. A sugary donut is so concentrated in sugar and starch that our pancreas reacts like we just ate a crate of oranges (an exaggeration). The pancreas releases an extra big shot of insulin to push sugar into the cells of your body. But the donut doesn't actually have that much sugar, so the excess insulin drives down the normal blood sugar. You're hungry for another donut in an hour. Junk food is just too darn concentrated for us.

Decades of research tells us that the foundation of a healthful diet is unprocessed food consisting of mostly plants, lean protein sources, and dietary oils from fruits and nuts. The addition of modest amounts of whole-grain foods provides additional fiber and nutrients. Something like this diet is our goal.

Exercise #1	**Energy Balance of the Body**
Exercise #2	**Energy in Food - Basic Calculations**
Exercise #3	**More About Nutrients**
Exercise #4	**Comparison of Dieting Strategies**

Exercise #1
Energy Balance of the Body

Energy is the ability to do work. The chemical energy in food is called potential energy because it has to be converted into work by the body's metabolism. The work performed by the body can be movement or it can be production of other organic molecules needed by the body.

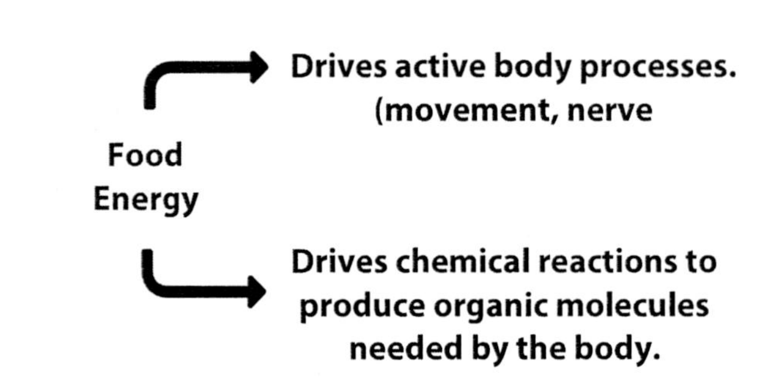

The ***energy balance equation*** of the body is a simple principle that explains weight changes. If energy input (food eaten) equals energy output (food metabolized for all body activities), then the body weight remains constant. If more food energy goes into the body than is used by the metabolic processes, your weight will increase (fat storage). Likewise, if less energy goes into the body than is used, body weight will decrease.

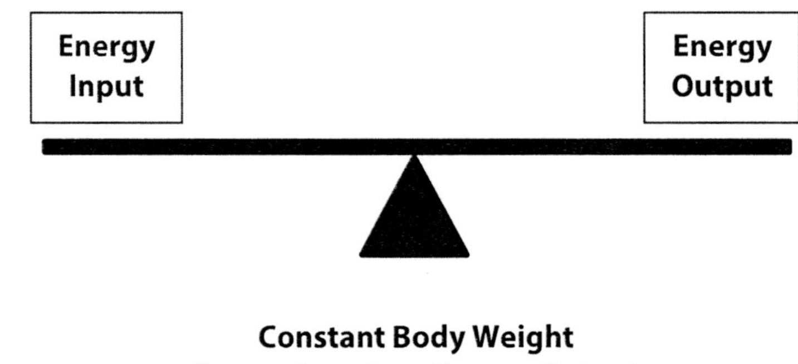

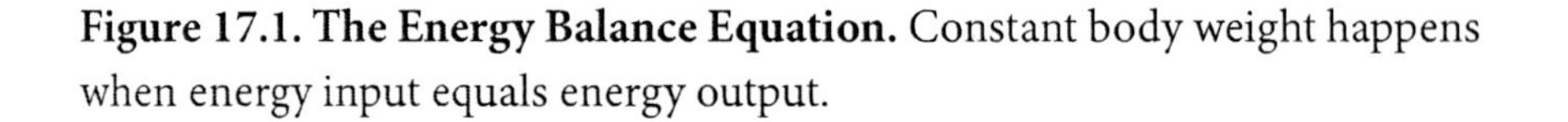

Figure 17.1. The Energy Balance Equation. Constant body weight happens when energy input equals energy output.

Energy Input

Food contains potential energy, and that energy is located in certain chemical bonds of the food molecule. These bonds are broken during metabolism, and energy is released as the food molecule is processed. The potential chemical energy in a food can be determined by measuring the amount of heat given off when that food is burned in a Calorie Chamber. One gram of sugar has a certain amount of potential chemical energy. If that amount of sugar is burned and we measure the heat given off, the amount of energy in one gram of sugar is revealed. The answer is about 4 Calories. A ***Calorie*** is a unit of heat energy. To be exact, it is the amount of heat required to raise the temperature of 1 liter of water by 1^0 C. One gram of sugar has 4 Calories of food value. That amount of energy is enough to raise the temperature of 1 liter of water by 4^0C. Another very different way to imagine one Calorie is to compare it to the amount of energy from a 100 watt light bulb for 40 seconds.

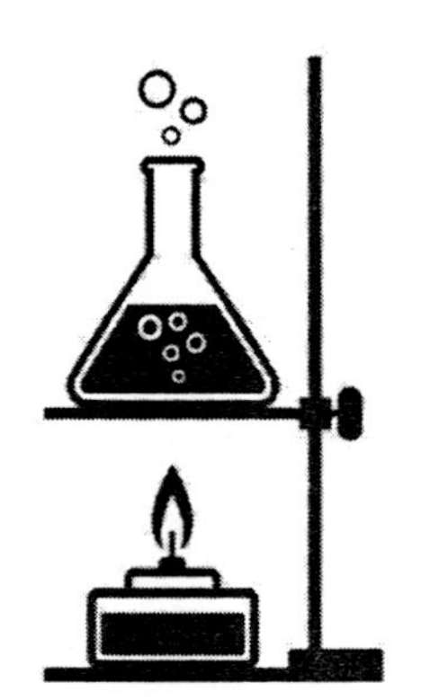

A Calorie is the amount of heat energy required to raise the temperature of 1 liter of water by 1°C.

Demonstration of Energy in Sugar

Your instructor will take the class outside to show you how much heat energy is released when sugar is burned. The amount of sugar in this demonstration is approximately what some people add to their coffee. A chemical has been mixed with the sugar that makes it burn faster. None of the released energy comes from that added compound.

? Question

Are you surprised at how much heat was released? Any comments?

Heat is released during metabolism, but organisms would combust if all of that energy were released at once. Cells release the energy a little bit at a time, and enzymes and hormones regulate the rate at which energy converts from potential chemical energy to body activities. The metabolic rate varies within the population because of the unique enzyme and hormone profile of each person. More on this later.

Food can be burned in a calorimeter to determine how much potential energy it has.

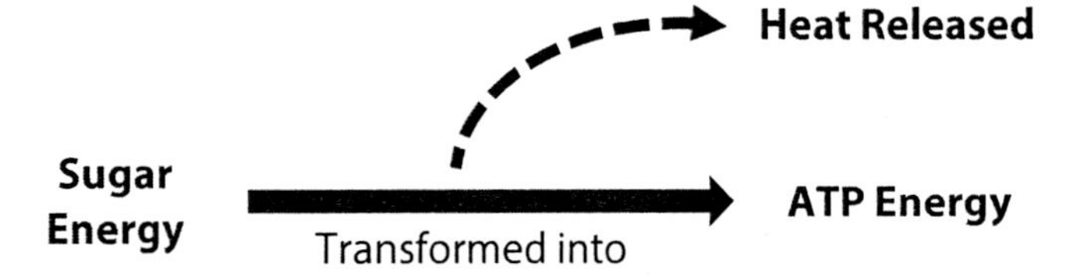

A first complication in calculating the energy input for humans is that some of our food is unusable for metabolism. The main component of plants is cellulose (fiber), and it isn't digested by humans because we don't have the enzymes to do so. And some of the digestible nutrients in our food aren't completely absorbed by the small intestine. Absorption depends on the concentration of digestive enzymes and the speed that food is moved through the digestive tract. That varies from person to person. And some people have an inherited ability to easily store fat. There is a nutrition opinion that says, "Don't count Calories, but determine the impact of Calories on your metabolism of fats?" Do you store fat easily, or do you burn fat? These nutritional advisors recommend that we change the balance of nutrients instead of counting Calories. The view we use in this lab is to focus on Calorie analysis as a foundational approach that certainly can be modified as you learn more about specific treatments for weight imbalance. All nutritional approaches require some consideration of the "counting Calories" method.

There are physiological differences among people. Although variations in physiology complicate the calculations, the values listed in Table 17.1 are accurate enough to begin an estimation of energy balance in your body.

How to Convert Grams into Calories.

Table 17.1. Approximate Calorie Value of Primary Nutrients.

1 g of carbohydrate ≈ **4** Calories of energy input
1 g of protein ≈ **4** Cal of energy input
1 g of fat ≈ **9** Cal of energy input
4000 excess input Calories ≈ 1 pound gain in fat
4000 deficit of Calories ≈ 1 pound loss in fat
* These estimates are "rounded off" for easier calculation.

Energy Output

Energy output is the term used to represent all of the energy expended to maintain the metabolic processes and activities of an organism. The most accurate and direct way of determining the energy output is to measure the ***amount of heat given off***. Physics tells us that heat is released whenever energy is transformed from one form into another (such as nutrient energy into physical work). A technical problem with using this method of analysis is that the person must be inside an insulated container surrounded by a known quantity of water. The temperature of the water increases as heat is released by the person. Although this procedure is very accurate, it is also expensive and difficult.

This is a very accurate method, but it is expensive and difficult.

Figure 17.1. Apparatus for measuring oxygen consumption. This method is easier to do than the heat production method.

The usual method of measuring energy output is the ***oxygen consumption technique***. If the oxygen requirement of a resting person is known, then that value can be compared to the increased amount of oxygen used during a particular physical activity. This approach is easier and less expensive than the heat method. Most general studies of energy expenditure are based on oxygen consumption.

? Question

1. If you were comparing a natural-food based diet to a processed-food diet, what would be typical?

   Natural food is almost completely digested and absorbed into your blood. Yes or No

   Processed food is high in fiber. Yes or No

2. The problem with a donut is not so much the amount of sugar and starch, but the __________________ of them.

3. Which hormone is released by the pancreas when we eat a food that is very high in carbohydrate (sugar and starch)?

4. List two kinds of work performed by the body.

5. Write the energy balance equation of the body.

6. How much heat energy is one Calorie?

7. Why does the "burning" method of analyzing some human foods provide an inaccurate Caloric estimate of how much energy we actually get from those foods?

8. One Calorie of fat provides more energy input than one Calorie of protein. (T or F)

9. List how many Calories of energy are provided to the body by 1 gram of each of the three primary nutrients.

   Carbohydrate =
   Fat =
   Protein =

10. Assume that a person requires 1600 Cal to maintain constant weight. If this person eats no food (water only), then how much fat can they lose in a week? (Assume that only fat is being metabolized, and that there is no metabolic slowdown during the fast.)

11. Why is the measure of heat production an accurate estimate of energy expended by the body?

12. Although the heat production method is very accurate, it is difficult. What is the usual method for measuring energy output by the body?

13. You will be using several foods as examples during this lab. Let's see how you evaluate them before doing the activities. Put a check mark if you think the food is high Calories, high protein, or high fat.

Food Example	High Calorie	High Protein	High Fat
Tuna Sandwich			
Milk (8oz)			
Afternoon Snack 1 cup of peanuts 1oz of cheese 6 crackers			
Cheeseburger			

14. If the normal resting energy output is 80 Calories per hour and the energy output during moderate exercise is 200 Calories per hour, then what is the energy output for the exercise?

Exercise #2
Energy in Foods - Basic Calculations

Internet Resource for Food Composition

The three primary nutrients (sometimes called ***macro-nutrients***) are carbohydrate, fat, and protein. We need to balance the ratio of these nutrients in our diet depending on the particular demands of our body. The general recommendation by governmental nutritional agencies is C=60%, F=20%, and P=20% of our Caloric intake. Furthermore, it is suggested that the carbohydrate not go above 65% or less than 45% over an extended period of time. Fat is to stay between 15-35%, and protein 10-35%. Complete ***Food Composition Tables*** can be found on the Internet (USDA National Nutrient Database for Standard Reference, Release 25 or newer) and in most nutrition books. Also, there are several sets of complete Food Tables in the lab room for your use.

Basic Calculations

There are three steps for calculating the nutrient composition of a particular food.

How Do You Calculate % of Nutrients in Food?

Step 1: Look up the grams of each nutrient. (Use Food Composition Tables.)
Step 2: Convert grams to Calories.
Step 3: Determine % of each nutrient.

Procedure

- We will use a tuna sandwich as an example for calculating the % of Calories for each nutrient in a food. A sandwich is a mixture of ingredients.

Step 1: The grams of each nutrient is determined from the Food Composition Table.

Tuna Sandwich	Nutrient Content		
	Carbohydrate	Protein	Fat
2 Slices of Bread	24 g	4 g	1.4 g
1 Tbsp. of Mayonnaise	trace	trace	11 g
Lettuce	trace	trace	trace
2 oz of Tuna	0 g	15 g	1 g
Totals	24 g	19 g	13.4 g

Step 2: Convert grams of nutrient into Calories using the Conversion Factors presented in Table 17.1. Do the multiplications now.

Tuna Sandwich	Caloric Conversion Factor	Caloric Value
24 g C	x 4 Cal per gram	
19 g P	x 4 Cal per gram	
13.4 g F	x 9 Cal per gram	

Total = _______ Cal

You should have calculated the total Calories to be 293? Did you get that?

Step 3: Calculate the % of each nutrient in the food using the following formula. Do that now.

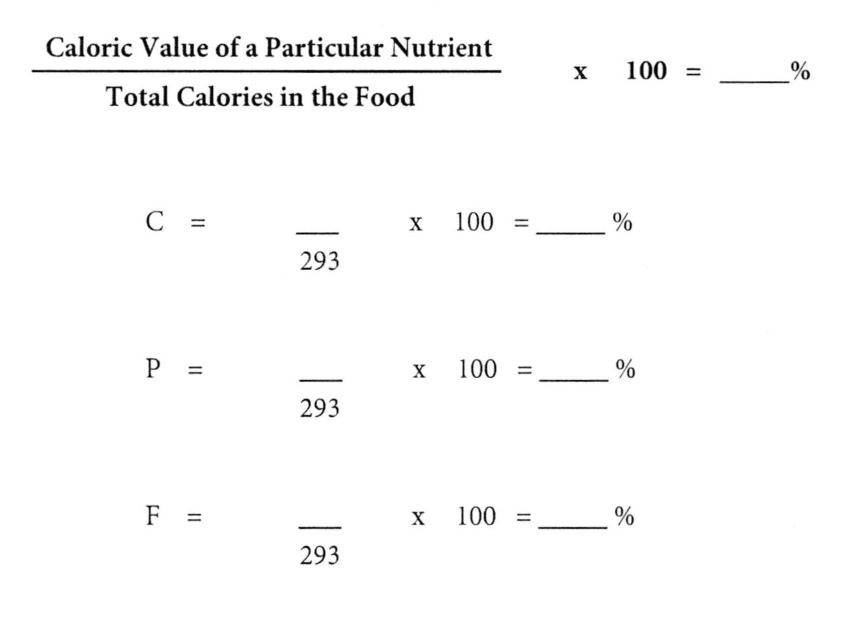

$$\frac{\textbf{Caloric Value of a Particular Nutrient}}{\textbf{Total Calories in the Food}} \times 100 = \_\_\_\_\%$$

$$C = \frac{\_\_\_}{293} \times 100 = \_\_\_\_\%$$

$$P = \frac{\_\_\_}{293} \times 100 = \_\_\_\_\%$$

$$F = \frac{\_\_\_}{293} \times 100 = \_\_\_\_\%$$

Procedure

- Let's see if you can calculate the Caloric percentages of each nutrient in 8oz of whole milk. This glass of milk would be approximately 12g of carbohydrate, 9g of protein, and 9g of fat.

Do the Calculation for Whole Milk.

Whole Milk (8oz)

Carbohydrate ______ %

Protein ______ %

Fat ______ %

? Question

1. Look at your answers to question #13 in Exercise #1. How would you describe these foods now that you have made the actual nutrient calculations? Are they high carbohydrate, protein, or fat?

   Tuna Sandwich = ______________

   Whole Milk = ______________

   After School Snack = ______________

   Cheeseburger = ______________

2. Calculate the Caloric % of fat in a 3 ounce piece of sirloin steak using the following data: P = 20 g; C = trace; and F = 27 g.

   ______ = total Calories

   ______ % fat

3. If you eat a 6 oz steak, what has changed?

   ______ = total Calories

   ______ % Fat

But what happens when you add a creamy dressing?

4. A green leafy salad is considered to be low in fat and low in Calories. What happens to the fat content when you add a tablespoon of salad oil or creamy dressing?

5. Whole milk is more accurately called 3% milk because it is 97% water. Reduced Fat Milk is 2%, and Low Fat is 1%. Are any of these actually low fat?

6. What is the actual fat % (based on Calories) for Low Fat Milk?

Exercise #3
More About Nutrients

Carbohydrates

```
   H   O
    \ //
     C
     |
 H - C - OH
     |
HO - C - H
     |
 H - C - OH
     |
 H - C - OH
     |
     C
   / | \
  H  H  OH
```

Glucose

Dietary carbohydrates are classified as simple or complex. ***Simple sugars*** consist of a single sugar like glucose or a few sugar molecules hooked together. You can metabolize all of the natural sugars, and they all have similar food value. ***Complex carbohydrates*** are hundreds of sugars hooked together to form a ***polysaccharide***—starch (in plants) and glycogen (in animals). Our digestive system breaks down the polysaccharides into simple sugars which are then absorbed and used for energy.

Polysaccharides —Digestion→ **Monosaccharides** —Absorption→ **Sugars in the blood**

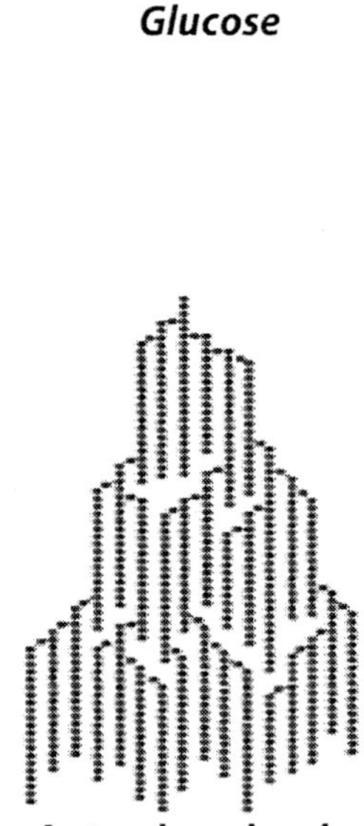

A starch molecule is hundreds of glucose molecules.

Carbohydrates are the body's immediate energy source, and most of our dietary requirement should come from fresh vegetables, fruits, and some whole grains. We typically eat refined grains rather than whole grains. Refined grains taste better and spoil slower, but they are not as nutritious as whole grains. And they usually have added fats to improve texture. Whole grains have more nutrients and fiber. But whole grains can be low in B vitamins compared to commercially refined grains, so that must be adjusted with another source for those vitamins. Nutrition can be very tricky.

New research indicates that whole grains, though an important part of most diets, should not be over-consumed with a false sense that anything "natural" is always ok. Medical studies in ***glycemic response*** (what actually happens to blood sugar when you eat specific foods) suggests that complex carbohydrates are sometimes absorbed as quickly as sugars. Diets very high in grains (or other carbohydrates) can contribute to overweight. The marbling of fat in a gourmet beef steak is a direct result in grain-fed cattle.

The "steak" got fat by being fed an excess of grain!

Carbohydrates are digested and absorbed faster than fats (Carbohydrates = 2-5 hours; Fats = 5-8 hours). Because fat takes longer to digest, we have a special hormonal mechanism (cholecystokinin) to slow stomach emptying when fat is in our food. Because of this hormone response, it can be beneficial to add a small amount of butter or sour cream to a baked potato. This would allow the digestion and absorption of the potato's large amount of carbohydrate to be moderated over a longer period of time than eating a plain potato. You would probably feel better. However, you would not want to add ice cream at the end of a large meal. Why not?

Fiber: The Indigestible Carbohydrate

Fiber (plant cellulose) in our diet is essential for good health even though it does not provide Calories. We lack the digestive enzymes necessary to break it down and derive nutrition from it. Health professionals recommend consuming 25-40 grams of fiber daily. Generally, fiber intake will be adequate when the bulk of the diet consists of vegetables, fruits, and legumes, and a few whole grains. Many processed foods are devoid of fiber unless it has been added.

Dietary fiber comes in two varieties: ***soluble*** and ***insoluble***. Soluble fiber (as found in legumes, oats, nuts, seeds, and many fruits) slows the absorption of carbohydrates and can decrease spikes in blood sugar. This reduces the risk of type-2 diabetes. Soluble fiber also lowers LDL (bad cholesterol) and decreases the risk of heart disease. Insoluble fiber (as found in many whole grains, nuts, seeds, and many vegetables) encourages favorable bacterial growth in the intestine. An added health benefit is that fiber decreases hunger and can help us from overeating.

Another important contribution of fiber is its ***fat-sequestering activity***. Fiber entraps some of the fat molecules in the colon. This ensures that fats get "swept out" of the body rather than staying in the colon where they can become potential carcinogenic metabolites as they are digested by bacteria. Fiber intake is associated with a decreased risk of colon cancer. However, it is possible to overindulge in fiber. Excess fiber in the diet can hinder the absorption of calcium, magnesium, zinc, and other important minerals.

Does Your Diet Have Enough Fiber?

There are methods for calculating the amount of fiber in your diet, but the simplest test to determine if you are eating enough fiber can be done in a three day experiment.*

Procedure

- Eat 1 cup of All Bran cereal in the morning.
- Eat 1 cup of All Bran cereal in the evening.
- If your bowel movements become softer* within three days, then your diet is inadequate in fiber.
- Try this simple experiment during the next three days.

*Gastroenterologists tell us that most people are partially constipated and don't know it. The cause is not enough fiber in the diet. They say that a proper bowel movement should easily fold when being flushed down the toilet.

? Question

1. Give an example of a simple sugar.

2. Is fructose (fruit sugar) easier to metabolize than glucose?

3. Give an example of a polysaccharide.

4. What is a glycemic response?

5. Cholecystokinin is a hormone released when you eat _____.

6. How would adding a small amount of butter or sour cream enhance the absorption of potato starch?

7. Sometimes a person adds a bowl of ice cream at the end of a large meal and feels extremely full for an uncomfortable 1-2 hours. Can you explain this?

8. What is the recommended daily fiber intake? (in grams)

9. Describe one specific explanation for how fiber in the diet might reduce the risk of colon cancer?

Protein

The World Health Organization considers protein to be the most critical macro-nutrient because it is in short supply, and much effort is expended to provide protein to everyone worldwide. Protein makes up most of the body's structure and all of the body's enzymes which control cell chemistry. At least 30 grams of complete protein are required daily for metabolic processes alone. When intake is less than this, muscle wasting, reduced immune response, and other degenerative processes occur. Government guideline for protein is 45–60 grams daily, but requirements for individuals can vary considerably. Decreased intake of protein may be indicated for those with kidney disease or gout, and increased intake is recommended for dieters, athletes, and burn victims.

With few exceptions, ***complete proteins*** are derived from animal-based foods, and they contain all of the amino acids in the proper ratios for sustaining human life. ***Incomplete proteins*** are plant-based foods but can be combined together to provide enough complete protein necessary for health. Proteins are built from amino acids. There are 21 amino acids necessary for human proteins, and 12 of them can be made during our own metabolism. Most people require at least nine others in their diet, and infants require another three. These are called ***essential amino acids*** because it is essential to have them in your diet. You've heard the phrase "Rice and beans, beans and rice". Look at Table 17.2 below, and you can see that each of these foods is lacking two essential amino acids. But together they are a complete food for protein. Note: Complementary essential amino acids must be eaten together in the same meal. They do not "wait around" for the complementary amino acid in the next meal!

Rice and Beans, Beans and Rice.

Table 17.2. Complementary Plant Foods for Complete Protein.

Plant Foods	Amino Acids in the Foods			
	Isoleucine	Lysine	Methionine	Tryptophan
Legumes	Yes	Yes	No	No
Grains	No	No	Yes	Yes
Together	Yes	Yes	Yes	Yes

Humans are quite individual in their ability to metabolize and benefit from protein in food. Because the protein content varies in vegetable sources, obtaining a full complement of essential amino acids is of particular concern for anyone with a mostly vegetarian diet, especially for vegans. The following Table 17.3 compares protein quality of various foods and is generally agreed on by most health professionals.

Table 17.3. Essential Amino Acid Rating of Various Foods.

PROTEIN SOURCE	RATING
Eggs (whites)	1
Milk—human	0.9
Soybean isolate	0.9
Meats (beef, pork, poultry, seafood)	0.7–0.9
Milk—dairy	0.7
Grains (wheat, corn, barley, oats)	0.4–0.7
Legumes	0.4–0.7
Potatoes	0.5
Rice	0.5

? Question

1. What is the recommended daily protein intake? (in grams)

2. What is a complete protein?

3. What is an essential amino acid?

4. What is one particular diet challenge facing strict vegetarians?

Dietary Fats (Lipids)

When visiting the Internet or reading magazine articles you can be overwhelmed by conflicting information about fats in our diet. Research evidence comes from two primary sources. First are the studies that correlate diseases with different diets. Second are the studies where people are put on special diets and certain blood measurements (maybe cholesterol) are monitored. Many unknowns remain after so much study, but there are some general recommendations.

What are Fats?

The chemical structure of fats and oils includes an alcohol molecule bonded to fatty acids and sometimes also with other organic molecules. In our blood the alcohol glycerol is bonded with three fatty acids, and the complex is called a ***triglyceride***. There are hundreds of variations on this general theme.

A	—	FATTY ACID
L		
C		
O	—	FATTY ACID
H		
O		
L	—	FATTY ACID

Fatty acids have their own complex and varied structure. One variation in structure is the number of hydrogen atoms and type of chemical bond between the carbon, hydrogen, and oxygen in the molecule. A fat is defined as being more saturated if there are many single-bond hydrogen atoms in the fatty acid. The unsaturated fat has fewer hydrogen atoms. There are three major classes of dietary fats: ***saturated***, ***monounsaturated*** (a little bit unsaturated), and ***polyunsaturated*** (very unsaturated).

Saturated Fatty Acid

O	H	H	H	H	H	H	H	H	H	H	H	H	H	H	H	H	
C	C	C	C	C	C	C	C	C	C	C	C	C	C	C	C	C	H
	H	H	H	H	H	H	H	H	H	H	H	H	H	H	H	H	

Unsaturated Fatty Acid

O	H	H			H	H	H	H	H	H	H	H			H	H	
C	C	C	C	C	C	C	C	C	C	C	C	C	C	C	C	C	H
	H	H	H	H	H	H			H			H	H	H	H	H	

Unfortunately, fats have acquired a great deal of controversy. It seems that many researchers disagree with each other, and some facts about fats may surprise you. Even though fats carry the most Calories (9 Calories per gram), they are necessary for good health. Cutting fat out of your diet is a misguided strategy for weight control. Not all fats are equal in health value. Some fats cause cellular damage, but other fats protect cells from damage. Beneficial fats are found in all three fat categories and so are damaging fats. How do we tell the difference?

First we were told to switch from saturated to polyunsaturated.

Bad Lipids?

One early observation about dietary fat was that people who ate high amounts of saturated fats developed a number of health problems. These disorders included heart disease, diabetes, cancer, and excess overweight. The overweight part was easily explained by the high Caloric value of saturated fat, but what about the rest? Health officials recommended that everyone switch to unsaturated fats and oils, which we did. Unexpectedly the health problems got worse. That began more detailed research about the subtleties of fat metabolism.

Polyunsaturated fats occur in high amounts in most processed foods. The food industry developed a special subcategory of polyunsaturated fat called ***trans-fat*** that is made by hydrogenating natural plant oils. These fats and oils are less expensive than the saturated fats they replaced. They are solid like saturated fats, don't spoil easily (no refrigeration), and can be reused many times for high temperature frying. But these fats cause free-radical damage and promote inflammatory reactions that contribute to cancer, heart disease, joint disease, and type-2 diabetes. Trans-hydrogenated fats raise LDLs (low-density lipoproteins or "bad" cholesterol) and they lower HDLs (high-density lipoproteins or "good" cholesterol). Imitation cheese, margarine, and solid vegetable fats used in frying and baking are three of the most notorious trans-hydrogenated fats. So, the next recommendation from health officials was to cut back in polyunsaturated fats, especially trans-fats, which Americans are now trying to do.

Then we were told to cut back on polyunsaturates.

Good Lipids

Saturated fats are not intrinsically bad for human health. Short-chain and medium-chain saturated fats, as found in butter fat, tropical oils, and human breast milk are essential to health. In fact, saturated fats are critical for many biological processes such as hormone production. Avoiding all saturated fats will increase susceptibility to many health problems. But we make plenty of saturated fats during our own metabolic processes, so the recommendation is to keep intake at 10% of Caloric intake for normal people and at 7% for people with cardiovascular disease. Researchers also discovered that two of the most saturated fats known—coconut and palm oils—are more beneficial to health than most other oils. Unlike many polyunsaturates, coconut and palm oils do not spoil quickly at room temperature, do not produce free-radicals when heated, and are very easy to digest. These oils are ideal for cooking and are not stored as body fat but are burned for fuel. Their properties make them useful in weight reduction diets and an excellent substitute for many polyunsaturates.

Modest amounts of polyunsaturated fats from whole nuts and seeds (sometimes called ***Omega 6 fats***) are also part of a healthful diet. Wheat germ and whole grains contain small amounts of these essential oils. However, these oils spoil very easily and create free radicals when heated above 150 °F. It is best to keep foods high in these oils refrigerated until eaten. In general, excess intake of Omega 6 fats is pro-inflammatory and probably contributes to disease. But none of the health problems occurred in people who restricted their intake to 2 grams per day. We can probably safely tolerate more than that, but how much more is the debate. Researchers are now focusing on the ratio of Omega 6 and Omega 3 fats.

Omega-3 fats from cold-water fish, krill, and flax seed contain both polyunsaturated and saturated fats. They are incredibly beneficial to health but are typically low in the average American diet. Omega-3 fats reduce inflammation and help ameliorate the pro-inflammatory effect of other polyunsaturates in the omega-6 class. Health officials recommend that we consume an amount equal to two meals of salmon per week. There are sources other than salmon such as the monounsaturated oils derived from fruits and nuts (olives, avocados, walnuts, macadamias, and almonds). These monounsaturated oils have been shown to reduce cholesterol and are the foundation of the often-praised Mediterranean diet.

Now we are told to balance the fats we consume.

The average American intake of fats is about 80 grams (mostly because of fast-foods) and exceeds the 40–60 grams recommended by most nutritional experts. Our typical diet provides an imbalance of polyunsaturated over monounsaturated fats and Omega 3 fats. This promotes inflammatory disease states—especially heart disease. Researchers have found that the proper ratio of the three classes of fats may be the key. Ideally, omega-6 to omega-3 ratios should be 1:1 in the diet. The omega-6 to omega-3 ratio in the American diet ranges from 10:1 to 30:1. Our government health officials now suggest that we not go beyond a 4:1 ratio. According to research worldwide, diets that maintain a low ratio produce the least diet related disease.

What can we conclude?

A healthful meal would include about 50-60% of Calories from carbohydrates. Most should be in the form of vegetables, fruits, and whole grains. This should be enough for both energy and fiber. If the fiber is not quite enough, then you should have a bowl of bran (like All Bran) each morning with breakfast. The nature of your bowel movements tell you directly if your fiber intake is adequate. The easiest way to lose weight starting with a typical American diet would be to cut carbohydrates. This is because the usual diet is too high in processed carbohydrate (combined with unhealthy fats). When weight reduction is the goal, a healthful diet might be closer to the 40-50% carbohydrate.

The protein content in each meal must be balanced. If you are eating eggs (white), low-fat milk products, soybean isolate, or meats, your diet will be balanced with essential amino acids. If you are a strict vegetarian, you must take more effort to balance your plant foods (especially over time). The amount of protein in a good meal would be at least 15% of Calories, but it should be higher (20+%) if you are dieting and maybe as high as 30% during periods of physical conditioning to prevent muscle loss and other protein deficiencies. But do not overindulge protein. A diet very high in protein will cause liver and kidney damage.

The fat intake in a good diet would be less than 30% of total Caloric intake. Saturated fats should be about 1/3 of all fats. Heart patients are sometimes recommended to reduce saturated fats to 7% of total Calories, but that is difficult to achieve depending on the main protein sources in your diet. You should have the goal of creating a balance of monosaturates plus Omega 3 fats about equal to the amount of healthy Omega 6 fats that you eat (from fruit and nuts). Certainly these recommendations will be improved with more research.

There are clinical reasons for deviating from the general population diet, and specialized diets have been developed for extreme physical training. There is room for some processed foods and excess Calorie foods in the diet of a healthy person. The secret is not to let those foods sabotage the benefits of a healthful diet.

? Question

1. What is a triglyceride?

2. What is the chemical difference between a saturated and unsaturated fat?

3. What health disorders were observed in people with high saturated fats in their diets?

4. People were directed to switch from (circle your choice)

Saturated to Unsaturated
or
Unsaturated to Saturated

5. As more Americans made the switch, what was the health outcome?

6. What are trans-fats?

7. What are two saturated fats that are suggested as replacements for trans-fats?

8. What is the recommended maximum % of saturated fats in our diet?

9. What is the recommended maximum % of total fats in our diet?

10. What kind of food is high in Omega 3 fats?

11. Describe a healthful diet.

Exercise #4
Comparison of Dieting Strategies

The first step in developing a diet strategy for a client is to determine whether they have a healthful diet. This should be the primary goal of any health professional before giving advice on how to lose weight, increase body conditioning, or start specialized physical training. If a diet change is necessary, there are two options: modify existing diet or take on a new diet.

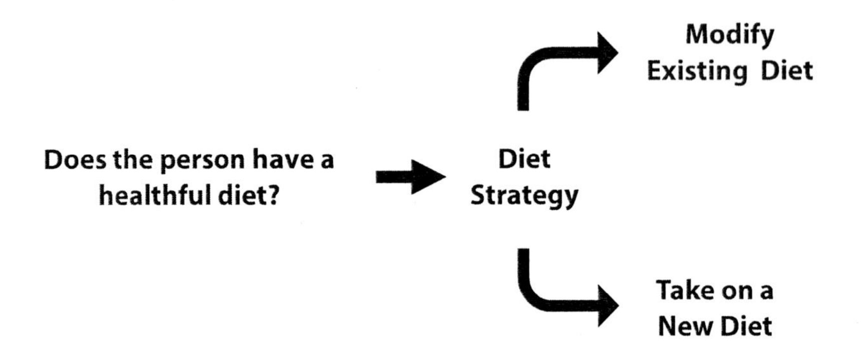

The second question in a diet strategy is, "Does the person have a problem with energy balance (body weight)?" Again, the research is very helpful here. If a person is in the process of gaining weight, you can best help them by starting with a focus on Calorie reduction in their diet. Once they begin losing weight because of the new diet, additional exercise (daily activity) should be added. A particularly favorable time to add exercise is when the person begins to plateau after early weight loss. You might ask why not start with an aggressive diet and exercise plan from day one. You can, but diet changes are a major challenge to initiate and maintain. Once a healthful diet is established, the person will experience significant improvements in how they feel. This can motivate the next step of increasing daily activity.

An important exception to this general approach comes with a patient who is very much overweight. Investigation has shown that people's activity level goes down dramatically when they have excess weight. They do not enjoy exercise! They must start moving toward a minimum normal activity level before putting serious effort towards Calorie reduction. Of course, this assumes that you have already assisted them to establish a healthful diet even if it remains too high in Calories.

Regardless of the particular approach you are using with a client, there will be a time for Calorie reduction. Thankfully, there are many helpful generalizations from weight loss studies. Reducing food intake by 500 Calories per day is considered to be the maximum change in diet without risking health consequences. Most nutritionists suggest starting with a more moderate 250 Cal reduction unless there is medical supervision. The number of days on a severe Calorie restriction is the critical question. If the person is on a 500 – 1,000 Calorie reduction for two weeks and are having no serious symptoms, then that would probably be fine. But reducing Calories that much over a longer time period requires dietary supervision to avoid malnutrition problems.

Health experts agree that ideal nutrient ratios in a healthful diet are: carbohydrate 50–60%, protein 15–25%, and fats 20–25%. In general, women require fewer Calories (perhaps 1500 compared to 2000 for men). If the Caloric Input is cut more than 500 below daily requirement during a weight reduction diet, then the percentage of protein and fat in the weight loss diet should be increased.

Table 17.4. Example Comparisons of Weight Loss and Weight Maintenance Diets.

	2000 Calorie Non-Weight Loss Diet for Men		
	Carbohydrate	**Protein**	**Fat**
Calories	1100	400	500
Grams	275	100	56
% in Diet	55%	20%	25%

	1200 Calorie Weight Loss Diet for Men		
	Carbohydrate	**Protein**	**Fat**
Calories	480	360	360
Grams	120	90	40
% in Diet	40%	30%	30%

Smaller portions is for the disciplined eater.

How Good are Popular Dieting Strategies?

Smaller Portions of Everything

This approach starts with a healthful diet and the person eats smaller portions of each food. This diet works well unless you cut more than 200-300 Calories per day. As Calories are cut, you must not reduce the required proteins and healthy fats. For the disciplined dieter this is a healthy way to lose weight slowly.

Special Prepared Meals

Special Prepared Meals work because they are expensive, and somebody else makes them.

The success of specially prepared meals for dieting is the result of two factors – expense and ease. You pay handsomely for the food and are motivated to stay on the diet longer and not violate it. Also, someone other than you has planned a reasonably good-tasting meal that is not high in Calories. However, these packaged foods are processed and do not compare to the quality of natural foods in a healthful diet. A conspicuous requirement of most of these diet plans is that you must provide your own produce, low-fat dairy, and lean meats. In a healthful diet, that is what you're supposed to be eating anyway! Liquid forms of these diets are different only in that they include the needed proteins and fats. They are a "this is all you need to eat" approach. The special prepared diets have helped many people to lose weight. They are not ideal healthful meals, but we can hope that competition between the companies producing them will lead to improved products in time.

Eat All You Want of This or That

There are many variations of this strategy to dieting. One method tells you to eat a certain amount of a particular food before eating your meals. For example, the ***Eat-Six-Grapefruits-Per-Day*** diet instructs you to eat two grapefruits before each of the three meals of the day. Though grapefruit does have some modest fat-burning properties, the idea behind eating so many grapefruits is more about filling you up with low-Calorie foods so that you eat less of higher-Calorie foods.

You have to want to eat it.

Another variation of this dieting approach states that you can eat all you want from a narrow list of foods. For example, the ***Eat-All-of-the-Hard-Boiled-Eggs-and-Oranges-You-Want*** diet. Again, this approach is more about displacing higher-Calorie foods. Versions of these dieting strategies can create nutritional imbalances that result in health problems. For example, excessive intake of citrus fruit can cause a loss of minerals through increased urination. Eating unlimited amounts of protein is hard on the kidneys and also can cause mineral losses. These diets are short-term in benefit, and they are not what people should use for long-term weight management.

? Question

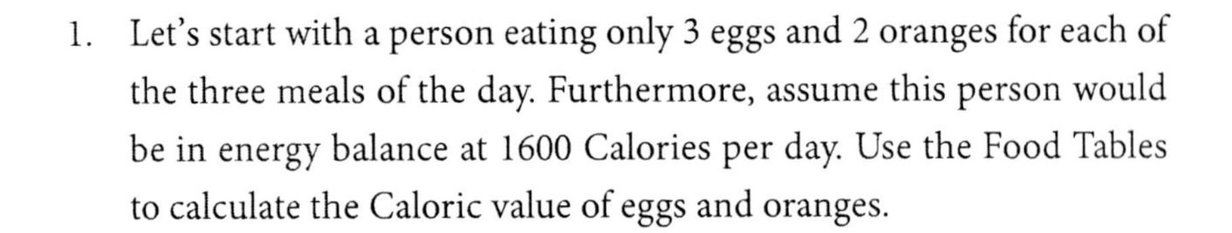

1. Let's start with a person eating only 3 eggs and 2 oranges for each of the three meals of the day. Furthermore, assume this person would be in energy balance at 1600 Calories per day. Use the Food Tables to calculate the Caloric value of eggs and oranges.

9 eggs = _______ Cal

6 oranges = _______ Cal

2. It takes about 4000 Calorie deficit to lose a pound of body fat. How many days would this person stay on the Eggs-and-Oranges diet in order to lose a pound? *

__________ days to lose one pound

* Note: All calculations in this lab's examples are assuming that the weight loss will be consistent over many days of dieting. Actually, there is a typical "surge-taper-plateau" effect, and weight loss does not follow these exact calculations. However, the general comparisons are a useful start.

Cut Something Out

Sometimes cutting one thing out of your diet can do the trick.

Another dieting approach is to eliminate something sweet or that has extra fats. This can be an effective strategy for those people with consistent eating habits. For example, let's consider the person who has a well-established habit of eating consistent meals and has noticed that he is four pounds heavier than two years ago. Each morning he eats the same breakfast.

1 bowl of oatmeal with 4 oz of low-fat milk
1 piece of toast
1 Tbsp. of butter
1 Tbsp. of jam
8 oz glass of whole milk

? Question

1. Use the Food Composition Table to determine the Caloric value of the whole milk and jam.

1 Tbsp. of jam = _____ Cal
8 oz of whole milk = _____ Cal

2. If only the jam and glass of whole milk were eliminated from his breakfast each morning, how many days would he be on the new diet in order to lose one pound of fat?

_____ days to lose one pound

A more extreme approach of cutting-something-out is exemplified by the *Pritikin-type* of diet during which you do away with most fat-containing foods. Though proponents have enjoyed success with cardiac patients, it should not be attempted outside a clinical setting because it can create nutritional imbalances. Eliminating fats from your diet has consequences including decreased absorption of fat-soluble vitamins, lower hormone production, and metabolic disorders. Eliminating most fats usually reduces protein intake and increases starch consumption to make up the Calorie deficit. Those interested only in weight reduction (and not heart-disease reversal) should avoid this diet or diets similar to it unless specifically prescribed by a medical professional.

Switch a Meal

Switch a taco for a hamburger?

Another version of cutting something out of the diet is the switching a meal method. Again, we start with someone who has a fairly consistent diet, and we will switch-out a meal. This person normally eats a Delux Hamburger with cheese (410 Cal) and adds a Mango Smoothie (290 Cal). He says he needs the convenience of a Fast Food but is willing to go to another restaurant if the meal is good tasting and he gets something sweet to drink like coke or smoothie. We are going to send him to Wahoo's and have him get a Grilled Fish Taco (209 Cal) and a serving of Brown Rice and Black Beans (359 Cal). We'll add a small Coke (90 Cal).

? Question

1. Have we changed the total Calories much?

   Fish Taco Meal = _____ Cal

   Cheeseburger Meal = _____ Cal

2. In what ways do you think the fish taco meal could be better than the hamburger meal?

   What about total fat content?

   What about saturated fat?

   Amount of protein?

   Anything else?

3. What is the primary accomplishment of making this switch? (Recall the First Rule of Diet Analysis.)

4. We are deliberately constraining our choices in this case because of what the client told us. If we only wanted to cut his Calories a bit by modifying his burger meal, what might we suggest?

The Subway Diet

The Subway Diet is an example of the "Pick a Meal" approach to weight loss.

The Subway Diet is a further extension of the "Switch a Meal" approach to weight-loss dieting. It could be called the "Pick a Meal" method. Pick a reasonably good Subway Meal and stick with it for a long time. There are many examples of dramatic weight loss success using this approach. In one case a college student weighed 425 lbs. while attending college. The student decided to lose weight by making a radical switch in diet. The old diet went as high as 10,000 Calories in a single day and was comprised of every high Calorie food. The switch was to a 6 inch Turkey Breast Sub before lunchtime and a 12 inch Veggie Sub at dinner. Each sandwich was accompanied with a diet coke and a baggie of baked potato chips. The student also walked about 4 miles extra each day during trips to the Subway store. That was an increase in exercise. The student lost 245 lbs. in about a year. This story is a powerful example of will and method. Let's examine the Subway Diet in more detail.

Subway Diet
Total Calories per day = 1020
Total Fat = 9g (about 2g was saturated)
Total Protein = 34g
Total Fiber = 20g

? Question

1. About what % of the Subway Diet is fat?

2. About what % is saturated fat?

3. How do you rate this student's fat intake?

4. About what % of this diet is protein?

5. How do you rate the protein intake?

6. Instead of a 6 inch Turkey Breast Sub and a 12 inch Veggie Sub, someone suggested that the student should have had a 12 inch Turkey Sub and a 6 inch Veggie. What do you think they were trying to improve?

7. How would you estimate the role of the extra walking in this example of weight loss?

Fasting won't work if you raid the refrigerator.

Fasting One Meal or Fasting One Day

This dieting strategy involves some degree of fasting. Perhaps the person will fast instead of lunch on one or two days of the week. Or they might fast one entire day every week or two. Both of these methods require that the person can feel ok and function normally in their lives when they fast. Most people can do this. It is an option to lower Caloric intake unless the person reacts to the fast by overeating after it.

? Question

1. About how many "fasts for lunch" would lead to a loss of 1 lb? (Make some assumptions.)

2. About how many one-day fasts would lead to a loss of 1 lb? (Make some assumptions.)

LAB

18

The Heart

How amazing is the heart! In the embryo it begins as a simple tube with two veins dumping into it and two arterial vessels leading out. Blood is being pumped by the human heart at three weeks after fertilization. Watch the YouTube 1951 film "Normal Development of the Heart" by Dr. Richard Blandau, University of Washington. The reality is more amazing than words can describe.

The heart, about the size of your fist, is located pretty much in the center of the chest (between the lungs) with a slight pointing to the left which makes it feel like it is on the left side. The "lub-dub" sound we hear when listening to the beat is the result of heart valves closing. It is an incredible pump. It can pump 35 million gallons of blood (3 supertankers) in a lifetime. Furthermore, the heart is capable of varying its output between 50ml and 250ml per beat and its contraction rate from 40 to 200 beats per minute (bpm). As a child your resting heartbeat was about 100 bpm, and as an adult the normal is near 70. When you are resting, the heart pumps about 5 liters of blood per minute. This is about the same rate as a slow flow of water from the bathroom faucet when you brush your teeth. During strenuous exercise your heart can pump 30 liters of blood per minute. An elite-trained heart can pump 40 liters per minute. This is about equal to the water flow when you fast-fill the bathtub.

The strength of your heartbeat is equal to a "hard squeeze" of a tennis ball. It beats 100,000 times in a day and about 2.5 billion times before it is completely worn out. Although a woman's heart beats 10% faster than a man's which would predict 10% fewer years, she outlives him by 10%. The heart defies simple description. You will refer to it as the center of your love for others – the seat of your character. It started you going at three weeks after conception and will provide life until your last breath. Now it's time to learn how the heart accomplishes these feats.

Exercise #1 **The Heart as a Pump**
Exercise #2 **The Heart Cycle**
Exercise #3 **Heart Sounds**
Exercise #4 **Heart Rate and Stroke Volume**
Exercise #5 **Blood Pressure**

Exercise #1
The Heart as a Pump

Two Pumps in One

The heart must control two separate circuits – pulmonary and systemic.

The heart is actually two pumps—a ***right pump*** and a ***left pump***. The right pump delivers blood to the lungs; this route is called the ***pulmonary circuit***. The left pump pushes blood to the rest of the body; this route is called the ***systemic circuit***.

Notice that Figure 18.1 is drawn as if the heart is facing you. This means that the right side of the heart is on the left side of the drawing. All anatomy diagrams are drawn in this view. Remember this whenever you look at a medical picture.

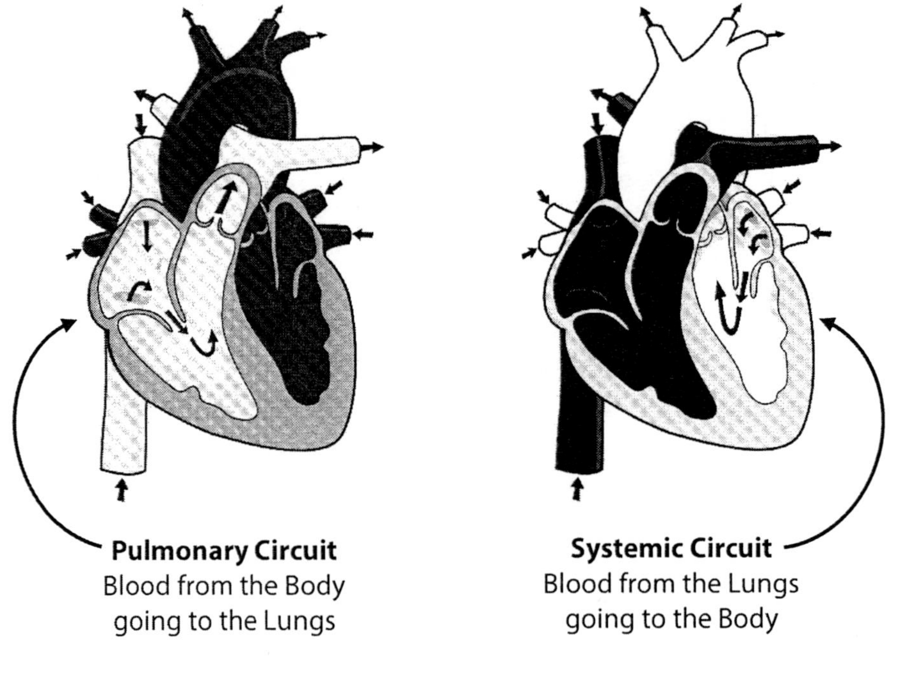

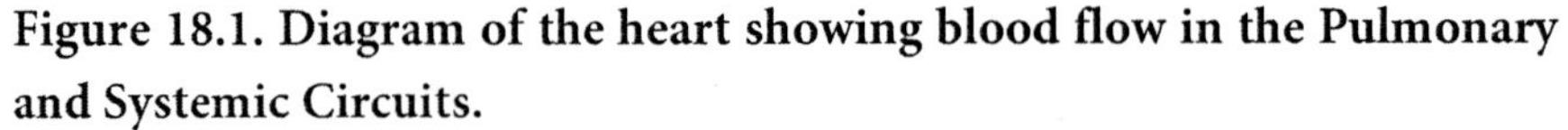

Figure 18.1. Diagram of the heart showing blood flow in the Pulmonary and Systemic Circuits.

Two Chambers per Pump

There are two chambers in each of the two heart pumps. The top one is a temporary storage chamber called the ***atrium***, and the bottom one is a pumping chamber called the ***ventricle***. Blood from the body tissues flows into the atrium of the right heart pump. This blood is then pushed through a valve and enters the right ventricle. The ventricle does the hard work of

Atria are storage chambers; Ventricles are pumping chambers.

pumping blood out of the heart. The right ventricle pumps blood to the lungs where it is ***oxygenated***. While the ventricle is pumping blood out of the heart, the atrium fills with blood entering the heart. This efficient design allows the atrium to quickly refill the emptied ventricle, resulting in a fast-pumping heart.

Oxygenated blood from the lungs enters the atrium of the left heart pump. This blood is then moved into the left ventricle which pumps the blood to all of the body tissues (except the lungs). The left pump has to work harder than the right pump.

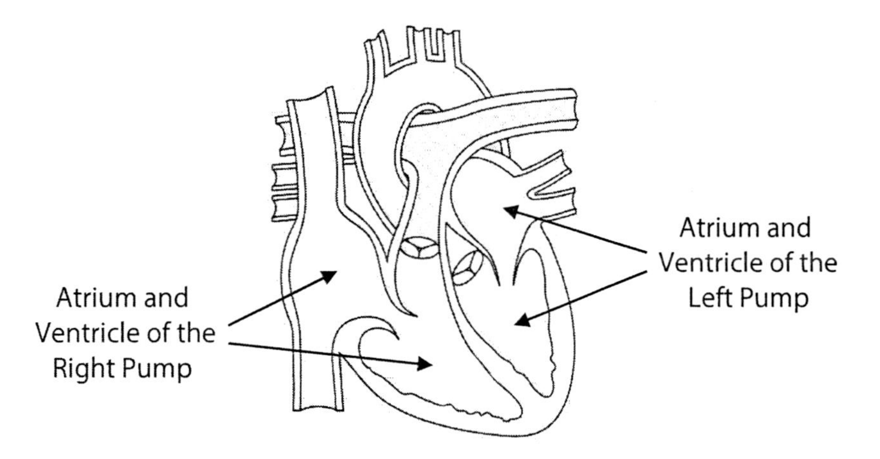

Figure 18.2. Diagram of the heart showing the Right Pump (Pulmonary) and the Left Pump (Systemic). Each pump has a storage chamber (atrium) and a pumping chamber (ventricle).

? Question

1. Which side of the heart is part of the pulmonary circuit?

2. Which side of the heart is part of the systemic circuit?

3. One congenital heart abnormality is a hole between the right and left ventricles. What two problems result?

4. Which chamber has to do the most work?
   Atrium or Ventricle

5. Which chamber has a thicker muscle wall?
   Atrium or Ventricle

6. Which pump has to do the most work?
   Right Ventricle or Left Ventricle

7. Which pump has a thicker muscle wall?
   Right Ventricle or Left Ventricle

8. The right heart pump moves blood to the ____________.

9. The left heart pump moves blood to the _________.

Heart Valves

A heart valve is designed to prevent blood from moving in the wrong direction.

Four heart valves are strategically located to prevent backflow as blood moves through the heart. These valves are like one-way doors—they only open in one direction. There is a chamber valve between each atrium and ventricle. These are called ***atrio-ventricular valves***, and they ensure that blood will not flow back into the atria when the ventricles contract. The medical name of the left atrio-ventricular valve is the ***mitral*** valve. The medical name of the right atrio-ventricular valve is ***tricuspid*** valve. To help you remember these names, think of the letter "l" and "m" as being next to each other in the alphabet. That means mitral = left side. Tricuspid is closer to the letter "r" (right side). Tricuspid = right side.

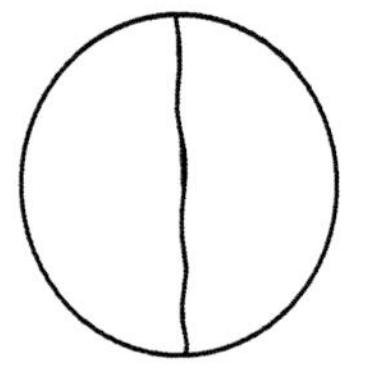

Bicuspid Valve (semi-lunar)

Blood is pushed out of the ventricles and into the two big arteries leaving the heart. The ***aorta*** receives blood from the left ventricle and directs it to the body. The ***pulmonary artery*** receives blood from the right ventricle and directs it to the lungs. There is a valve at the beginning of each of these arteries. They have a half-moon shape and are sometimes called ***semi-lunar valves***. Otherwise, the medical term refers to the vessel name (i.e. aortic valve for the aorta valve, and pulmonary valve for the pulmonary artery). The artery valves prevent backflow into the heart once blood has been pumped into the arteries. Together the four heart valves ensure that blood moves in only one direction through the heart circuit.

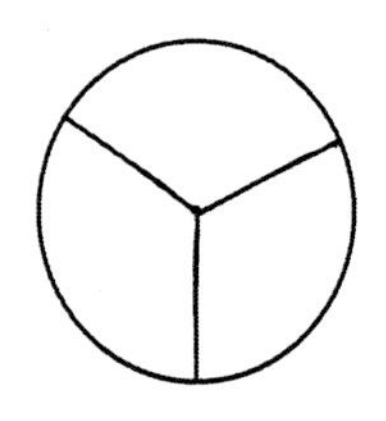

Tricuspid Valve

A heart valve is designed to plug an opening when blood moves in the wrong direction. Think of a valve as being something like a parachute that is attached to the wall of the heart or artery. If blood moves in the wrong direction, the "parachute" (valve) fills with blood and expands to plug the opening. When the blood moves in the correct direction, the valve collapses like an upside-down parachute. This allows the blood to easily pass through the valve.

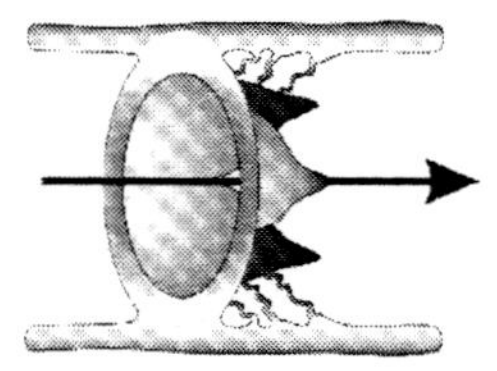

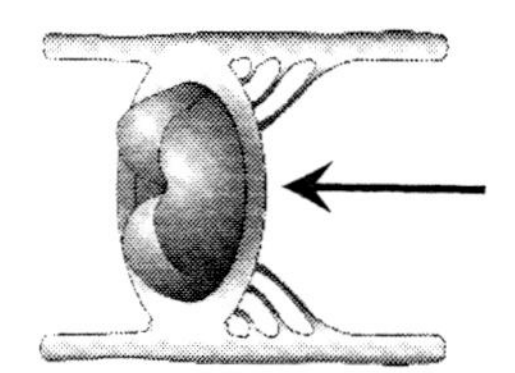

Figure 18.3. Structure of a heart or vessel valve. Valves are somewhat like a parachute that fills with blood when the blood moves in the wrong direction.

The heart valves are very flexible so that they can easily fill with blood. Special cords, called ***chordea tendineae***, attach the valve to the heart wall. These cords operate similar to the ropes of a parachute, and poets refer to them as our "heart strings".

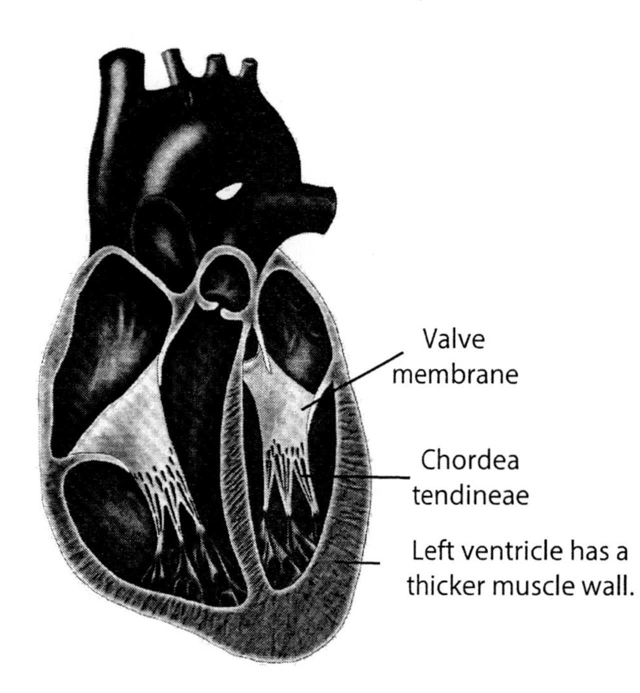

Figure 18.4. Section of the heart showing the membrane structure of valves and string-like anchors for them.

? Question

1. The atrio-ventricular valves are between the ___________ and the ___________.

2. Is the mitral valve on the right or left side of the heart?

3. There are two other valves in the heart. Where are they located?

4. What would happen to blood flow if one of the valve "cords" broke? Be specific.

5. What happens to blood flow if a valve opening is narrowed by disease scarring?

6. If there is a moderate heart valve problem, what would the heart do to compensate?

7. What would you then expect to happen to the size (thickness) of the muscle on the affected side?

8. On which side of the heart would a moderate heart valve problem have more consequence to your health? Explain your answer.

Examination of the Sheep Heart

Procedure

- Go to the dissection table and observe the sheep heart.
- See if you can identify the four chambers of the heart. Remember: one of the ventricles should have a thicker muscle wall. Which one? _______________ Find it to get oriented.
- Find a heart valve, and feel the valve to determine its flexibility. Can you find the valve "cords"?
- Show your instructor when you can identify all of these structures.
- Draw a simple sketch of the dissected heart. This will remind you of what you saw in case you are tested on it later.

Sheep Heart

Exercise #2
The Heart Cycle

Conduction of the Heart Impulse

Heart muscle exhibits an ***intrinsic rhythmic contraction***. This means that a piece of it will contract on its own. Different parts of the heart muscle have different rates of contraction. In general, all parts of the heart are relatively slow in contraction rate compared to that of the pacemaker of the heart, and because they are slower they are controlled by it. The pacemaker is the ***sino-atrial node*** located in the upper part of the right atrium where the superior vena cava brings blood into the heart from the arms and head.

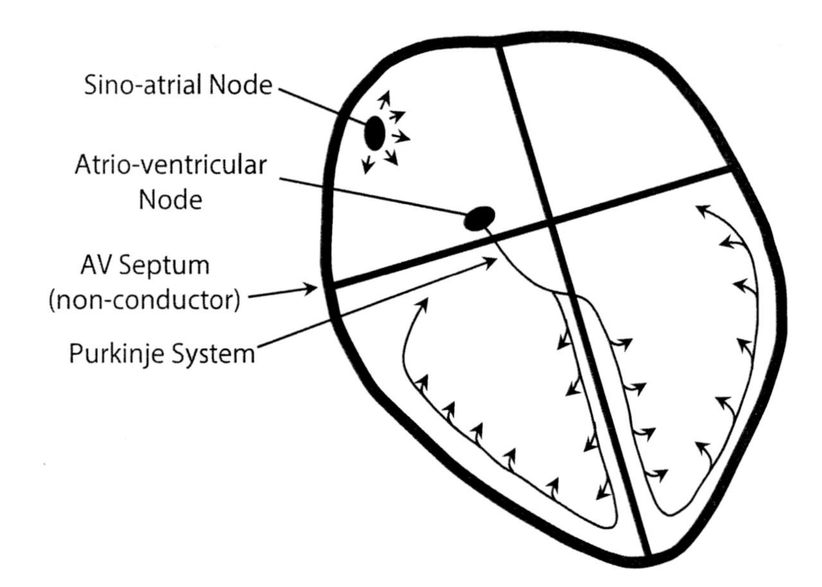

Figure 18.5. The Conduction System for the heart impulse.

A period of time after a muscle has contracted and during which it is impossible to stimulate that muscle to contract again is called the ***refractory period***. Think of it being something like "firing" a gun and then "cocking" it to fire again. You can't fire the gun until it has been re-cocked. All muscle, including heart muscle, has a recovery time during which the contractile mechanism is returning to a "ready to go again" condition.

The normal refractory period in human heart muscle is about 0.3 seconds. ***Normal conduction time*** for the heart impulse to travel throughout the entire heart during a heartbeat is about 0.06 seconds. If the heart impulse starts at the SA node (which it normally does), it will travel throughout the heart and then stop because the heart muscle is still in refractory period. Normally, conduction time is much shorter than refractory period, and heartbeats are separated by a short resting period. Medical problems arise when anything slows conduction time.

The ***AV septum*** is connective tissue and does not conduct the heart impulse. It would stop the impulse with the contraction of the atria, but there is a special "delay center" called the ***atrio-ventricular node*** that carries the signal into the ventricles. This cluster of specialized muscle cells has a very slow transmission rate (about 10X slower than normal heart cells). This is a "delay" in the transmission of the heart impulse. Once the impulse leaves the AV node, it enters the ***Purkinje System*** (another specialized set of muscle cells). Purkinje fibers conduct the impulse very rapidly (almost 10X faster than normal heart muscle). The heart impulse conduction system operates to create two separate contractions in a single heartbeat – first, contraction of the atria, followed by a delay, and then contraction of the ventricles.

The normal ***cardiac cycle*** can be summarized as:

1. Initiation of heart impulse in the SA node.
2. Spread of impulse through atria.
3. Both atria contract together.
4. Heart impulse is stopped by the AV septum (non-conductor).
5. AV node has been activated, but the spread is very slow which delays the heart impulse before it enters the ventricles.
6. Heart impulse enters the Purkinje System which then conducts very fast throughout both ventricles.
7. Both ventricles contract together.
8. Heart impulse stops because of the refractory period.
9. The next heartbeat is again initiated by the SA node about 1 second after the previous impulse.

Heart rate and strength of contraction are controlled by the autonomic nervous system – ***sympathetic nerves*** increase the heart output, and ***parasympathetic nerves*** decrease the heart output. There are important medical implications of this control process discussed later.

Electrocardiogram (EKG)

The ***electrocardiogram*** (called ***EKG*** or ECG) is a recording of the small electrical currents produced by the traveling heart impulse and the contracting and relaxing heart muscle. These electrical patterns indicate whether there is a normal or abnormal functioning of the heart. A normal EKG is shown in Figure 18.6.

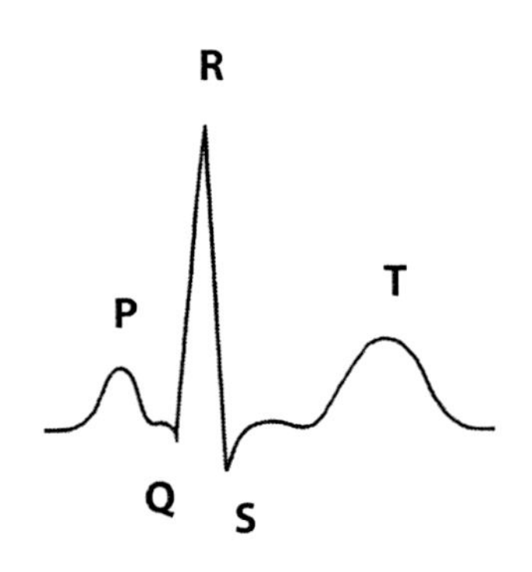

Figure 18.6. Normal EKG. The P wave is the pacemaker; QRS wave is conduction through the ventricles; and T wave is recovery of the ventricles.

The ***P wave*** is a recording of the electrical activity in the atria, and it is especially important in diagnosing problems in the heart's natural pacemaker.

The ***QRS wave*** is a recording of the conducting current in the ventricles. This current travels very rapidly along the Purkinje System. The result is that all muscle cells in both ventricles are stimulated at the same time, causing these chambers to contract quickly and strongly.

The ***T wave*** occurs just after the ventricles contract, and it is a recording of the normal recovery phase of the ventricles. This is a period when the muscle cells perform various biochemical reactions that prepare them for the next contraction.

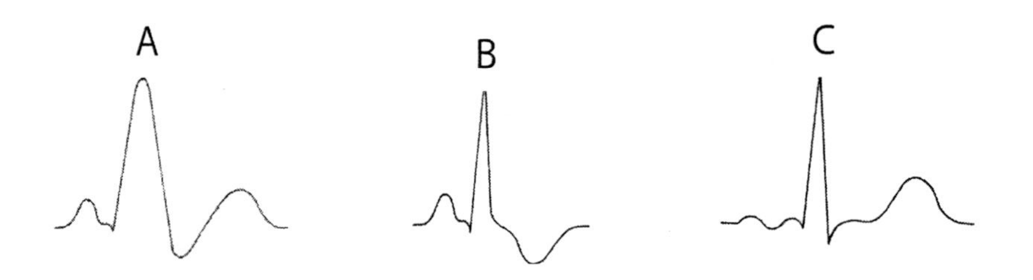

? Question

1. A person who drinks a lot of coffee complains that he has irregular heart rate. Which one of the above EKG waves could reflect this problem? Explain your answer.

2. A person has a greatly enlarged heart from the overwork created by years of high blood pressure. Which abnormal EKG wave could reflect this problem? ____________ Explain your answer.

3. A person with poor coronary circulation to the heart muscle has some heart injury. Which abnormal EKG wave could reflect this problem? _____ Explain your answer.

4. What is the name of the pacemaker of the heart?

5. Where is the pacemaker located?

6. What is the refractory period?

7. In a normal beating heart the impulse conduction time is longer than the refractory period? (yes or no)

8. What is the general function of the atrio-ventricular node?

9. Where are the Purkinje fibers?

Circulation to the Heart Muscle

Heart muscle works very hard, and it must be supplied with oxygen and nutrients just like any other part of the body. The vessels that supply blood to the heart muscle are called ***coronary arteries***. Two important circulation patterns can be seen in Figure 18.7. The right atrium is fed only by the right coronary artery, and the left atrium is supplied only by the left coronary artery. However, each coronary artery supplies blood to parts of both ventricles. Partial blockage to one of the coronary vessels could affect the atrium on one side of the heart more than the ventricle on that side. However, the ventricles do more work than the atria and must be supplied with more blood. This makes each clinical case a bit different.

Another important aspect of coronary arteries is the connection between arteries. Connections between arteries are called ***collateral circulation***. These connections are alternate routes of blood flow to tissue if one path is blocked. Some parts of the heart have no collateral circulation. Other parts have only very small-diameter collateral vessels because they are mostly unused. Also, there are different amounts of collateral vessels among people. Can you find a large collateral vessel in Figure 18.7? Color that vessel. Hint: It is called the ***posterior circumflex***.

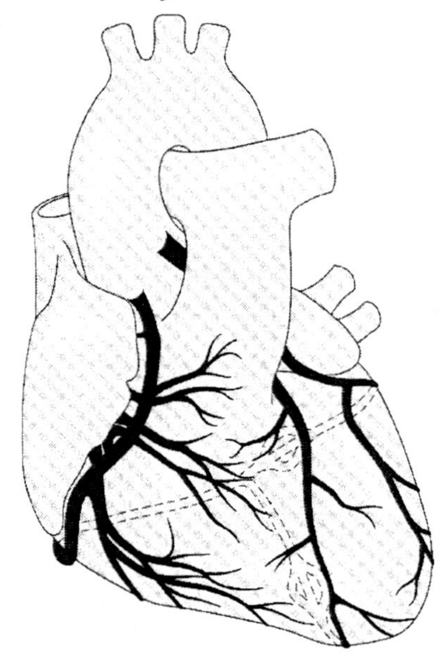

Figure 18.7. Collateral circulation in the coronary arteries.

One of the medical paradoxes happens when a healthy person dies suddenly with the first heart attack. Another less healthy person has a series of small heart attacks over many years and lives a long life. Some of this is explained by differences in collateral circulation. After the first small heart attack, a collateral vessel can be slowly stretched by more blood flowing through it. This can provide a degree of protection from a large heart attack in the future.

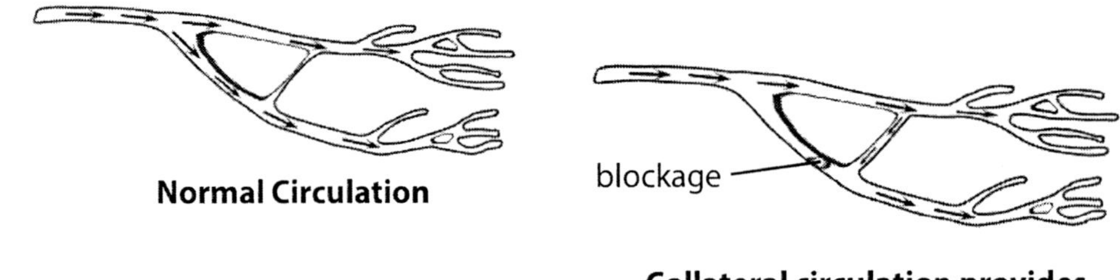

Figure 18.8. Collateral circulation. A collateral vessel may be unused until there is a blockage that forces blood to flow through it.

Use of Stents in Coronary Vessels

Stents are medical devices that are used to open partially blocked arteries. Once positioned in an area of blockage, the stent is expanded by an inflatable bladder which is then removed leaving the stent in place.

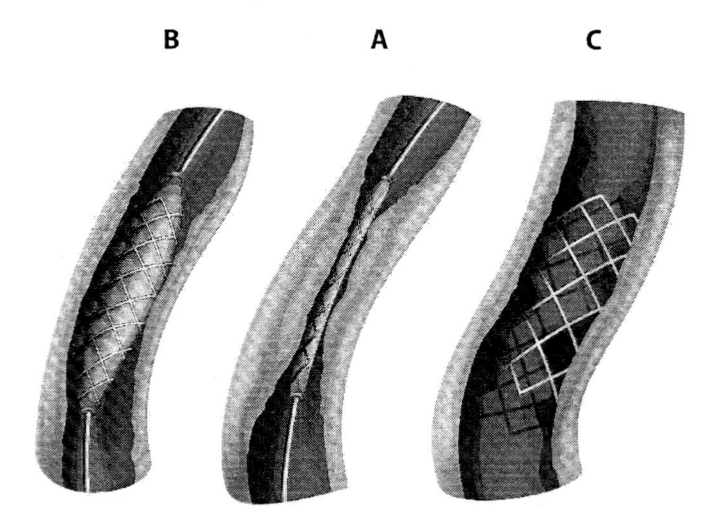

Figure 18.9. Arterial stents. In (A) the stent is positioned; (B) bladder is inflated; and (C) bladder is removed leaving the expanded stent in place.

? Question

1. A patient is told that she has a narrowing of the right coronary artery. Which chamber of her heart is going to be affected the most by this disorder?

2. Which chamber would also be somewhat affected?

3. A group of patients were told that they had plugged arteries in their hearts. In addition, they had all suffered a similar size of heart attack. All of these patients survived. Some of them had parts of their injured hearts return to normal after several months. The other patients had no such luck. Explain these differences in terms of coronary circulation.

4. A heart attack on which side of the heart would probably cause the most serious immediate risk to the person?

Abnormal Holes in the Heart

The heart in a normal fetus has a hole between the right and left atria. This opening allows fetal blood to partially bypass the lung circuit since the lungs aren't used to supply oxygen during life in the womb. Usually, the atrial hole closes shortly after birth. A birth defect results if this hole does not grow closed. Another birth defect occurs when there is an abnormal opening between the right and left ventricles. Both of these abnormalities in heart development are called ***septal defects***. Septal defects allow mixing of oxygenated and deoxygenated blood. This reduces the efficiency of the heart. Also, an opening between ventricles means that the left side can not pump at the needed high pressure. The right and left ventricles will have the same pressure. Septal defects can have serious health implications if left uncorrected.

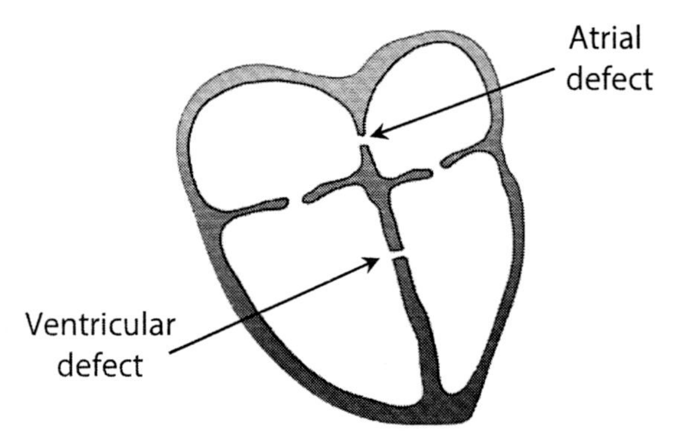

Figure 18.10. Septal defects of the heart. All are significant problems, but some must be corrected sooner than others.

? Question

1. What important molecule is carried by blood entering the left heart pump and is not in the blood entering the right heart pump?

2. If the fetal hole between the right and left atria does not close, what happens to the blood in these two chambers?

3. Which ventricle operates under the most pressure (does the most work)?

4. What would happen to the pressure in the two ventricles if there was a hole between them?

5. What would the heart do to compensate for the pressure problem created by a ventricle hole? Hint: People with this abnormal hole must have it repaired while they are young, or they won't live long.

Exercise #3
Heart Sounds

The heart sound is often described as "**lub-dub**." You might think that the two parts of this sound come from the actual contractions of the upper and lower chambers of the heart. That is not correct. These sounds are more closely associated with the closing of the heart valves. The first sound, ***lub***, happens when the blood vibrates just after the ***chamber valves close*** between the atria and ventricles. Heart sounds are created by fluid vibration waves rather than a physical closing sound of the valves. The second sound, ***dub***, occurs just after the heart ***artery valves close***. With the aid of a stethoscope, a physician can hear these heart sounds and determine if there has been damage to any of the valves.

Heart Sounds on the Internet

Procedure

- Go on the Internet and find examples of heart sounds. Try a search for "Demonstration: Heart Sounds & Murmurs" posted by the Department of Medicine (University of Washington). http://depts.washington.edu/physdx/heart/demo.html

Listen to the following heart sounds:

What you hear depends on where you place the stethoscope.

- **Normal**
  Can you hear two distinct sounds? _____ The first sound happens when the valves between the atria and ventricles close. The second sound happens with the closing of the artery valves after blood leaves the ventricles.

- **Mitral Regurgitation**
  What is the main difference in this sound?

  This is a mitral valve defect, so the sound should be heard during the lub or the dub? _____

- **Benign Murmur**
  How is this benign murmur different from the sound of the mitral valve defect?

- **Pericardial Rub**
  Can you hear the difference between a murmur and a rub? Describe the difference.

- **Extra Heart Sounds**
  When does the extra sound happen in this example? (circle your choice)

  During lub sound (S_1)
  During dub sound (S_2)
  After dub sound (S_2)

Listening to Your Own Heart Beat

Procedure

- Get a stethoscope.
- Clean the earpieces of the stethoscope with a cotton ball soaked in alcohol. Always repeat this procedure whenever another person uses the stethoscope.
- Carefully fit the stethoscope earpieces in your ears so that they are comfortable and point slightly forward in the ear passage. (Your ear passage points forward before it turns inwards to the ear drum.)
- Move the diaphragm of the stethoscope around the left side of your chest starting at the lower center notch of your rib cage.

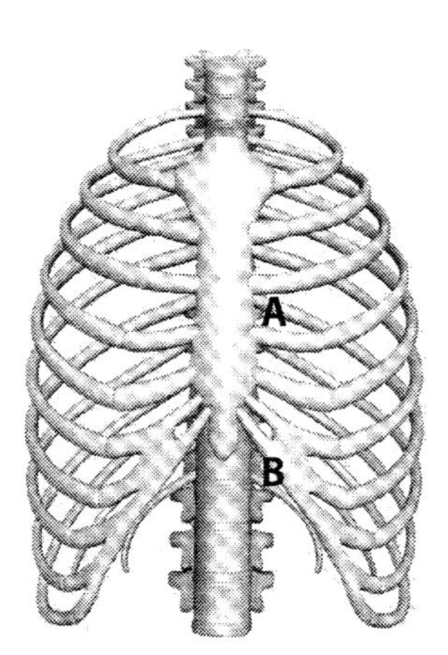

As the stethoscope is moved around the heart area, you will hear the "lub" sound better at some places and the "dub" sound better at other places. An experienced physician or nurse can position the stethoscope to hear each heart valve and determine whether there is an abnormal sound. Abnormal sounds indicate possible valve damage or other circulation problems.

The "lub" sound of the heartbeat resonates downward from the chamber valves, so you hear it best at the bottom of the heart. In which position of the stethoscope (**A** or **B**) do you hear mainly a "lub" sound?______ This would indicate that you are located nearer the bottom of the heart where the sound of chamber valves is loudest.

Most people expect that the two big heart arteries would exit from the bottom of the ventricles, but they don't. These arteries come out of the top of the ventricles and arch upwards above the heart. Refer to the previous diagrams of the heart, and notice the location of the two heart arteries (Figures 18.1 and 18.2). The "dub" sound of the heartbeat resonates upward which is the direction that the heart arteries leave the heart. In which position of the stethoscope (**A** or **B**) do you hear primarily a "dub" sound?_____ This would indicate that you are located near the top of the heart where the heart artery valves make their sounds.

There are other clinical procedures for listening to patient heart sounds. Some of these procedures include positioning the patient onto their left side or right side, or having the patient try to push out air while the throat is closed, or squatting, or quickly standing, or clenching fists. These are methods for changing specific chamber pressures during the cardiac cycle, and they can help with specific diagnoses of heart problems by listening to heart sounds.

Exercise #4
Heart Rate and Stroke Volume

There are two ways of increasing heart output. The first is to pump faster (heart rate), and the second is to pump more blood per beat (***stroke volume***). Both methods allow us to recover from an accumulating oxygen debt during exercise. Heart output usually shows an emphasis on one or the other method depending on the nature of that particular heart. For example, people with a larger than average heart have a greater range in stroke volume possible per heartbeat. The smaller heart does not have this same range in stroke volume, and it must adapt to the output demand by increasing the heart rate. Your heart will favor either stroke volume or heart rate, and this can be determined by results from a Step Test. Perhaps your lab instructor will have you perform this test if time permits. If your heart quickly returns to the resting heart rate but your blood pressure remains above normal for a longer time after exercise, then your heart probably favors stroke volume. If your heart rate is fast during exercise and the blood pressure is about normal when heart rate recovers, then your heart may favor heart rate as the method of recovery after exercise.

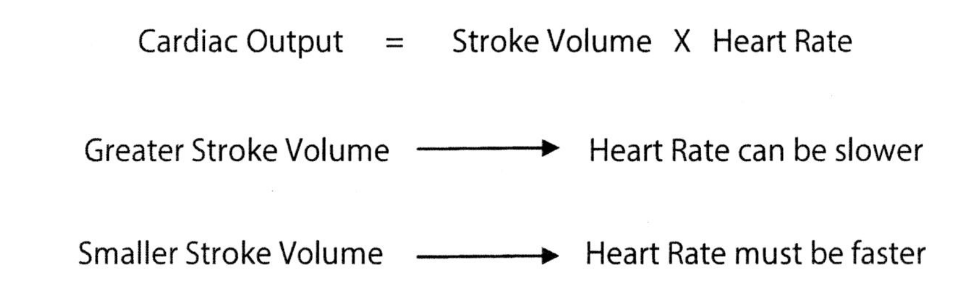

Daily training will increase the size and strength of the heart. Consider the following differences in stroke volumes.

Table 18.1. Comparison of Stroke Volumes in Conditioned and Unconditioned Subjects.

Degree of Conditioning in the Subject	Representative Stroke Volumes (ml)	
	While Resting	During Exercise
Unconditioned	60	120
Conditioned	80	150
Elite-Conditioned	100	200+

? Question

1. Define stroke volume.

2. Compare stroke volumes of the unconditioned with the elite conditioned (Table 18.1). About what % more cardiac output is occurring in the elite conditioned?

3. When the unconditioned runs 1 mile, how far has the elite conditioned person run with the same number of heartbeats?
_____ miles

4. If we assume that an average person needs to pump 5 liters of blood per minute during resting conditions, then how many heartbeats are needed by each of the subjects?

   Unconditioned person needs _____ heartbeats in a minute.
   Conditioned person needs _____ heartbeats in a minute.
   Elite Conditioned person needs _____ heartbeats in a minute.

5. (Note: When you finish the lab activity on "Heart Rate and Longevity", come back to this and the next question.) Based on the Internet Resting Heart Rate Longevity Calculator, how many extra years would you get by strengthening your heart from the unconditioned to the conditioned stroke volume? _____ years

6. How many more years by going from conditioned to elite conditioned? _____ years

The Pulse Wave

The ***aorta*** is the large artery leaving the heart and supplying blood to all other arteries. Each heart contraction forces a volume of blood (50–200 ml) into the aorta. The aorta is "ballooned" out by that blood. The elastic artery wall snaps back an instant after it has been stretched. This recoil causes the next segment of the aorta to balloon out and snap back. The alternate expansion and recoil of the aorta wall "pulses" outward from the heart to the other arteries of the body. You can feel these waves passing by whenever you press a finger on an artery. The pulse wave is not moving blood but is something like ripples spreading from a rock thrown into a pond. Its strength is determined by the strength of the heartbeat and the elasticity of the artery wall. If the beat is strong and the artery walls are very stretchy (a healthy condition), then the pulse wave will be strong.

Pulse Wave

Radial Pulse

The resting heart rate of most people is somewhere between 60 and 80 beats per minute (bpm). These contractions can be counted by listening to the heart beat or by counting the number of pulse waves that pass by a spot on an artery. Counting these waves by touch is how you measure the ***pulse***. The ***carotid pulse*** is felt when you press your fingers against the side of your throat. The ***radial pulse*** is felt when you press your fingers on the thumb side of your upward-turned wrist.

Procedure

- Use the stethoscope to count your heartbeats for 30 seconds.

  Stethoscope Heart Rate = _____ beats per minute.

- Count both your radial pulse waves and carotid pulse waves for 30 seconds.

  Radial Pulse = _____ waves per minute.

  Carotid Pulse = _____ waves per minute.

? Question

1. Which had the stronger pulse waves? (circle your choice)

   Carotid or Radial

2. Which is closer to the heart?

   Carotid or Radial

3. Which would produce a stronger pulse wave?

   Smaller heart or Bigger heart

4. Arteriosclerosis hardens the artery walls with scar tissue. If arteries have been partially injured by arteriosclerosis and the arterial wall is less flexible than normal, would the pulse be stronger or weaker than normal?

5. If a person has arteriosclerosis, what happens to the blood pressure near the end of the arteries? Hint: Is some of the energy of the heart contraction "used up" by the pulse wave? (circle your choice)

   It is the same as normal
   It is higher than normal
   It is lower than normal

6. What would happen to the health of those small arterioles and capillaries affected by the condition posed in question #5?

7. Half of the people have a smaller-than-average sized heart and half have a larger-than-average sized heart. In which group would you expect the heart rate to be faster? Explain your answer.

8. In general, females have a higher heart rate than males. What explanation can you give for this difference?

Calculating Maximum Heart Rate

Most interval exercise programs tell you to perform high-intensity movements in order to elevate your heart rate to 70-85 % of maximum heart rate. This requires us to know our own maximum heart rate. It is not healthy to push your heart to absolute maximum unless your survival is at stake. So, it is best to make a calculation. Use the following formula:

Maximum Heart Rate = the number 210 minus (½ your age, + 5% of your body weight, + another 4 if you are male)

This formula tells us that a 20 year old male who weighs 150 lbs would have a calculated maximum heart rate of 210 minus 10, and then another minus 7, and finally another minus 4. His maximum heart rate is 189.

Calculate Your Maximum Heart Rate. _______
Now calculate 70% of Your Maximum Heart Rate. _______
Then Calculate 85% of Your Maximum Heart Rate. _______

Heart Rate and Longevity

Here comes the longevity judge.

Longevity means long life or "holding up" for a long time. It is very difficult to measure how well something is holding up, so scientists use life expectancy which is a statistical description of how many people live how many years. Modern longevity studies analyze data from large samples of people who have been evaluated on many aspects of their life – both medical and other. These people are followed through time until they get various illnesses or eventually die.

We know that there are serious problems with using correlations, because correlation is not causation. Causation is much more difficult to demonstrate. Correlated factors can overlap. For example, the type of diet may overlap physical activity. So, if you find that people with a certain diet live 5 years longer on average, and you discover that people with 30 minutes of strenuous daily exercise live 5 years longer, then you might assume that people with both habits would gain 10 years of life expectancy. It doesn't work out that way with correlations. The same generality applies to negative factors that take away from life expectancy.

How reliable is the current use of resting heart rate to predict longevity, and what is this calculation based on? Research in comparative physiology has demonstrated that the average mammal heart beats about 1.5 billion times before it wears out. Although there are exceptions to this heart longevity rule, it seems to be generally true whether you're a mouse or an elephant. A mouse's heart beats about 10 times faster than an elephant's heart, and a mouse lives about one tenth as long. Modern humans score above most other species for longevity. This may be because we are smarter and can avoid more hardships than the average mammal. However, we also have a limit. The accepted estimate is somewhere around 2.5 billion beats—if we are lucky enough to survive disaster and illness. How you spend these heartbeats is determined by what you do.

When does running more shorten my life?

Let's assume that the longevity rule (2.5 billion heart beats) is generally true for humans. The example we start with is a woman who already does enough daily activity to keep her heart healthy. She asks the question, "If I train in a very strenuous sport for 4 more hours a day beyond my normal activity, then how much might I shorten my life by doing this extra sport? " Assume that her normal heart rate of 70 is elevated to 120 during the heavy training.

? Question

1. If this athlete does an extra four hours of hard training beyond her normal active life, how many extra heartbeats has she used in that day? ____ extra heartbeats per day of training. (This is the difference between the heart beating at 70 and 120 for the four hours of training.)

2. If her normal heart rate is 70 and there are 1440 minutes in a day, then she normally uses ______________ heartbeats in a day. If an extra 100800 heartbeats represents one more lost day of longevity, then how many days of extra sport training does it take to shorten her life by one day? __________ (Hint: Your calculation will involve the difference between normal and training heart rate.)

3. How many years of this extra sport training would it take to shorten her life by one year? (You don't have to do a lot of multiplication here. It is the same ratio as the previous question.)

There were a number of assumptions in this example that we need to critique. First, we assumed that this person was already doing the perfect amount of exercise to keep her heart in prime condition. We assumed that additional sport training was not going to improve her heart, only use up more longevity heartbeats. That could be wrong. What if this extra training resulted in the heart getting even stronger so that her normal resting heart rate fell from 70 to 60 bpm? That slower heartbeat would mean that it is beating 14400 fewer beats per day than it was before the sport training program.

? Question

1. How many years of beating at this new improved and reduced rate of 60 bpm would add a year of longevity to her heart?

Before we ascribe too much importance to a higher heart rate, remember that females generally live 10% longer than males even though females have a 10% higher resting heart rate. Obviously, other important factors affect longevity.

2. Smoking elevates the heart rate about 10% above normal; so does drinking 3-4 cups of coffee per day. If you were a smoker or a coffee drinker for 40 years, how many years of longevity might be lost due to the increased heart rate?

Could coffee shorten my life?

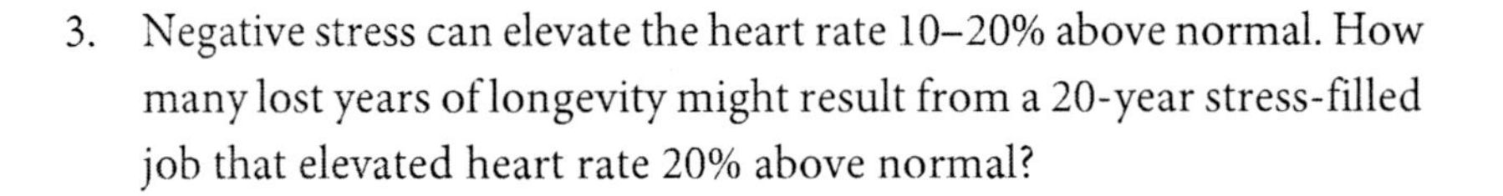

3. Negative stress can elevate the heart rate 10–20% above normal. How many lost years of longevity might result from a 20-year stress-filled job that elevated heart rate 20% above normal?

Resting heart rate is a useful predictor of live expectancy. There are longevity calculators on the Internet that will quickly predict how old a person is expected to live based only on the resting heart rate. The resting heart rate is measured in the morning before you get out of bed. This will be lower than what you would measure in lab today.

Longevity Calculator on the Internet

Procedure

- Go on the Internet and find a longevity calculator based on resting heart rate. Determine the calculated lifespan based on the following:

  50 bpm = ___ years old at death
  60 bpm = ___ years old at death
  70 bpm = ___ years old at death
  80 bpm = ___ years old at death
  90 bpm = ___ years old at death

- At the beginning of Exercise #4 (Question #4) you calculated the resting heart rate based on unconditioned, conditioned, and elite-conditioned. Use the Internet calculator to project the life expectancies of these three groups based on their resting heart rates.

  Conditioned _____
  Unconditioned _____
  Elite-Conditioned _____

- Answer Questions #5 and #6 in that same group of questions.

? Question

1. List two implications of using resting heart rates as a health tool in your chosen profession.

2. List two limitations of this approach.

Diving Reflex

Diving Reflex: Heart Rate slows when you put your face in cold water.

Popular magazines report that heart rate decreases when our face is submerged in cold water for 15 or more seconds. This reaction is called the ***diving reflex***. In diving mammals, like seals or dolphin, it is a major reaction with very dramatic changes in overall body physiology. The diving reflex is fun to investigate and is good practice with experimental design. Your job now is to design experiments that would test this idea. Also, test whether this reaction is caused by cold water only, or would it happen in normal temperature water? And, could it just be holding your breath that initiates the reflex?

There are plastic tubs in the lab that can be used for face immersion. Clean the container before you start, and use clean water for each subject.

Setup 1 (Testing whether temperature of the water is important.)

Description of experimental design:

Results:

Setup 2 (Testing whether it is holding your breath or face in the water that matters.)

Description of experimental design:

Results:

? Question

1. What do your results say about the stimulus for the diving reflex?

2. About how much change (from normal) in heart rate did you observe?

Exercise #5
Blood Pressure

Control of Blood Pressure

There are both nervous and hormonal controls of blood pressure.

Blood pressure is primarily the result of cardiac output and constriction of blood vessels. Anything that increases cardiac output or constricts blood vessels will raise blood pressure. During contraction of the heart the pressure is somewhat higher than during relaxation. When the heart relaxes, the vessel walls "snap back" which helps to maintain blood pressure and keep the blood moving.

There are two basic control processes of blood pressure – autonomic nervous system and biochemical/hormonal mechanisms. These control processes target the heart and the blood vessels, and they also target organs like the kidney which regulates salt and water in the body.

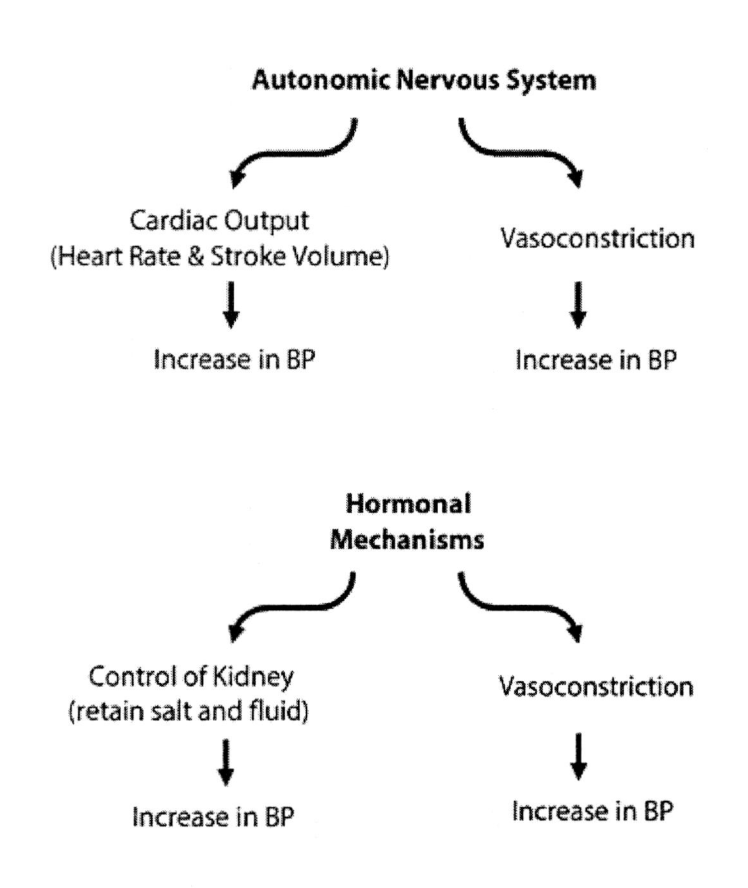

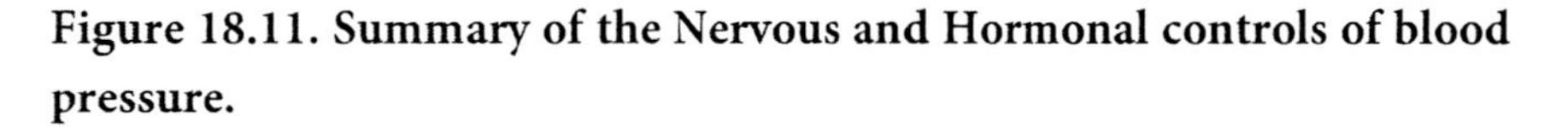

Figure 18.11. Summary of the Nervous and Hormonal controls of blood pressure.

Measuring Blood Pressure

Knowing how to measure your blood pressure is one of the best health-maintenance tools you can learn. Blood pressure is created by the contraction of the left ventricle. As you would expect, the pressure is highest when the chamber contracts. Pressure during heart contraction is called the ***systolic pressure***. This causes the vessel walls to be stretched like an expanding balloon if the vessel walls are healthy. When the ventricle relaxes the blood pressure drops. However, instead of dropping to zero, the blood pressure is partially maintained by the recoil of artery walls that have been stretched during systole The lower artery pressure during the relaxation of the ventricle is called the ***diastolic pressure***.

Figure 18.12. Pulse Wave in body arteries. The elasticity of artery walls moderates the pressure rise during heart systole, and the recoil of the artery wall moderates the pressure drop during diastole.

Medical books state that the normal resting blood pressure is 120 (mmHg) over 80, although normal depends on sex, age, size, and position of the person (sitting, etc.). The 120 refers to the systolic pressure, and the 80 refers to the diastolic pressure. The method we will use to measure blood pressure is fairly simple. It is easier to show you how to measure blood pressure than to explain all of the details in writing. Therefore, most of the instructions will come from your lab teacher. Read through the general steps of the procedure and the hints that follow before you begin practicing.

If this is your first time, team up with someone who already knows BP technique.

Materials Needed

- ✓ blood pressure cuff
- ✓ stethoscope

Procedure

- There should be several students in the lab who already know how to measure blood pressure. If this is your first time, pair up with one of them.
- Clean the stethoscope earpieces with a wipe of alcohol.

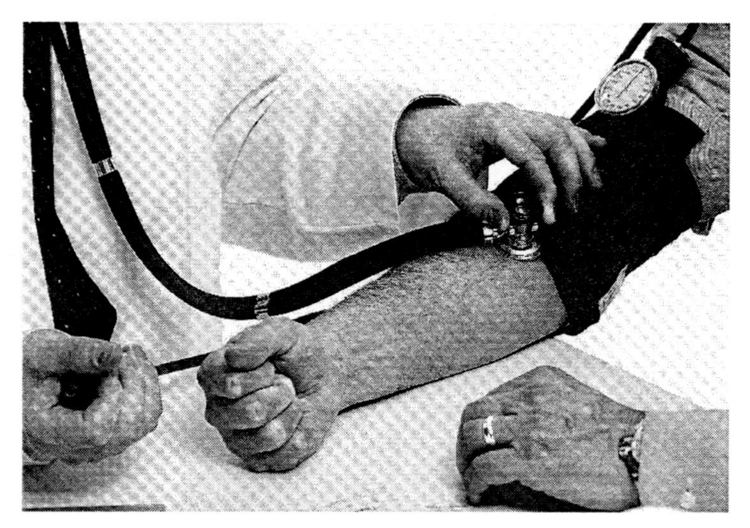

Figure 18.13. Proper technique for measuring blood pressure.

- Choose the correct cuff size for your arm, and fasten it around your upper arm. Place the stethoscope diaphragm over the brachial artery (inside bend of elbow).
- Position the arm of the sitting subject to be at about the level of the heart.
- Pump the cuff full of air until all blood is stopped in the brachial artery. The thumping heart sound that you hear through the stethoscope will fade and disappear as the cuff pressure is pumped above the systolic pressure.
- Release the pressure on the cuff by slowly opening the valve. Listen to the brachial artery with the stethoscope.

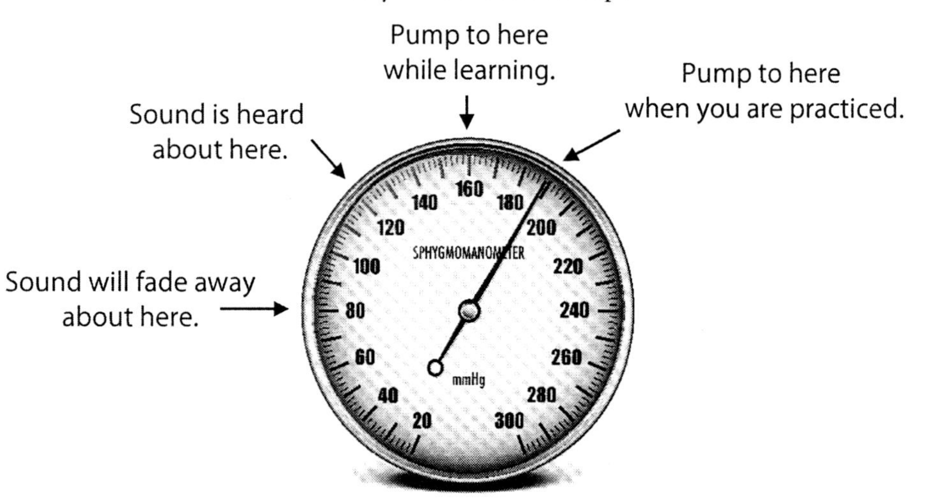

Figure 18.14. What to expect when watching the pressure gauge on the blood pressure equipment.

- When you first hear a "thumping" sound, read the gauge on the pressure cuff. This is the systolic pressure. The blood is just starting to squirt past the cuff during the contraction of the heart. That is the "thumping" sound.
- Continue to release the pressure on the cuff until the thumping sound disappears. Read the pressure gauge. This is the diastolic pressure. The blood is now flowing past the cuff during both the contraction and relaxation of the ventricle. The sound disappears when the flow changes from a pulsating squirt to a constant flow.

Hints for Measuring BP

- Don't pump the pressure cuff over 150 until you've practiced the technique several times.
- Important! Don't keep pressure on the arm for more than 30 seconds.
- Let your arm rest for at least 2 minutes after each reading before taking another measurement. This is especially important while you are learning the technique.
- Take your time. Learn this procedure well. It is important that you know how to measure your own blood pressure. Checking your blood pressure every few months provides you with a thorough understanding of your normal physiology. When an abnormal change occurs, you can seek medical advice.

Your blood pressure while sitting. _____

Your blood pressure while standing. _____

Your blood pressure while lying down. _____

With the blood pressure cuff still on your arm but not pumped up, run in place for 30 seconds. Measure your blood pressure immediately after exercise:

Your blood pressure immediately after exercise is _____

In Exercise #4 you learned about stroke volume and heart rate related to the size of your heart. If you continue timing after exercise until your heart rate returns to normal and then measure blood pressure, a remaining higher pressure would indicate that your heart is favoring stroke volume.

Are you favoring heart rate or stroke volume?

Cold Pressor Test

This test is used to determine the effect that a sensory stimulus (cold) has on blood pressure. The normal reflex response to a cold stimulus is a slight increase in blood pressure (both systolic and diastolic). In a normal individual, the systolic pressure will rise no more than 10 mm Hg, but the increase in a hyper-reactive individual may be 30 to 40 mm Hg.

Don't freeze the hand. Just get it cold.

Procedure

- The subject should be seated comfortably with the BP cuff attached.
- Immerse the person's free hand in ice water (approximately 5°C) to a depth well over the wrist.
- After waiting 30 seconds, measure the blood pressure.

Your normal resting blood pressure is _______.

Your blood pressure after cold water is _______.

Are you a normal or a hyper-reactive individual based on the Cold Pressor Test? _______________

What are possible implications of the Cold Pressor Test? There is some evidence that people showing a hyper-reaction to a cold stimulus have a greater chance of developing high blood pressure later in life. They may have inherited a more reactive nervous system - favored for survival in hunter-gatherer times but easily overstimulated by our modern chaotic life. We just don't know. Before giving too much importance to the results of the cold water test, remember that lack of exercise, high salt intake, unhealthy diet, and stressful situations are known causes of high blood pressure and are mostly under our control. A much stronger predictor of future high blood pressure (hypertension) is a current high pressure under normal resting conditions.

White Coat Syndrome

This reaction happens to some patients whenever they are in a clinical setting. It is called ***white coat syndrome*** because you become hypertensive whenever you see a "white coat". It has been studied to determine whether these patients are actually hypertensive or not.

White Coat Syndrome – What could possibly be responsible for it?

There is a strong correlation between long-term high blood pressure and a number of serious ailments. But people with white coat syndrome show less correlation with serious ailments than those with demonstrated high blood pressure. A series of questions remain. Are these people more reactive and perhaps a bit more vulnerable to high blood pressure than others? How useful is it to follow those with white coat syndrome but have normal pressures at home? More than 20% of patients with a consistent record of high blood pressure in the doctor's office have normal blood pressure during the rest of the day. Studies have compared clinical recordings with home measurements and 24 hour monitors. Interestingly, the expensive monitors proved no more useful than home self-measurements. Some researchers are investigating blood pressure while patients sleep as a better way to detect true hypertension. With better and cheaper monitors there will be more accurate diagnosis.

Poll your class to see how many have elevated blood pressure at the doctor's office, but are normal at other times.

Possible White Coat Syndrome in this class = ____ %.

? Question

1. List the two control processes of blood pressure.

2. What are the two factors controlled by the autonomic nervous system that change blood pressure?

3. What are the two factors controlled by hormonal mechanisms that change blood pressure?

4. Define systolic pressure.

5. Define diastolic pressure.

6. What is the Cold Pressor Test?

7. What is the White Coat Syndrome?

LAB

19

Senses and Perception

We must be aware of our environment and make adjustments to its demands. The senses provide the necessary information, and our perception provides an evaluation of that information through the experiences of trial and error. Working together, these two processes have defined the success of the human species. Today's lab is designed to demonstrate some of the sensory and perceptual mechanisms of the nervous system. Our senses depend on specialized receptors located in various parts of the body. These receptors are activated by only one kind of stimulus (sound, touch, light, chemicals, etc.). The information from one receptor is kept separate from that sent by another sense organ. In the brain, particular areas are specialized for processing and interpreting the information pertaining to each of the senses.

Exercise #1 **Touch**
Exercise #2 **Temperature Sensation**
Exercise #3 **Hearing**
Exercise #4 **Smell**
Exercise #5 **Taste**
Exercise #6 **Vision**
Exercise #7 **Reflexes**

Exercise #1
Touch

A string-line that loops over on the top of your head from ear to ear approximately separates the motor cortex of the frontal lobes from the touch areas of the parietal lobes. The ***motor area*** controls movement of individual muscles and patterned movement of groups of muscles. The ***touch area*** localizes which parts of the body are sending signals, and it interprets the meaning of those signals.

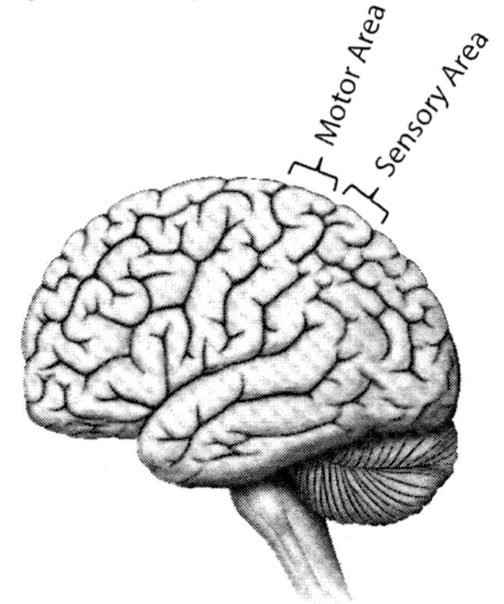

Figure 19.1. Location of the sensory area in the human brain.

In this Exercise you will determine the density of touch receptors in several areas of your body. Using that information, you can test the idea that the areas with greater sensitivity have more touch receptors.

Materials

- ✓ horse hair
- ✓ metric ruler
- ✓ compass or two-point discriminator
- ✓ fine hair brush

Procedure

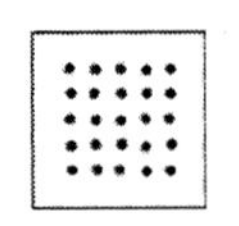

Mark off a 1 cm square.

- Be sure to perform all the tests on each person in your lab group.
- Mark off a 1 cm x 1 cm square on your fingertip, back, and another area of your body that you would like to test.
- Close your eyes while your lab partner lightly touches you 25 times with the horse hair. The touching should be done in a grid-like pattern that covers all of the square you have marked.

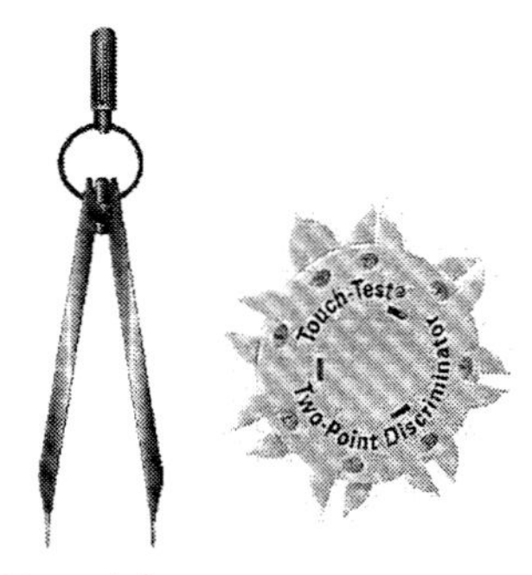

Use either a compass or two-point discriminator.

- Each time you feel the touch of the horse hair, say so. Record the number of positive responses in the Data Table 19.1.
- Next, use the points of a compass or two-point discriminator to lightly stimulate the subject's skin in the area of the marked boxes. The compass points must be blunt and not poke through the skin. *File the points if necessary.* Start with the points close together, then increase their distance apart until the subject definitely feels two distinct points. Be sure that the two points are applied simultaneously each time, and retest to see if there is error due to imagination.
- Measure the distance between the two compass points when the subject clearly perceives two points. This is called ***two-point discrimination.*** Record the results for each area of the body that you mapped for touch receptor density. Include the results from everyone in your lab group.
- Another clinical tool to detect damage in sensory pathways is to brush an area of the skin with a very fine soft brush. Different areas of the body have different sensitivities, but an experienced clinician can determine if a particular patient has lost some of the normal sensitivity. Practice with this technique on each of the areas of the body that you tested for two-point discrimination. Make sure the subject closes their eyes.

Lightly touch with a fine brush.

Table 19.1. Results of Touch Discrimination Experiment. Record results for everyone in your lab group.

Area of the Body	# of Positive Responses during 25 "touches" in 1 cm^2	Two-Point Discrimination (in cm)
Fingertip		
Back		
Other ________		

? Question

1. Which test area had the greatest density of touch receptors?

2. Which test area had the best two-point discrimination?

3. How is two-point discrimination related to density of touch receptors?

4. When you have an itch somewhere on your back, why does it take so much scratching before you finally find it?

5. Did you discover any particular value for using the fine brush test? Describe.

Exercise #2
Temperature Sensation

During this Exercise you will determine whether your body detects the actual temperature or the change in temperature.

Materials

- ✓ a large beaker of cold water (10^0 C)
- ✓ a large beaker of hot water (50^0 C)
- ✓ a large beaker of 30^0 C water

Procedure

- If the water beakers are already set up at the demonstration table, then check and adjust the temperatures using water from the hot plate or ice cubes in order to maintain the three temperature conditions listed above.
- Place the index finger of one hand into the cold water, and put the index finger of the other hand into the hot water for 15 seconds.
- After 15 seconds, quickly place both fingers into the 30 °C water.
- Record the sensations.

Cold-water Finger feels _____.
Hot-water Finger feels _____.

? Question

1. What seems to be the most important factor related to your perception of skin temperature?

   actual temperature   or   change in temperature

2. If you wanted your clients to swim in a pool that was 76^0F, what kind of shower would you have them rinse off with before entering the pool?

Exercise #3
Hearing

The ear is divided into three parts: the outer, middle, and inner ear. When sound waves enter the ear, the ***eardrum*** (between the outer and middle ear) is shaken, and special small bones attached to the eardrum vibrate. These ***middle ear bones*** transmit the sound vibrations to a small membrane (oval window) opening into the inner ear. The vibrations of the oval window create oscillations in the fluid of the ***inner ear*** (cochlea). This moving fluid activates the ***auditory nerves*** leading to the brain. The ***auditory cortex*** of the brain is specialized for interpreting sounds.

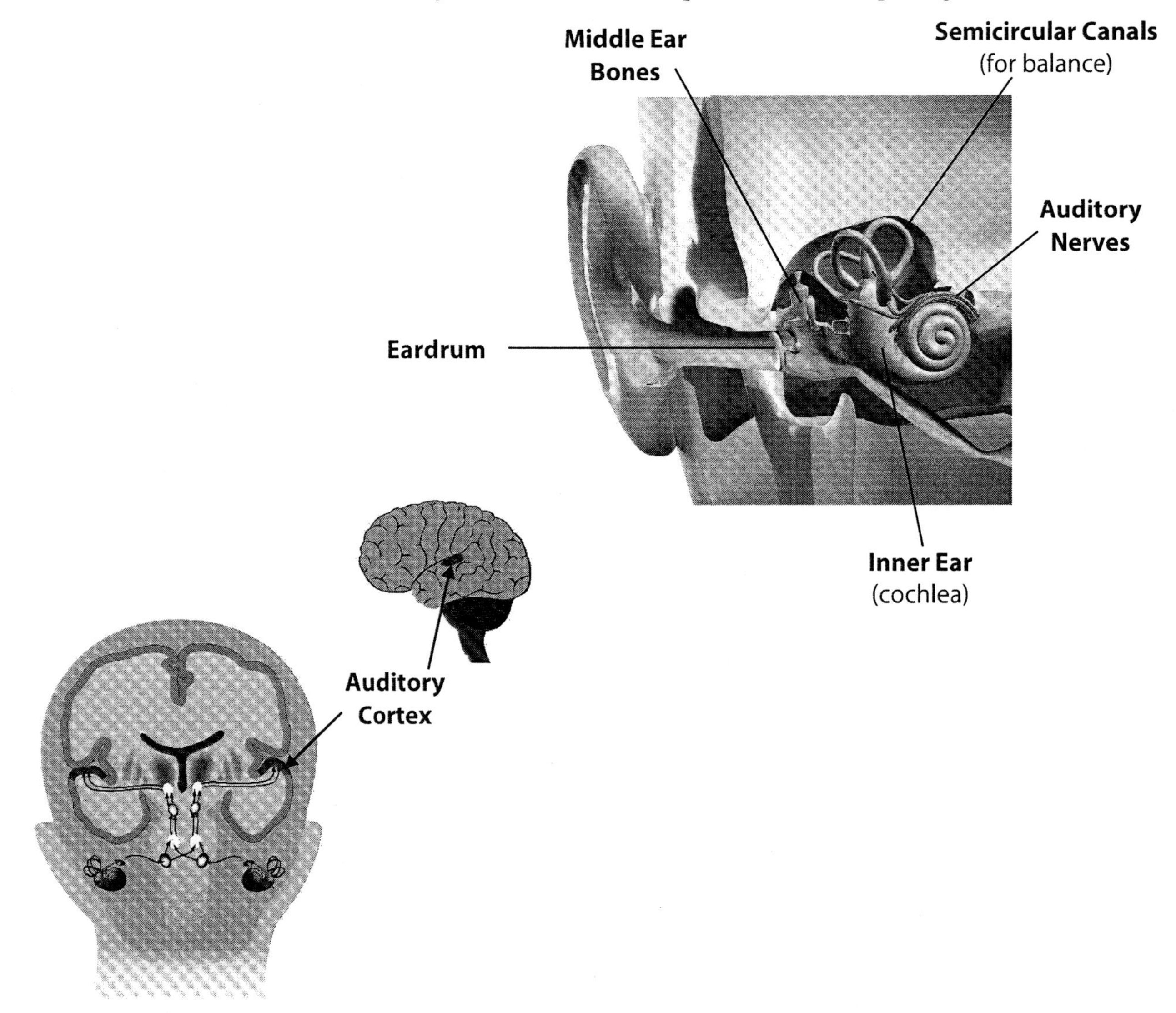

Figure 19.2. Anatomy of the Hearing Apparatus. Sound waves vibrate the eardrum which transmits those vibrations through the middle ear bones into the inner ear (cochlea) where the actual nerve receptors are located. Signals from the cochlea are sent to the auditory cortex where sounds are identified and interpreted.

Materials

- ✓ cotton for ear plugs.
- ✓ set of tuning forks.
- ✓ meter stick.

Procedure

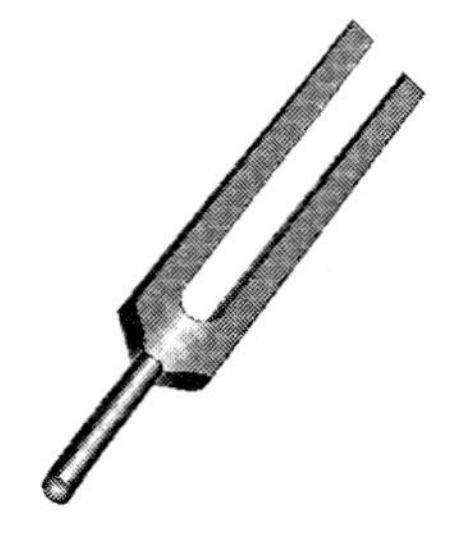

Be consistent when striking the tuning fork.

- Do the first test in a quiet room. Have the subject close one ear with cotton and close his eyes. Strike the tuning fork against the table and hold it in line with the open ear. Move the tuning fork away from the ear until the subject just loses the ability to hear it. Measure that distance. Repeat the test again to validate your first measurement. Record the hearing distance for both ears (Table 19.2). Be sure to strike the tuning fork with equal force each time you do the test.
- Perform the same test with each of the six tuning forks of different tones to determine if you have hearing loss in any of the six ranges. If one of your ears has a hearing loss at a particular tone range, then do the next test.
- This next test should not be performed in a quiet room. It will determine whether your hearing loss is due to a problem in the middle ear or inner ear. Place the handle of a vibrating tuning fork on the midline of the subject's forehead. A person with normal hearing will localize the sound as if it were coming from the midline. If one ear has defective middle-ear function (ear bones), then the sound will be heard much better in the defective ear when the tuning fork is touching the forehead. If there has been damage to the auditory nerve, then touching the tuning fork to the forehead won't improve hearing in the defective ear.

Table 19.2. Results of Hearing Tests.

Sound Frequency (cycles per second)	**Farthest Distance sound is heard from the Left Ear**	**Farthest Distance sound is heard from the Right Ear**
128 cps tuning fork		
256 cps tuning fork		
512 cps tuning fork		
1024 cps tuning fork		
2048 cps tuning fork		
4096 cps tuning fork		

? Question

1. Middle ear damage often comes from serious ear infections during childhood. Check your class for anyone diagnosed with middle ear damage by our lab tests, and ask them if they had childhood ear infections.

2. Inner ear damage in young people is often because of exposure to very loud sounds. Check your class for anyone with inner ear damage (nerve), and ask them if they have been exposed to loud sounds.

3. What is the best design for a hearing aid if a patient has hearing loss in a particular sound range?

4. What would be the design of a cheaper hearing aid?

5. Which person would more likely be a candidate for surgery?

   Middle ear damage
   or
   Inner ear damage

Exercise #4
Smell

It has been observed that one's nose is never so happy as when it is thrust into the affairs of another, from which some physiologists have drawn the inference that the nose is devoid of the sense of smell.

Ambrose Bierce

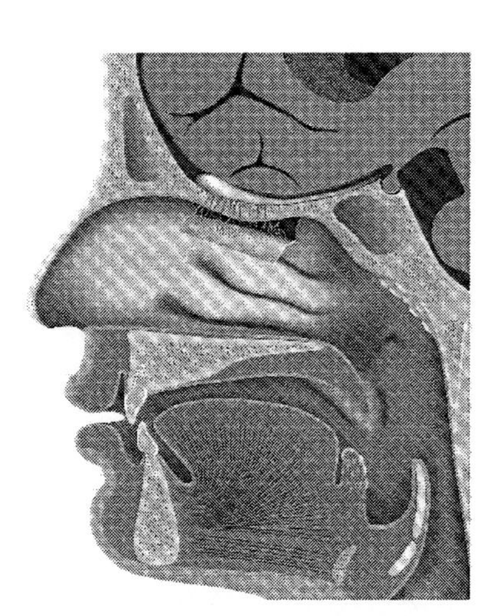

There are more than 30 million smell receptors that line the top of the nasal cavity.

Taste studies show that most of food flavor comes from smell rather than taste itself. Some researchers suggest that the evolutionary specialization of the mammalian forebrain began with the sense of smell because there are so many smells in the environment. Although the complete role of this sense is not understood, it seems more closely linked to emotional memories than to the conscious activities of our brains, and it can change in different conditions. As you experiment with the various odors in the smell experiment, describe the type of emotional reaction and memories that you have to each.

As air enters the nasal cavities, it swirls around and aromatic molecules contact specific nerve receptors in the roof of the nasal cavity. There are 30+ million receptors in your nose, and nerve cells are probably dedicated to particular odors. Each aroma has a chemical binding with specific nerve endings. Evidence suggests that learning is involved during the "wiring" of the smell neurons, and memory may be essential because Alzheimer's patients have no apparent smelling ability. Much of the neuron circuitry is not "hard wired" as other senses are, and it can be modified and improved somewhat during our life. Smell receptors are the only nerve cells that replace themselves every few months. This may be necessary because the nerve endings are directly exposed to the air. Smell may be the most adaptable sense.

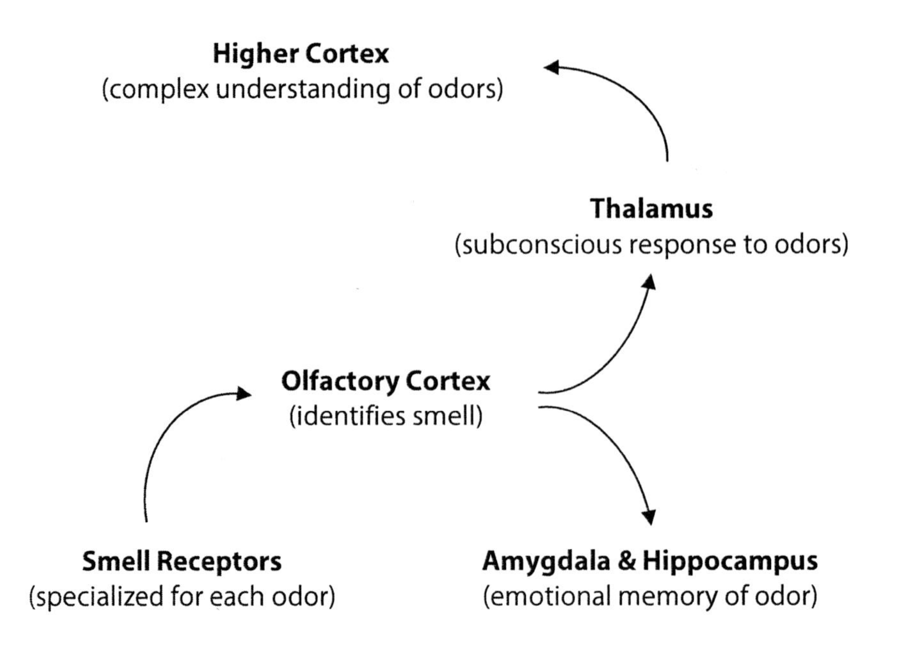

Figure 19.3. The Smell Sensory System and a general description of the detection and interpretation of smells.

Materials

✓ a smell kit

Procedure

- Close your eyes. Have your lab partner pass an open odor vial about 3" under your nose for a couple of seconds. Repeat the test if necessary.
- First determine if you can smell an odor. Then determine if you can correctly identify the smell. Finally, describe any special memories associated with the smell.
- Record the results of your test in Table 19.3.

Table 19.3. Data Table for Odor Recognition.

Sample	Can detect a smell	Can identify the smell	What are any memories associated with the smell?
1			
2			
3			
4			
5			
6			
7			
8			
9			
10			
	Total =	Total =	

What memories are associated with each smell?

? Question

1. There are two abilities being tested. One is to determine if you can detect an aroma. What was the average number of odors that students could detect but not necessarily identify?

2. What was the average number of odors that students could correctly identify?

3. Repeat the experiment after you have some more practice with the smells. Did your recognition improve?

4. How many of the smells were associated with emotional memories?

5. List three examples of how specific smells might be used to sell you a product.

Exercise #5
Taste

The tongue has at least four different taste receptors (salty, sweet, bitter, and sour). However, the taste of many chemicals is also influenced by your interpretation of their smell. In this Exercise you will examine several aspects of your ability to taste.

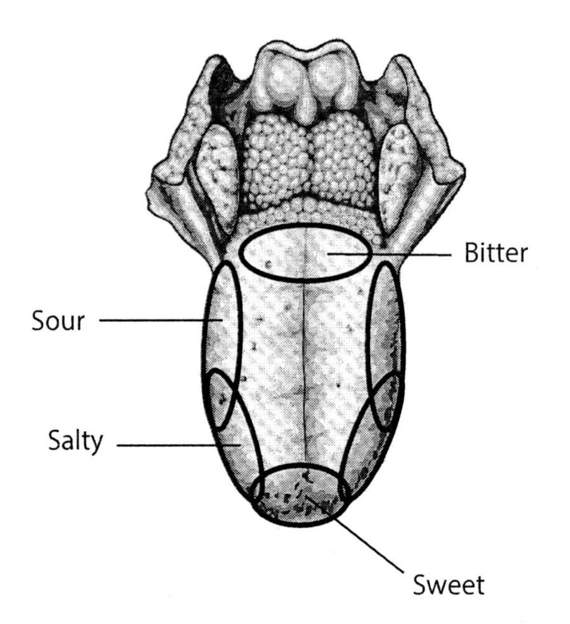

Figure 19.4. Location of four taste areas on the tongue.

Genetically Determined Taste

Your ability to taste some chemicals is determined by whether or not you have inherited the gene controlling the taste response to that particular substance. This is an important lesson for you to remember when certain foods don't taste bitter to you but other people complain about them (especially your children). They may have the gene to taste it, and you may not.

Materials

- ✓ special taste papers for PTC and thiourea

Procedure

- Take one of the taste papers, and touch it to your tongue. You will immediately know if you are a taster!
- Do the same test with the other taste paper.
- Put the used taste papers in the trash.

? Question

1. Are you a taster for thiourea?
2. Are you a taster for PTC?
3. If you are a non-taster and you want to be a high-class chef, what might you do to compensate for this genetic limitation?

Some have a more refined sense of taste than the rest of us!

Sugar Taste Threshold

This experiment will determine differences in sugar tasting threshold among students, and the results might help us to understand why some people prefer more sugar in their foods.

Procedure

- Go to the demonstration table and determine your sugar ***taste threshold*** (minimum percent of sugar that you can taste).
- Dip a strip of tasting paper into each solution, and record whether or not you can detect a sweet taste. Discard the used taste papers.
- After testing all of the solutions, go to the front desk for the key to the sugar concentration of each solution.
- Record your results in the summary chart on the front chalkboard, and compile the class results below.

Table 19.4. Data Table for Sugar Taste Threshold.

Taste Threshold	Sugar Concentration				
	0.25%	0.5%	1%	2%	5%
# of Students Who Could Taste					
# of Students Who Could Not Taste					

? Question

1. What was your sugar tasting threshold?

2. What was the lowest percent of sugar tasted by your classmates?

3. What was the highest threshold in the class for tasting sugar?

4. Do the people with a high taste threshold also like more sugar in their food? (You could determine this by asking your classmates how much sugar they add to their coffee.)

Estimating the Number of Taste Buds on Your Tongue

The ability to taste depends on many factors, but one of them is the density of taste receptors on your tongue. It is quite easy to make an estimate for your own tongue. The process involves putting a colored fluid (either food coloring or blue sucker) on your tongue and counting taste buds within a small circle. The taste buds will appear uncolored (pink) on the blue tongue. Each taste bud has many receptor cells (100 or so). You will be categorized as average, supertaster, or low taster based on the number of taste buds.

Materials

- ✓ paper towels
- ✓ paper-hole reinforcers (circle used for counting)
- ✓ blue sucker
- ✓ magnifying glass
- ✓ bright light or flashlight

Procedure

- Everyone wash their hands before starting.
- The subject is to suck on a blue sucker for a few minutes until the tongue is blue. (There are other instructions if you are using blue food coloring. Ask your instructor.)
- Use a piece of paper towel to slightly dry the tongue.
- Place one of the paper-hole reinforcers on the subjects tongue.
- The taste buds are slightly raised and are not stained blue as much as the rest of the tongue.

- Use a flashlight and magnifying glass to count the number of taste buds.
- Help the subject remove the paper circle with a paper towel.
- Repeat this procedure for each person in your lab group.
- Record your results in the summary chart on the front chalkboard, and compile the class results in Table 19.5.

Table 19.5. Tasting Ability Based on Density of Taste Buds.

Taster Category (# taste buds)	**# of Students in Each Category**
Low Taster ≤ 15	
Average Taster ≈ 16-29	
Super Taster ≥ 30	

? Question

1. How many taste buds did you count in your tongue sample area?

2. What is your taster category?

3. Analyze the class data to determine if there is a correlation between your genetically determined taste test, sugar tasting threshold, and density of taste buds. Use Table 19.6.

Table 19.6. Relationship Between Taste Bud Density and the Ability to Taste Sugar and PTC (or Thiourea).

Taster Category	**# of Students who could taste 0.25% or 0.5% sugar**	**# of Students who were tasters of PTC or Thiourea**
Low Taster		
Average Taster		
Super Taster		

Exercise #6
Vision

Human beings are primarily visual animals. It is our dominant sense for relating to the environment. There is a lot of scientific literature on visual perception, and we encourage you to investigate this information when you have time to do so. Most human behavior is strongly influenced by visual perception, and what we don't know can affect us without our knowledge.

Preferred Eye

This test is designed to reveal which one of your eyes is used for certain visual functions. Your preferred eye is the one your brain chooses to use when both eyes can see the same object.

Which is your preferred eye?

Procedure

- Pick an object that's about 30 feet away. Make a circle with your thumb and first finger of both hands.
- Straighten and raise your arms from your waist to a position where the circle surrounds the object. Keep your head and feet positioned straight ahead.
- Without further movement, close one eye. Then open the closed eye, and close your other eye.
- Which one of your eyes has the same view as the view with both eyes open? This is your preferred eye. When you close your preferred eye, the distant object will move out of the circle formed by your hands.

? Question

1. Which eye is your preferred eye?

2. If you are left-eyed, what problem will you have in shooting a rifle?

3. Why should you use your preferred eye when looking through a monocular microscope?

Eye with Best Vision

Procedure

- Use the classroom eye chart to determine which of your eyes has the best vision (without glasses).

? Question

1. Which of your eyes has the best vision?
2. Is this the same eye that is your preferred eye?
3. Are you left or right handed?
4. Talk with other lab students, and discover whether the eye with the best vision is always the same one as the preferred eye or handedness.

   Results

Eye with Best Depth Perception

There are fairly simple ways of determining which of your eyes has the best depth perception. If the testing equipment is available in lab, then determine the depth perception for each of your eyes, and tabulate the class data like that in Table 19.7. If the depth perception equipment is unavailable, then use the information in Table 19.7 to answer the questions.

Table 19.7. Relationship Between Handedness and Vision Tests.

Vision Tests	13 Left-Handed People	11 Right-Handed People
Eye with Best Depth Perception	9 left eye 3 right eye 1 same in both eyes	2 left eye 8 right eye 1 same in both eyes
Preferred Eye	7 left eye 6 right eye	5 left eye 6 right eye
Eye with Best Vision	1 left eye 1 right eye 11 same in both eyes	2 left eye 1 right eye 8 same in both eyes

? Question

1. The preferred hand (whether left-handed or right-handed) is most closely associated with . . . (circle your choice)

Eye with best depth perception
or
Preferred Eye
or
Eye with best vision

Exercise #7
Reflexes

Which reflex personality type are you – calm or quick reacting?

The brain is capable of sending both ***fascilatory*** and ***inhibitory*** signals to the reflex centers in the spinal cord. The balance of these opposite effects determines the quickness and magnitude of reflexes and one aspect of your personality. When you are excited or threatened or preparing for competition, the balance shifts to faster and stronger reflexes and you are more excitable. There are differences in the normal disposition of the nervous reflex system among people. You can see this difference by watching how reactive a person is under normal conditions (very calm vs. quick reacting).

A reflex test while reading can be used to determine which reflex type you are. Concentrating on reading reduces the effect your brain normally has on reflex centers. If reading reduces your reflex response, then normally your brain must be stimulating reflexes (you are a quick-reacting person). If reading increases your reflex response, then your brain normally inhibits reflexes (calm-reacting). This test is not 100% determinative because your brain's reflex emphasis does change under various circumstances, and this lab class could be one of them. No person is 100% one type or the other all of the time.

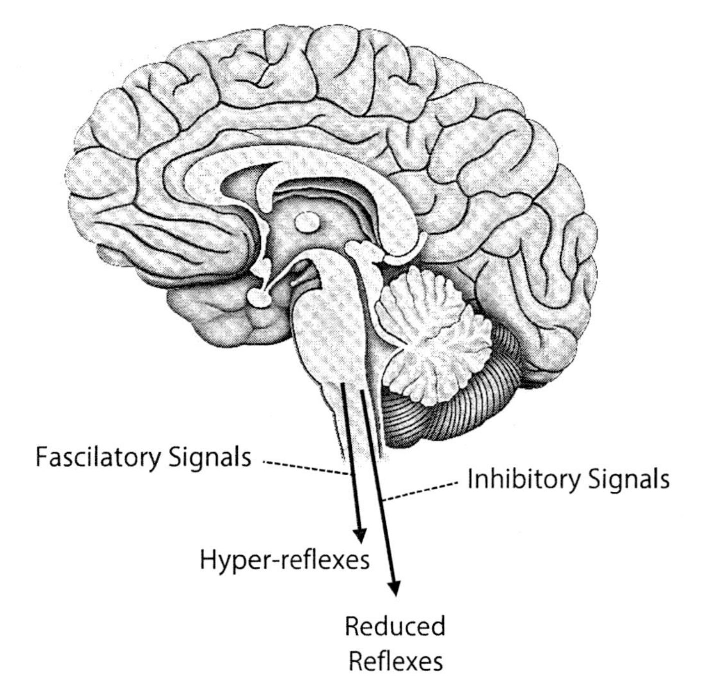

Figure 19.5. Excitation and inhibition of the spinal reflexes. Pathways leading out of the brainstem are capable of shifting the reactivity of the reflex centers in the spinal cord.

Materials

- ✓ patellar hammer
- ✓ meter stick

Procedure

- Before beginning the test, ask your lab partner to evaluate whether you are the calm or quick-reacting type.

Lab Partner's Opinion:

Your Opinion:

- Sit on a table so that your legs hang freely over the edge. Have your lab partner hit the patellar ligament (just below the knee) with the reflex hammer. Don't hit too hard. This may take some practice. Measure the amount of leg movement several times to get an average estimate of the reflex intensity.

  Normal Reflex =

- Next, read from a textbook while your lab partner measures the amount of reflex leg movement. Is the reflex more intense or less intense during the reading conditions?

  Reflex while reading =

? Question

1. Under normal circumstances, did your brain activate or inhibit your spinal reflexes?

2. So, which reflex type are you?

3. Does this agree with how you evaluated yourself before the test?

4. Compare your conclusions with those of other students in the class. What did you discover?

LAB

20

Human Evolution

Strickly speaking of course we know nothing about prehistoric man, for the simple reason that he was prehistoric.
-G.K. Chesterton

We are naturally curious about who was here before us, and our families and societies pass down written and oral stories about the past. Discoveries by anthropologists and molecular biologists have added more to our understanding about what may have happened before written history. This new evidence comes from four main sources: 1) cultural artifacts, 2) fossils, 3) comparisons of human traits with other primates, and 4) DNA similarities among living humans and other primates. We are now able to present a partial scientific story about human evolution. During this lab you will learn some of the methods for developing that history.

Exercise #1 **Trails of the Cross-Country Hikers**
Exercise #2 **The Mitochondrial DNA Clock**
Exercise #3 **Time Trails of Homo sapiens**
Exercise #4 **The Story of the Early Hominids**

Exercise #1
Trails of the Cross-Country Hikers

This first Exercise will help you to develop a way of thinking about human evolution. The information we give you may seem inadequate, but work through the problem and you will be able to draw a map of the different paths taken by five students, and you will discover that you can determine when they separated from each other and the trails they took. Your time map will look something like Figure 20.1 with branches and arrows that represent when the different hikers separated from each other.

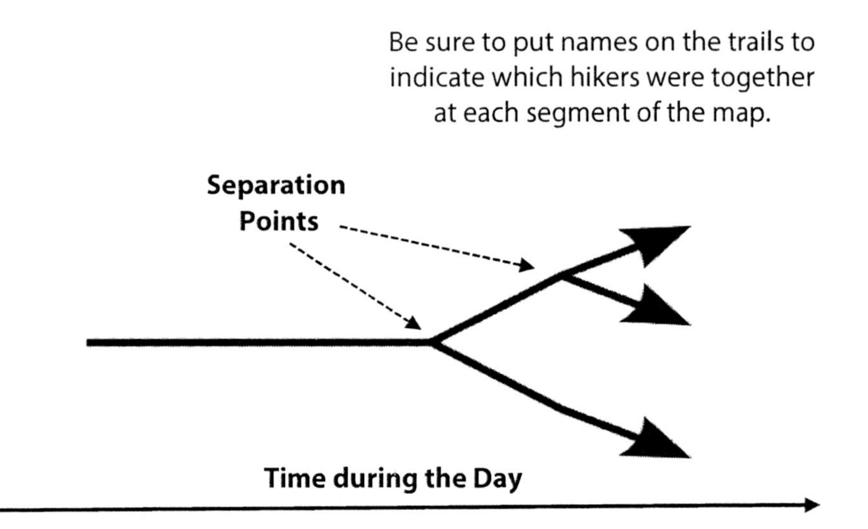

Figure 20.1. Example of a time map for the hiking path of the student hikers. This map is based on known associations of those students during the day.

Materials

- ✓ Refer to Figure 20.6 at the end of this lab titled, "Trails of the Cross-Country Hikers."
- ✓ A five-card set of names. You can use these name cards as an aid to physically keep track of the hikers during your analysis of the problem.

Procedure

- Work in groups of 3 or 4 students.
- Your group may need some quiet space to do this Exercise, so don't hesitate to work outside or in another room if necessary.

Cross Country Hikers

The hikers are Bill, Hector, Julie, Tom, and Maria.

Five members of the Cross-Country Running Team decided to have a rugged day of fun last Saturday. They drove out a long dirt road and parked the car at a grove of trees that was 12 hours of hard hiking from the freeway. All they had to do was to walk west and they were guaranteed to reach the Highway 8 Freeway where they could hitch a ride back to their parked car. Your task is to figure out the general paths that were taken by the individual people using the limited information provided below, and then you are to draw those paths in Figure 20.6.

Information

- The five hikers started out together, but they divided into smaller groups and took different paths as they went along.
- You are given clues about the hikers. The clues tell you when particular hikers separated from each other and took different paths. You will have to use those clues to figure out and draw their approximate paths to the highway.
- The hikers' names are: Bill, Hector, Julie, Tom, and Maria.
- They all began to walk together at about 6:00am. All of the hikers reached Highway 8 at around 6:00pm, but they did not necessarily arrive together.
- Tom and Maria arrived at Highway 8 together.
- The last time Hector was with Julie was 5 hours before he reached the highway. (This clue does not tell you whether or not Hector or Julie were with anyone else when they separated from each other. You will have to figure that out—read on.)
- The last time Julie was with Tom was 8 hours before she arrived at the highway. (Again, this clue does not tell you whether or not Julie or Tom were with anyone else when they separated from each other.)
- The last time Bill and Hector were together was 10 hours away from the highway. (Was Hector or Bill with anyone else at the time they last saw each other? Again, this clue does not tell you.)

Procedure

- Make a practice map to work out the problem, then draw your final map in Figure 20.6.
- You will find it helpful to use name cards to keep track of the hikers. Lay them on the table, and move them along the trails as you work through the clues.
- Draw the map of the trails taken by the five hikers. Show when they split up during the day and how they were grouped when they arrived at the highway at 6:00pm.
- Check with your instructor to make sure that you get the correct answer before going on. The next Exercise is going to use the same kind of thinking as in this hiker map.

Exercise #2
The Mitochondrial DNA Clock

The Mitochondrial Clock is not a real clock, but it does keep time.

There is an evolutionary "clock" made of human DNA that will tell us how long it has been since any group of humans has been separated from any other group of humans. Think about it this way: When humans live together, they mate and blend their genes over many generations of time. This creates common traits in human DNA. Now, what happens if one group splits into two separate groups that migrate far away from each other, and they never get a chance to interbreed with each other again? As you would expect, the two groups are no longer mixing genes (and DNA), and over time they will begin to look somewhat different from each other.

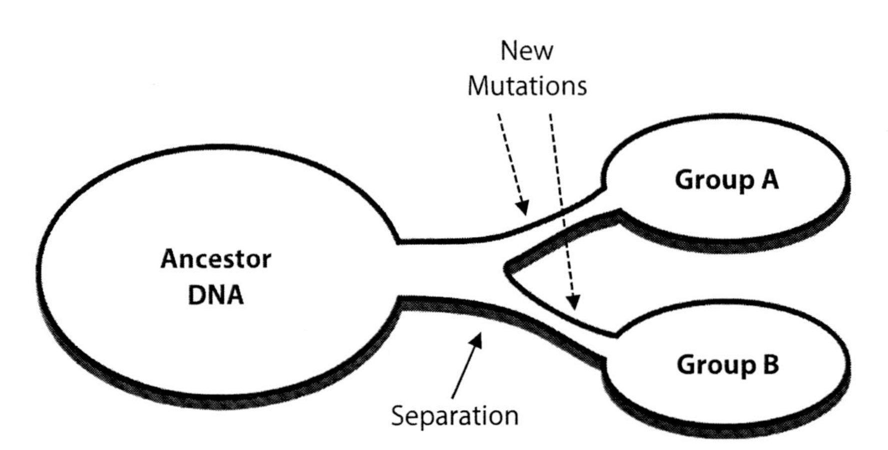

Figure 20.2. If groups of people are separated for long periods of time, they begin to collect different DNA mutations in their populations. The more time of separation, the more differences will collect in the two groups.

Mitochondrial DNA

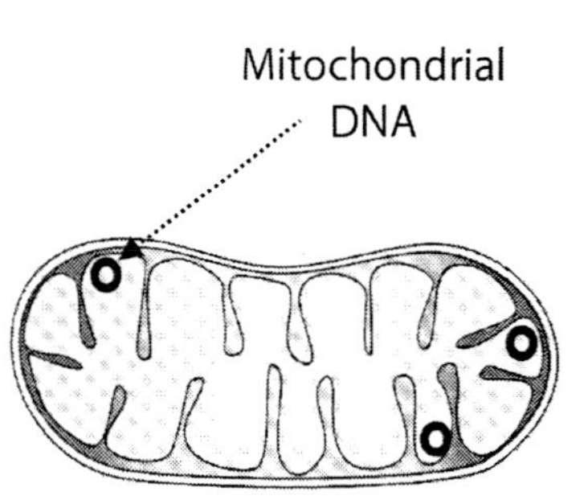

Mitochondrion

You have learned that the hereditary DNA is carried by 23 pairs of chromosomes in humans. But all organisms, including humans, have a small pieces of DNA in their mitochondria. This DNA is involved with the reproduction of mitochondria. It does not mix with DNA in the cell nucleus during sexual reproduction. The only way it gets passed onto the next generation is with the mitochondria in the mother's egg. The father's sperm does not give mitochondria to the egg during fertilization. This strange situation means that mitochondrial DNA is passed on by the mother and stays the same unless it is changed by "new" mutations. These mutations occur randomly as a natural process of living on this planet.

By comparing DNA patterns, chemists can detect any new mutations that have been added to the small piece of original mitochondrial DNA that was passed on from mother to child through thousands of human generations. If a group of humans splits off from a common ancestral group, then both groups begin to accumulate different mutations from each other. Each group is genetically separated from the other group. This starts at the point where they no longer interbreed.

The easiest way to show that two groups have separated in the past is to count the number of new mutations found in the DNA of one group and not found in the DNA of the other group. The greater the number of different mutations, the longer the amount of time that the two groups have been separated from each other. (Remember: Mutations take a lot of time to happen. If the DNA of the two groups is quite similar, then they have not been separated very long.)

Comparative DNA Analysis

Diagrammed below is a comparison of the mitochondrial DNA of three geographically separate human racial groups. Each vertical bar represents a mutation.

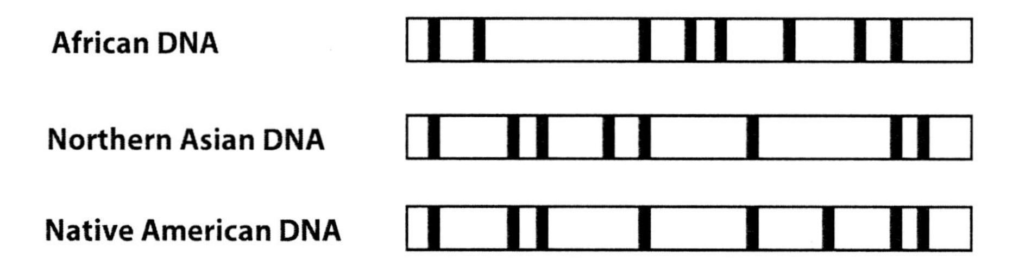

? Question

1. Which two groups are the most similar?

2. Which one of the three groups is the most different from the other two groups?

3. So, which group has been separated from the others for the longest time?

4. Which two groups haven't been separated from each other for very long?

The "Clock"

It takes about 500,000 years for 1% of the mitochondrial DNA to be changed by mutation.

Biologists have studied many different species, including humans, and have estimated that it takes about 500,000 years for 1% of the mitochondrial DNA to be changed by mutation. This estimated mutation rate is currently being debated, but we will base our calculations on it. Perhaps more research will clarify our understanding of evolutionary timelines. This mutation rate is a kind of "clock." For years, researchers have gone around the world collecting mitochondrial DNA samples. They have counted the total number of small mutations that have occurred in the humans of today. Each one of those mutations represents an amount of "time" on the ***mitochondrial clock***.

A representative sample of different human geographical/racial DNA has been collected and analyzed. When all of the different mutations were counted and the human groups were compared, scientists found that only 0.4% of the DNA of modern humans is different within all groups. If all modern humans evolved from a common group in the past, then all of the mitochondrial DNA differences started when the groups separated from the ancestor group. The common ancestor group wouldn't necessarily be the original humans, but they would be the common ancestor group of all humans today. If you want to estimate how long it has been since two racial or geographical groups have been separated, you must:

- Use the figure for the % of genetic difference between the two groups (as determined by genetic researchers).
- Multiply the % of genetic difference by 500,000 years. (It takes 500,000 years for 1% of the mitochondrial DNA to be mutated.)

- Practice Example: If Group A and Group B have 0.1% genetic difference, then it has been about __________ years since they separated and have not had the opportunity to interbreed.

500,000 years x 0.1% genetic difference = 50,000 years

Answer: 50,000 years

? Question

1. How many years have modern humans been on this planet? Remember: It takes 500,000 years for 1% of the DNA to be mutated.

2. Let's see if you can estimate how long two geographical/racial groups have been separated from each other.

   Group A, the "Altitudinals," have lived in the high mountain ranges of Nepal for longer than anyone can remember. Group B, the "Basinals," have lived on the flat river delta plains of Southern India for hundreds of centuries according to their written history.

   A genetic researcher has measured a 0.05% genetic difference in the mitochondrial DNA of Group A and Group B.

   When did Groups A and B separate?

   How long would you say it has been since these two groups lived together and had the opportunity to interbreed? In other words, how long ago was it that the Altitudinals and the Basinals split company, one moving to the highlands and the other moving to the lowlands?

   __________ years (Show your work.)

It's time to move on to Exercise #3, the "Time Trails of Homo Sapiens", and test your understanding of the "Mitochondrial Clock". If you are having trouble understanding the mitochondrial clock, then now is the time to get help from your instructor.

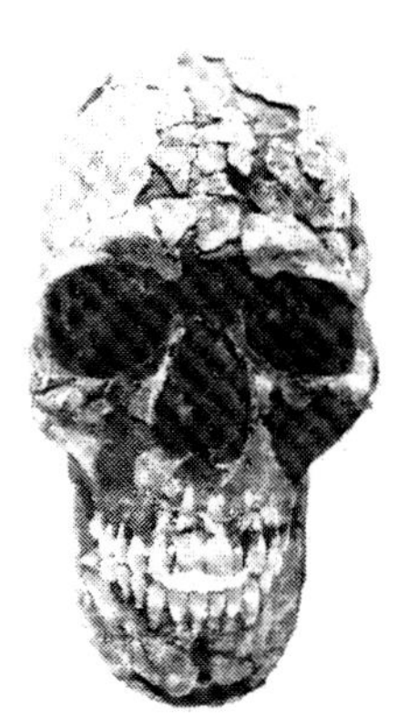

This human skull is about 100,000 years old, and it is very similar to a modern skull.

Exercise #3
Time Trails of the Homo sapiens

Archaeologists have found skeletal evidence of a modern-type human (looks like people today) who lived in Africa about 100,000 years ago. There is a question as to exactly how long modern humans have been here and exactly where they originated. This Exercise will give you a partial answer to these questions.

Modern human types are known as *Homo sapiens*. They are somewhat lighter in build with a larger brain size than earlier human fossils. Evidence also suggests that they probably lived much like the hunter-gatherer peoples of today. If you were to bring some of these skeletal remains back to life, put some modern clothes on them, and put them on a bus, you could not tell them apart from anyone else on that bus.

What is the "trail" that humans took as they spread around the world?

The migration of humans can be traced just like the trails of the cross-country hikers.

Figuring out the answers to this question is somewhat like doing the "Trails of the Cross-Country Hikers" problem. Here, you will follow five geographical/racial groups instead of five hikers. You must calculate when these groups separated from each other and relate those times of separation to a world map. From that information, you will be able to tell a story of modern human origin and migration.

? Question

1. When genetic researchers compare the DNA of Northern Asian populations versus Native American populations, they find a 0.07% difference. The "Mitochondrial Clock" formula tells us that there has been about _______________ years since the separation of these two groups.

2. When comparing the DNA of Northern Asian and European populations, researchers discover a 0.10% difference. The "Mitochondrial Clock" formula tells us that there has been about _______________ years since the separation of these groups.

3. When comparing the DNA of the Indonesian peoples to either the European group or the Northern Asians, researchers find a 0.12% difference. The "Mitochondrial Clock" formula tells us that there has been about ______________ years since separation of these three groups.

4. When researchers compare the DNA of African populations to any other group, they find a 0.20% difference. The "Mitochondrial Clock" formula tells us that there has been about ______________ years since the separation of Africans from all other groups.

Materials

- ✓ Refer to Figure 20.7 "Time Trails of *Homo sapiens*" at the end of this lab.
- ✓ Refer to Figure 20.8 "World Map-Migrations of Homo sapiens" at the end of this lab.

Procedure

- Fill out the map of the "Time Trails of *Homo sapiens*" in Figure 20.7. Use the "years since separation" information you just calculated on the previous page for the different geographical/racial groups.
- Show when each group split off from the ancestor group of humans (just like you did for the Cross-Country Hikers).
- Convert the time of separation map in Figure 20.7 into a geographical map in Figure 20.8. Draw your best interpretation of the trail taken by humans as they separated and spread out on this planet. Show where the ancestor population started, and show the sequences of where they separated and where they went after that. There may be more than one possible trail based on the limited genetic information presented in this Exercise.

Here is a table of the general anthropological evidence of the earliest settlements of modern humans that have been excavated so far.

Region	Time of the Earliest Settlements
Americas	Less than 35,000 years ago
Indonesia	50,000 years ago
Europe	35,000 years ago
Asia	60,000 years ago
North Africa	100,000 years ago

? Question

How does your story based on DNA comparisons, illustrated by the Migration Map in Figure 20.8, agree with the anthropological evidence?

Exercise #4
The Story of the Early Hominids

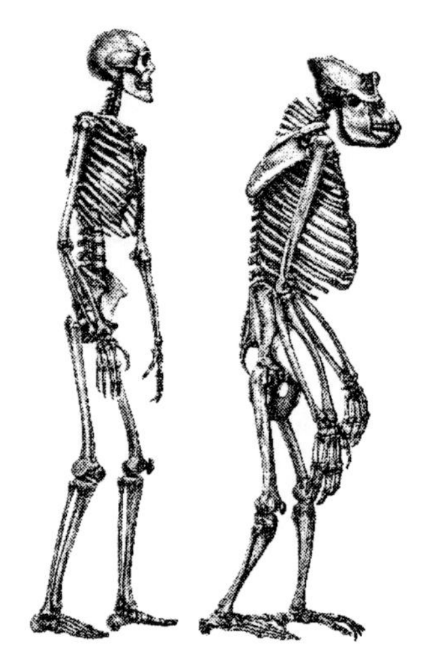

Figure 20.3. Comparison of the human and gorilla skeletons. At first, anthropologists thought we were most closely related to gorillas. However, our skeletons are quite different.

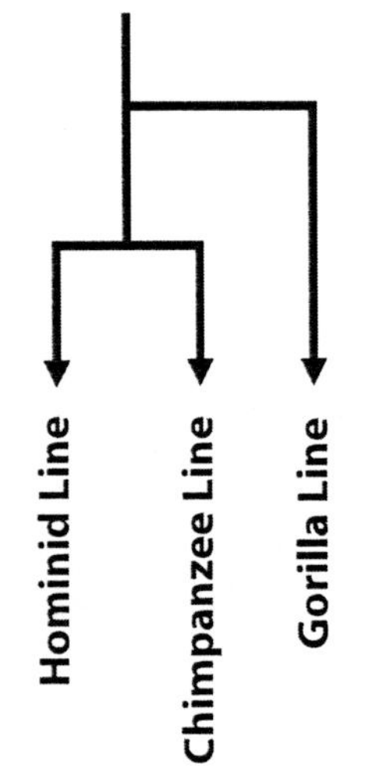

The story of human evolution is very old indeed.

Comparison of human and gorilla DNA evidence suggests that the gorilla line branched off the human ancestral line about 7 million years ago. Then approximately 6 million years ago the chimpanzee and human lines separated. The details of the last six million years of hominid evolution comes from the limited fossils that have been discovered. These are the only clues to our origin and evolution.

- **Earliest hominid**
  Two fossils were vying for earliest hominid, but one of them has been declared an ape by the anthropologists, leaving ***Orrorin tegenensis*** as the oldest discovery in our lineage. The bones of this hominid suggest that it was about four feet tall and bipedal. The fossils were found in Kenya (east Africa) and date to more than 6 million years ago.

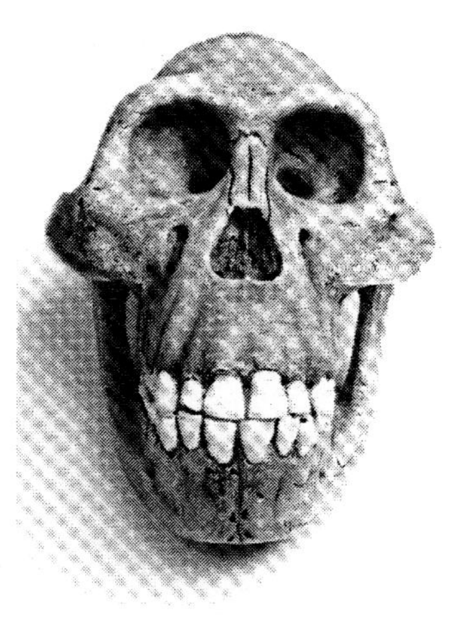

Australopithecus afarensis

- <u>**Other early hominids**</u>

Originally all early hominid fossils were called Australopithecines. New research has subdivided them into ten or more species within three separate genuses. The oldest is called ***Ardipithecus*** (about 6 million years ago). This genus is followed by several species of ***Australopithecus*** (2–4 million years ago) that are lighter-boned. A branch off *Australopithecus* apparently led to a heavy-boned species named ***Paranthropus***.

All of these hominids were smaller than modern humans, standing about four feet high and walking slightly bent over, and they had smaller brains. Skeletal comparisons suggest there were different lineages and niches. One or more of these early hominids (probably *Australopithecus afarensis*) are in the ancestral line leading to *Homo habilis* and *Homo erectus*.

- <u>*Homo habilis*</u>

The *Homo habilis* group was a tool maker. They had somewhat larger brains than Australopithecines, stood more erect, and were slightly bigger in size. Compared to *Homo erectus*, who also made tools, the *Homo habilis* group had a smaller brain. These fossils occur between 1.5 and 2.5 million years ago.

- <u>*Homo erectus*</u>

The original fossil groups of *Homo erectus* were designated as Peking Man, Java Man, and Heidelberg Man. They stood fully erect (for which they are named), and they had the same general height and brain size as modern man. The earliest fossils of *Homo erectus* (1.8 million years ago in Africa) are sometimes referred to by the name *Homo ergaster*. They have been found with more artifacts than earlier hominids which tells us something about how they lived. Apparently, they were very much like *Homo sapiens* but may have had less ability to speak because of some throat structure differences. This is currently being debated. There is evidence suggesting that *Homo erectus* may have included more subgroups, and their lineage and migration out of Africa might be a much more complex story than we currently understand. Some of them, including Neanderthals, survived until 30,000 years ago, and they may have interbred to a small extent with some *Homo sapiens*. (Researchers have estimated that 60% of modern humans carry some Neanderthal genes.) The *Homo erectus* group started to die out about 200,000 years ago at the same time as *Homo sapiens* appeared and migrated around the world.

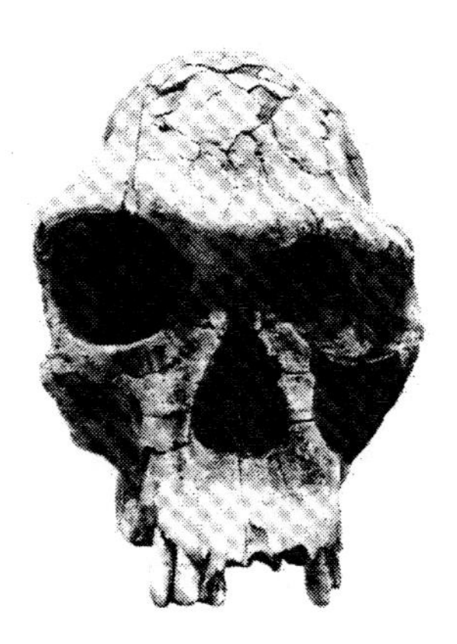

Homo habilis

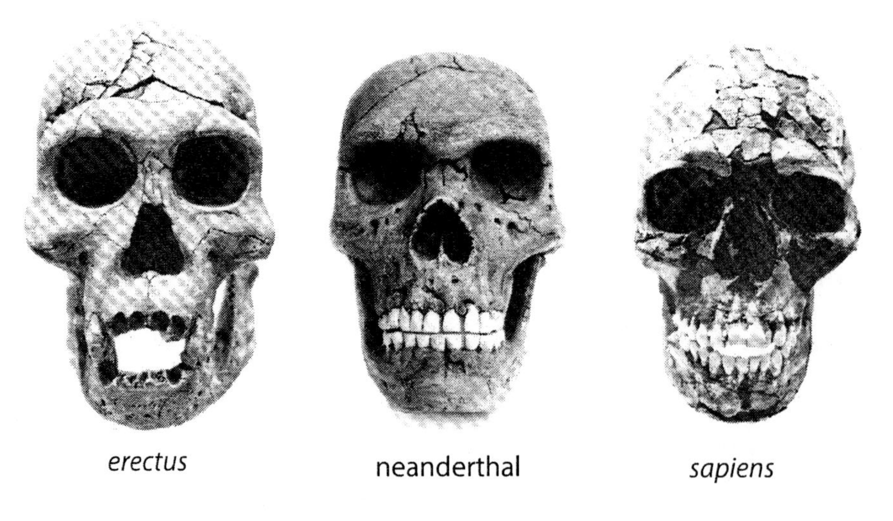

Figure 20.4. Comparison of hominid skulls. *Homo sapiens* is quite different from *Homo habilis* 2.5 mya), but there is more similarity with *Homo erectus* (1.5 mya) and Neanderthals.

Procedure

- Using the information provided in Table 20.1 complete the Figure 20.5 Time Map of known hominid species. Then complete the Figure 20.8 Geographical map by putting a dot on the map for each of the hominid species. See if the dots suggest a story of how these human-like species might be related to each other and where they developed and migrated.

Table 20.1. Age and Location of Fossils from Five Hominid Genuses.

Fossils	Age of Fossils	Where Found
Orrorin tegenensis	6.2 million years	Africa
Ardipithecus ramidus	5.5 million years	Africa
Australopithecus anamensis	4.2 million years	Africa
Australopithecus afarensis	3.5 million years	Africa
Australopithecus africanus	3.0 million years	Africa
Homo habilis	2.5 million years	Africa
Homo erectus	1.8 million years	Africa
Homo erectus (Peking)	500,000 years	China
Homo erectus (Java)	1.5 million years	Southeast Asia
Homo erectus (Heidelberg)	800,000 years	Germany

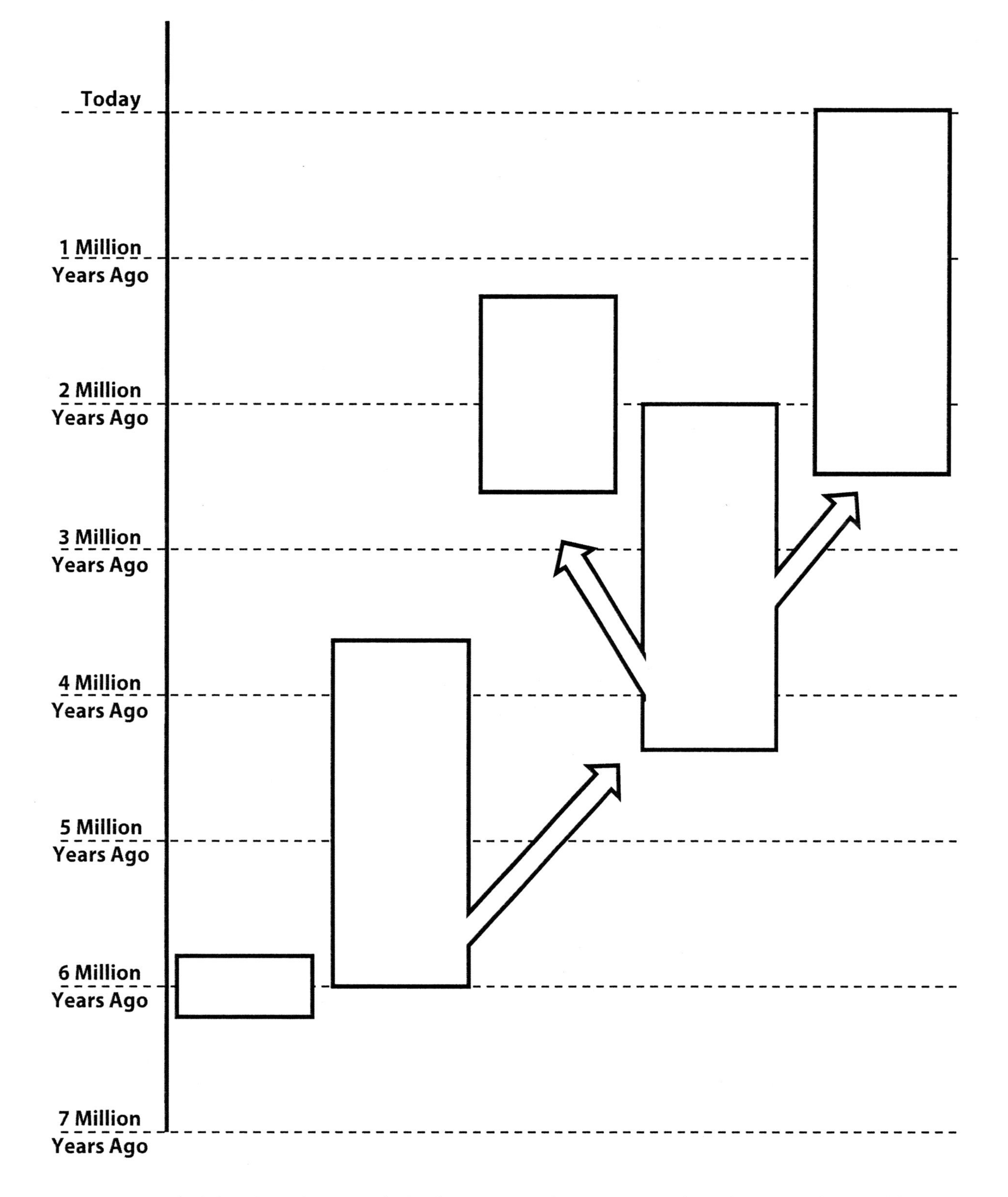

Figure 20.5. Label the above boxes with the five Hominid genuses. Use the information from Table 20.1.

1. How long ago did the human and chimpanzee lines separate?

2. How old is the oldest known hominid?

3. Based on fossil evidence, where did the early hominids originate?

4. A branch off the Australopithecines led to a heavy-boned type that is not considered a direct ancestral group to us. The name of that group is ____________________.

5. How long ago did *Homo habilis* live?

   Where did they live?

6. Which group first migrated out of the continent of their origin?

   How long ago?

   How far did that group spread?

7. Was there another group that came out of the continent of origin after the group in Question 6?

   Which species was that group?

   When did the migration of the new species happen?

   Indicate that event on your map in Figure 20.7.

   Show your completed maps to your instructor.

If you think that all of this is hard to figure out, remember that many researchers have spent their careers just to give you the small amount of information that you are working with in this lab. There is a great deal more to know that will make the picture more accurate. Anthropology and Archeology are other fields of study in the natural sciences. If you are enjoying what you are learning in this lab, consider studying more in those fields.

Trails of the Cross-Country Hikers

(Use for Exercise #1.)

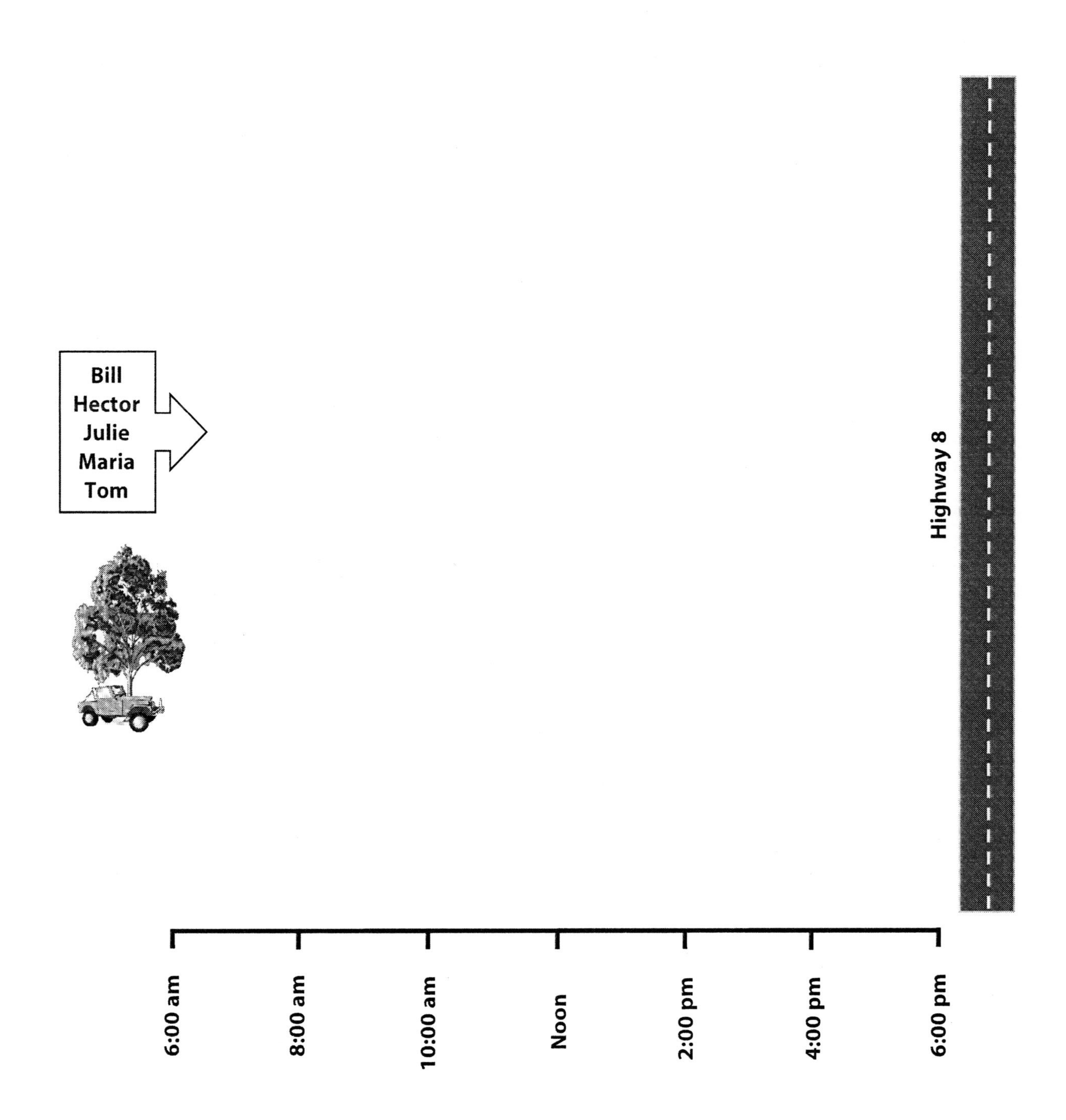

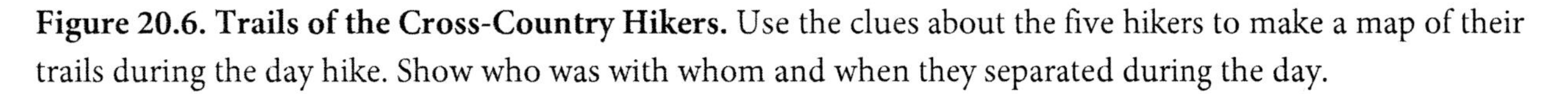

Figure 20.6. Trails of the Cross-Country Hikers. Use the clues about the five hikers to make a map of their trails during the day hike. Show who was with whom and when they separated during the day.

Time Trails of *Homo sapiens*
(Use for Exercise #3.)

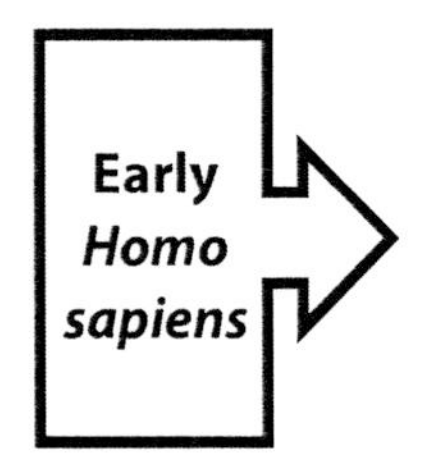

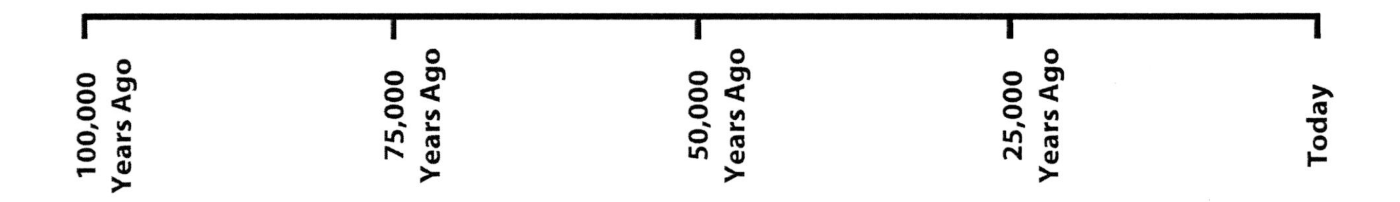

Figure 20.7. Time Trails of Homo sapiens spreading around the world. Use the mitochondrial DNA information about the five geographical/racial groups that you calculated in Exercise #3. Make a time map to show when the various human groups separated from each other.

World Map of Migrations of Modern Humans
(Use for Exercise #3)

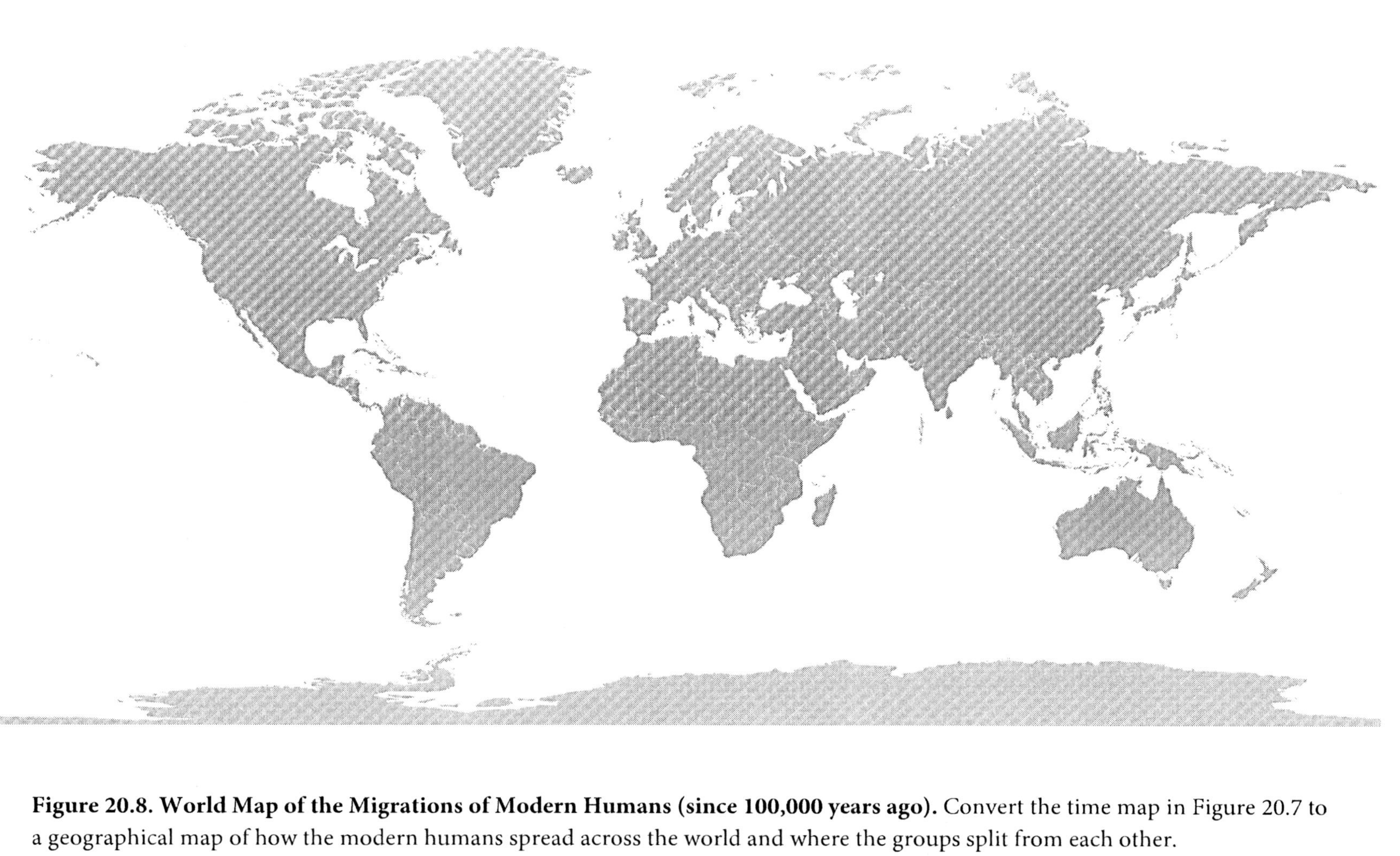

Figure 20.8. World Map of the Migrations of Modern Humans (since 100,000 years ago). Convert the time map in Figure 20.7 to a geographical map of how the modern humans spread across the world and where the groups split from each other.

Figure 20.9. World Map of the Story of the Early Hominids. Use the information from Figure 20.5 and Table 20.1. Put a dot on the map for each of the hominid species (label them), and see if the dots suggest a story of hominid evolution.

LAB

21

Embryology

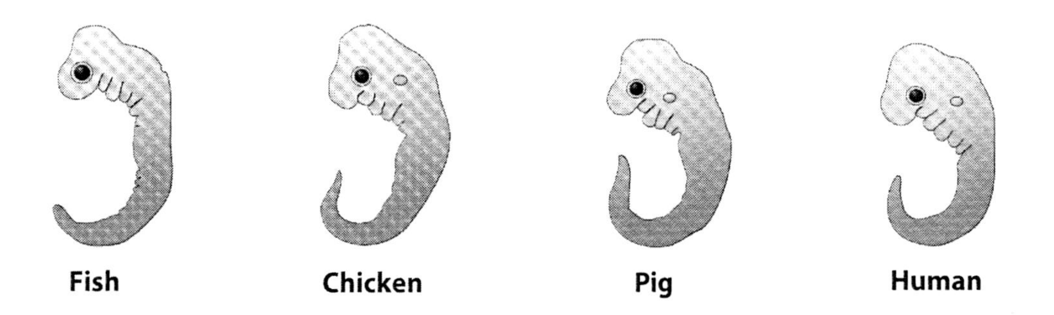

Embryology is the study of early developmental phases in plants and animals. An ***embryo*** is any stage after the egg becomes fertilized but before the developing organism looks like the adult form (about 8 weeks for humans). In advanced vertebrates, any stage after the embryo and before birth is called a ***fetus***. A century ago, there were two contrasting theories about development. One view, called ***preformation***, held that a miniature version of the adult existed in the egg and this grew into the adult form. This theory is incorrect. The second theory, called ***epigenesis***, stated that new structures and body systems are continually being created during embryonic development. This description is correct. All vertebrates begin as a single cell but develop amazing degrees of complexity by the time they are born.

Scientific research has revealed two very important embryological concepts. First, the study of embryos shows how simple life-forms could have evolved into complex life-forms. Developing traits provide biologists with the opportunity to observe how genetic mechanisms operate to create an organism. The second concept is that there are developmental processes common to all vertebrates. This means that studying the embryos of other organisms can help us to understand development in human embryos. In today's lab, you will learn the general terminology and descriptions of embryonic stages in the embryos of various organisms at similar stages found in human development.

Exercise #1 **Ontogeny Recapitulates Phylogeny**
Exercise #2 **Eggs, Sperm, and Fertilization**
Exercise #3 **Methods of Feeding the Embryo**
Exercise #4 **Stages of Development**
Exercise #5 **Disruption of Normal Development**

Exercise #1
Ontogeny Recapitulates Phylogeny

The embryonic development of an animal provides clues to its evolutionary history.

The title of this Exercise has been a recurring TV quiz-show question. The quote is from Ernst Haeckel, a German scientist at the turn of the 20TH century. He thought that embryonic development (***ontogeny***) replayed the entire evolutionary history (***phylogeny***) of a species. In other words, watching the development of a frog would be like seeing a movie of the evolution of vertebrates that lead to the frog. Since Haeckel's time, science has discovered that both evolution and developmental phases are far more complicated than his original idea suggested.

The human embryo does proceed through various levels of complexity similar to those of earlier vertebrates. There is certainly as much change in the human embryo from conception to birth as there is change in the fossil record from one-celled organisms to complex land vertebrates. And the situation faced by the human embryo is similar to the challenges presented to vertebrate groups that evolved from a water environment to dry land conditions. But, we do not first become a fish, then become a frog, followed by a reptile, and finally a mammal. The developmental "movie" is much more blurred than that.

? Question

1. Define embryo.

2. Define fetus.

3. What week of development marks the beginning of the fetus stage in humans?

4. What is meant by "ontogeny recapitulates phylogeny"?

5. Is it true that the human embryo first looks like a fish, a frog, and then a reptile before reaching the fetal stage? Explain your answer.

Procedure

- Read the brief descriptions of human embryonic stages in Table 21.1, and then refer to the "Ontogeny Recapitulates Phylogeny" chart in Table 21.2. Write the name of the appropriate human embryonic stage that compares in complexity to organisms in evolutionary history.

Table 21.1. Description of Human Embryonic Stages.

Human Embryonic Stages	
Name	**Description**
Zygote	Diploid cell formed by the union of egg and sperm.
Morula (3-4 days)	Small solid ball of identical cells.
Blastocyst (1 week)	The heart begins differentiating into tissues.
25 Day Embryo	The heart begins to beat, but there are no organ systems yet (i.e. Circulatory, etc.).
4 Week Embryo	The notochord has formed. This is the very beginning of a skeletal system which develops in later weeks.
4-5 Week Embryo	Organ systems are established including a primitive type of kidney called the mesonephros. This kidney degenerates and is replaced by a more complex one.
5-6 Week Embryo	Organ systems continue rapid development. A new type of kidney appears called the metanephros. It eventually grows into the adult kidney.
6-8 Week Embryo	There is rapid development of a brain lobe called the telencephalon which grows into the cerebrum in later weeks.
8 Week Fetus	This stage is now called the fetus because it looks like the adult form.

There is as much change in the human embryo from conception to birth as there is change from one-celled organisms to complex vertebrates.

Table 21.2. Ontogeny Recapitulates Phylogeny - Human Development. Fill in the human developmental stage that is comparable to each of the listed events in the evolution of life.

Ontogeny (development stages)	**Recapitulates** (repeat)	**Phylogeny** (evolutionary history)
______________		Pig and Primate Embryo (developed cerebrum)
		↑
______________		Chick Embryo (has dry-land kidney called metanephros)
		↑
______________		Frog Embryo (has aquatic kidney called mesonephros)
		↑
______________		Early Vertebrates (beginning of internal skeleton)
		↑
______________		Flatworms and Roundworms (definite organs present)
		↑
______________		Jellyfish Types (cells organize into tissues)
		↑
______________		Colonial Protozoa (cluster of identical cells)
		↑
______________		Eukaryotic Cells (unicellular with organelles)
		↑
(no comparable stage in humans)		Prokaryotic Cells (no cell organelles)

Exercise #2
Eggs, Sperm, and Fertilization

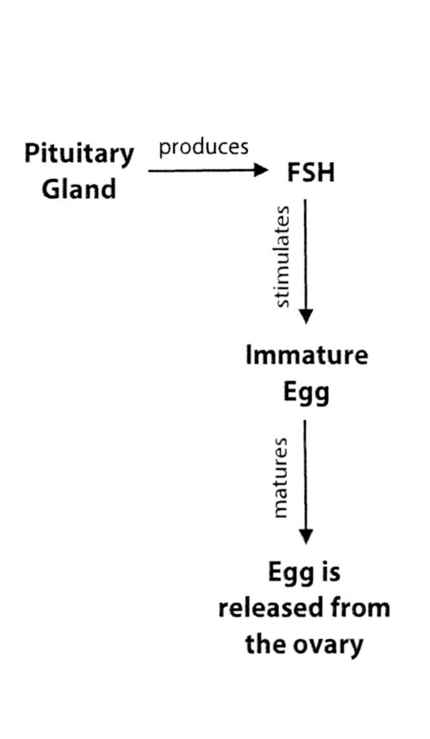

By the fifth week of human embryonic development, a female embryo has pre-egg making cells multiplying in her ovaries. These pre-egg cells begin meiosis and stop at a cell stage called the ***oocyte***. Each oocyte is surrounded by a small cluster of cells called the ***follicle***. There are about a million oocytes in the ovary of a female baby at birth. These immature eggs are stopped at Prophase I of meiosis until she reaches puberty. Eggs have a significant vulnerability because the chromosomes of oocytes are exposed to many environmental chemicals during both pregnancy and childhood.

The oocytes mature into eggs beginning at puberty. Usually one egg is released (called ***ovulation***) each lunar month. This monthly process, called the ***menstrual cycle***, is initiated by follicle stimulating hormone (***FSH***), which is produced by the pituitary gland (one of the glands of the endocrine system). As the egg matures, the cells surrounding it divide and grow. Most of this growing cellular mass remains in the ovary after the egg is released, and it is called the ***corpus luteum***. There is a fatty substance stored in the corpus luteum that contains the hormones ***estrogen*** and ***progesterone***. These hormones are released into the bloodstream and stimulate the growth of the inner uterine wall (***endometrium***) for possible implantation by an embryo. Estrogen also has an inhibitory feedback effect on the pituitary gland. It stops the production of another egg cell. Most birth control pills operate by mimicking this mechanism.

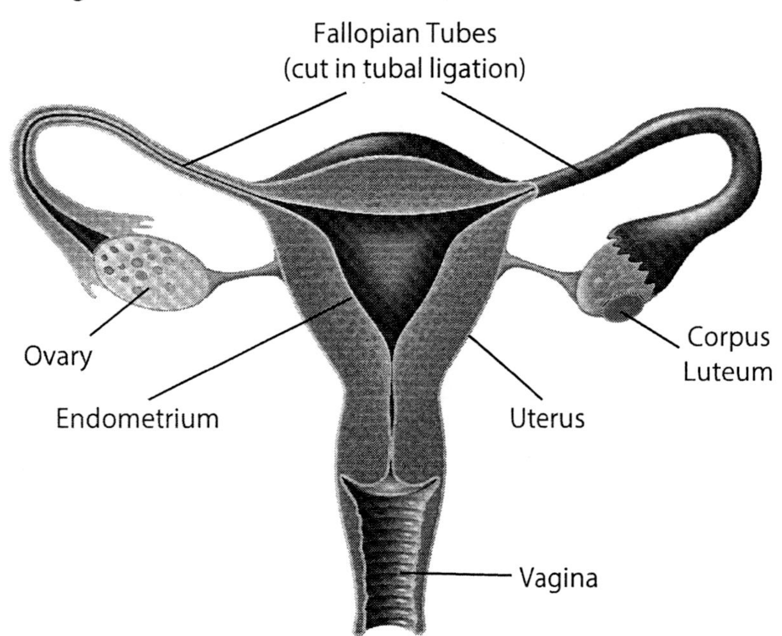

Figure 21.1. Anatomy of the female reproductive system.

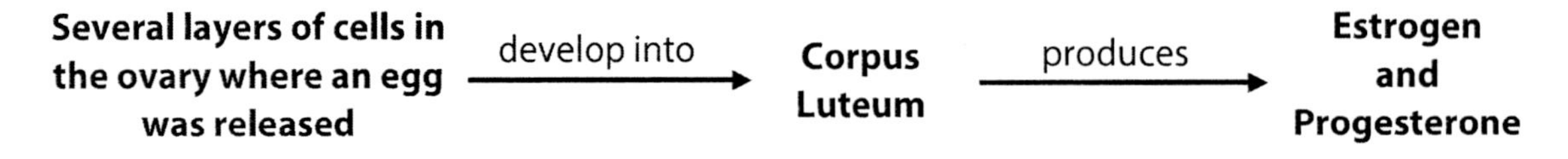

Figure 21.2. Production of estrogen and progesterone in the female. The monthly production of these two hormones occurs in the corpus luteum which is the spot in the ovary where an egg has been released.

Menstrual Cycle

- **Day 1:** Menstrual flow begins from previous month's cycle. One egg begins to mature in one of the ovaries for release in the next cycle.

- **Day 14–15:** Ovulation (Ovary releases an egg). The corpus luteum produces hormones (estrogen and progesterone) that stimulate the endometrium of the uterus to grow and store nutrients.

- **Day 28:** Next Menstruation: If the egg isn't fertilized, then the uterus lining is shed with the menstrual flow.

? Question

1. Health officials warn: "The eggs of your daughters can be damaged by certain chemicals you are exposed to during pregnancy." How can the eggs of your daughter be damaged when she won't start menstruating until she is 12 or 13 years old?

2. Which endocrine gland produces a hormone that stimulates an egg to mature?

3. Where does the corpus luteum form, and what hormones does it produce?

4. What day of the menstrual cycle does ovulation usually happen?

5. The menstrual flow is a shedding of the _______________ lining.

Materials

- ✓ a compound microscope
- ✓ three microscope slides: Immature Egg, Mature Egg, and Corpus Luteum

Procedure

- Work with two other lab groups. Put a different slide on each microscope, and use low magnification to get an overview of the sectioned ovaries.
- The immature eggs are bigger than most other cells in the ovary. A maturing egg cell is even larger and has a ring of cells surrounding it. The spherical mass of cells is called the ***follicle***. The ***corpus luteum*** is even larger than the follicle, and is a solid mass of cells without an egg inside. (The egg has been released.)
- Look back and forth among the three slides until you can see the difference in the structures.
- Draw each structural stage in enough detail so that you can find it again (perhaps on a test) with the help of your picture. Return the slides when you are finished.

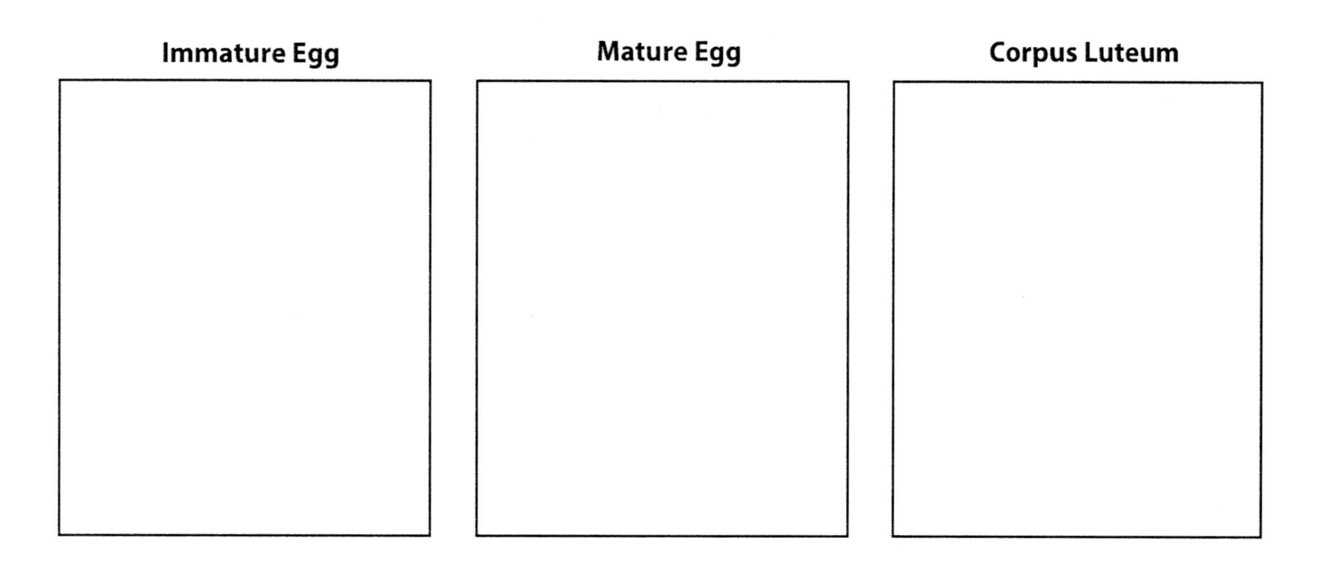

Sperm

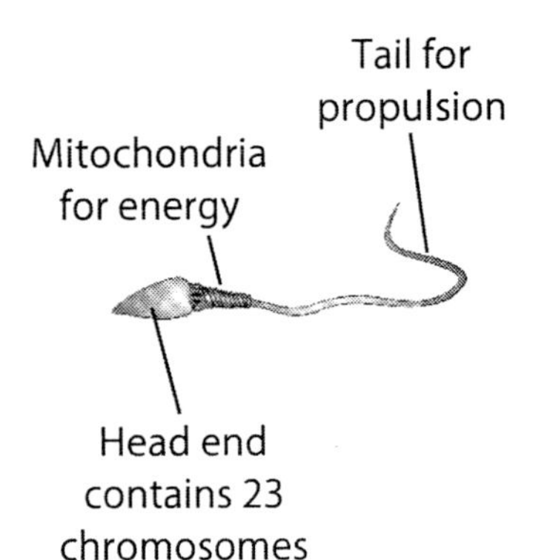

Human Sperm

The reproductive system of the human male embryo is noticeably differentiated by the eighth week. However, the male fetus does not begin meiotic division of sex cells as happens in the female fetus. The female fetus has immature eggs that are stopped at Prophase I of meiosis. The male doesn't begin sperm production until puberty.

It takes about three weeks for sperm to be produced. If a male is exposed to damaging chemicals, only the current "batch" of sperm is affected. After a month or so the sperm production could be back to normal. Of course, chromosome damage does happen in the male gametes as well as in the female. However, consider the short time sperm chromosomes are exposed to potential damage by environmental factors as compared to egg chromosomes. Sperm are produced (200–500 million daily) along the inside of an extensive system of tubes, called ***seminiferous tubules***, in the testes. They are stored in enlarged tubes called the ***epididymis***.

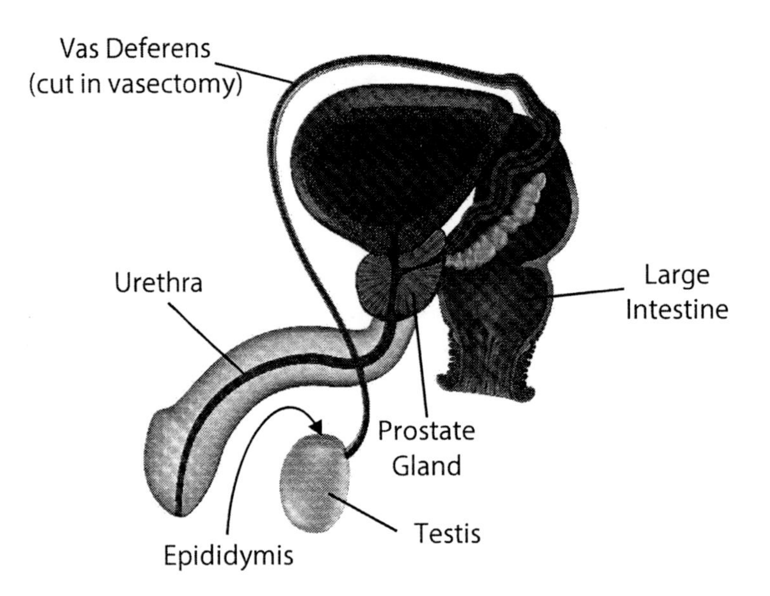

Figure 21.3. Anatomy of the male reproductive system.

? Question

1. Explain how sperm are less vulnerable to chromosome damage than eggs.

2. Where specifically are sperm produced?

3. Where are they stored?

Materials

✓ two microscope slides: Testis and Human Sperm

Procedure

- Examine the cross-section of a testis and locate the inner layers of the seminiferous tubules which are lined with various stages of sperm production.
- Draw a picture of the testis. Label the seminiferous tubules and sperm.
- Examine the slide of human sperm. Identify the head, mitochondrial region, and tail. Make a quick sketch and label it.
- Return the slides when you are finished.

Testis

Human Sperm

Fertilization

The human egg is released from the ovary with a layer of follicle cells around it. Sperm cannot fertilize the egg until this layer of cells has been removed. Each sperm cell produces only a small amount of enzyme that dissolves the jelly-like glue holding the small cells to the egg. Much of this enzyme is necessary to separate the follicle layer from the outside of the egg. Eventually, enough of the protective layer surrounding the egg is dissolved by many sperm cells. Then one of the sperm penetrates the egg. Once that happens, the egg membrane reacts by forming a new layer of jelly around the egg, thereby preventing any more sperm from entering it. If more than one sperm fertilizes an egg, the resulting embryo dies.

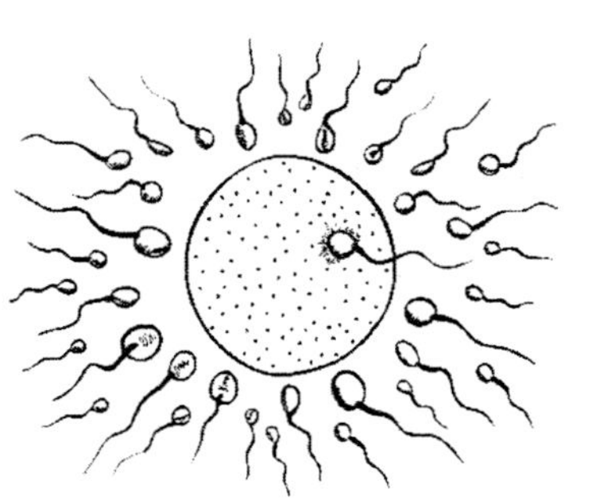

Although only one sperm fertilizes the egg, many sperm are required for the fertilization process.

Fertilization usually occurs as the egg moves through the fallopian tube towards the uterus. The embryo continues to develop for about one week before it implants in the wall of the uterus. After implantation a hormone, called chorionic gonadotropic hormone (**CGTH**), is released from the growing embryo. This hormone signals the corpus luteum (in the ovary) to produce large amounts of estrogen which then stimulates the mother's reproductive system to prepare for pregnancy. Estrogen also prevents new eggs from developing during pregnancy. This is why estrogen can be used for birth control. By the way, CGTH appears in the mother's urine about a week after implantation, and the urine pregnancy test detects whether it is there.

? Question

1. It is necessary for a male to have a high sperm count to be fertile. If only one sperm is necessary to fertilize the egg, then what function is served by the other sperm?

2. What event signals the mother's reproductive system to prepare for pregnancy?

3. What prevents more than one sperm from fertilizing the egg?

4. Describe how the pregnancy test works.

Sea Urchin Fertilization

If your instructor has been able to get mature sea urchins, then this is the time to watch fertilization actually happen.

Materials

✓ a depression slide and coverslip

Procedure

- Your instructor will inject the sea urchins with a salt solution that stimulates the release of their gametes. The urchins are then placed upside-down in a beaker of sea water. The gametes will flow out of the animal.
- Sperm looks milky. The eggs are granular and usually have a slight pink or yellow color.

- Put a drop of the eggs into a depression slide (no coverslip). Carefully place the slide on your microscope and examine the unfertilized eggs using the 10x objective lens.
- The sperm must be diluted with sea water because too many sperm causes abnormal fertilization. If the sperm collection beaker looks slightly milky, then dilute the sperm more. Add a drop of diluted sperm to your egg slide. Cover the slide with a coverslip, and immediately observe the events under the microscope. A ***fertilization membrane*** usually forms within 2 minutes.
- When you've seen the membrane form, put this slide aside and recheck it every 30 minutes. Don't leave the slide on the microscope with the light on because the fertilized eggs will soon overheat. If everything goes well, the first ***cleavage*** (division) should happen in about an hour. Watch for it!
- Draw pictures of the unfertilized egg, fertilization membrane, and cleavage.

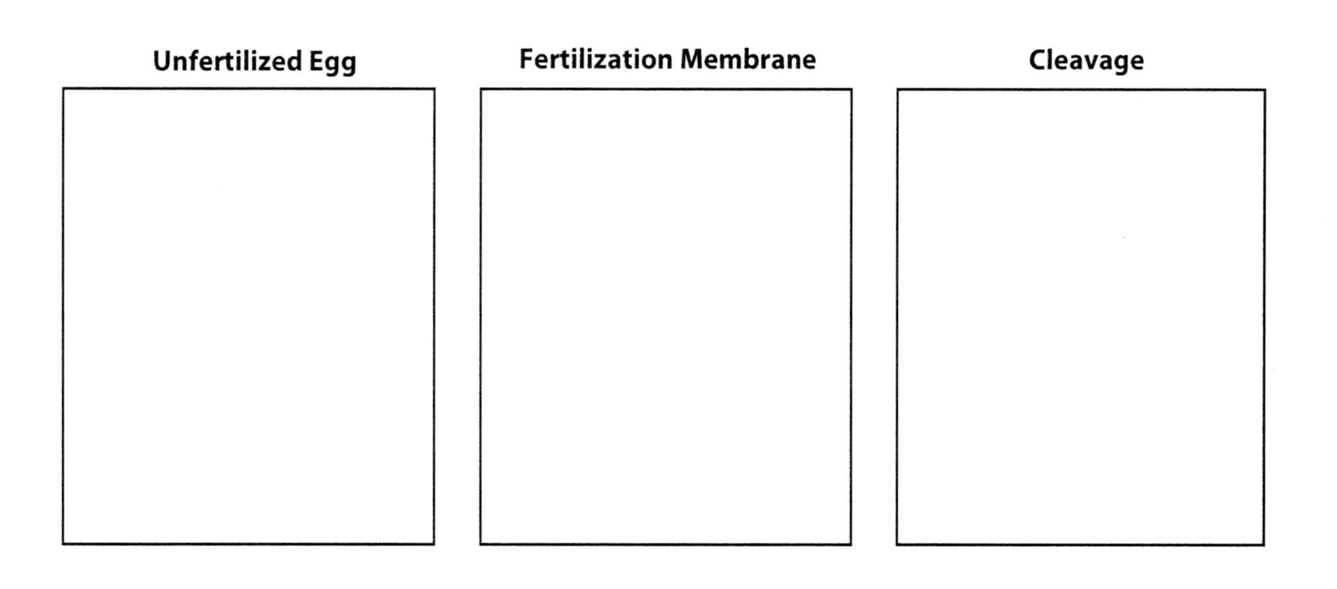

Exercise #3
Methods of Feeding the Embryo

There are two strategies for embryonic development: fast, with the embryo soon on its own; or slow, with the embryo fed by yolk or placenta.

Quick-Development
Self-Feeding

Invertebrates usually provide no special care to their embryos. The only source of food for the first stages of development is available in the cytoplasm of their eggs. After fertilization, there is rapid cell division and development to allow the embryo to feed on its own. As one example, the sea urchin embryo develops into a self-feeding stage in 24 hours or less. There are many small pieces of organic matter and micro-organisms in water for young embryos to eat. You will see sea urchin embryos during Exercise #4.

Yolk

Nutrition in the form of ***yolk*** is a way of giving more food to the embryo. Fish are examples of organisms that have small amounts of yolk in the egg. Birds and reptiles have much more yolk in their eggs. Larger amounts of yolk provide a longer time for development. This is especially necessary for the more advanced land vertebrates.

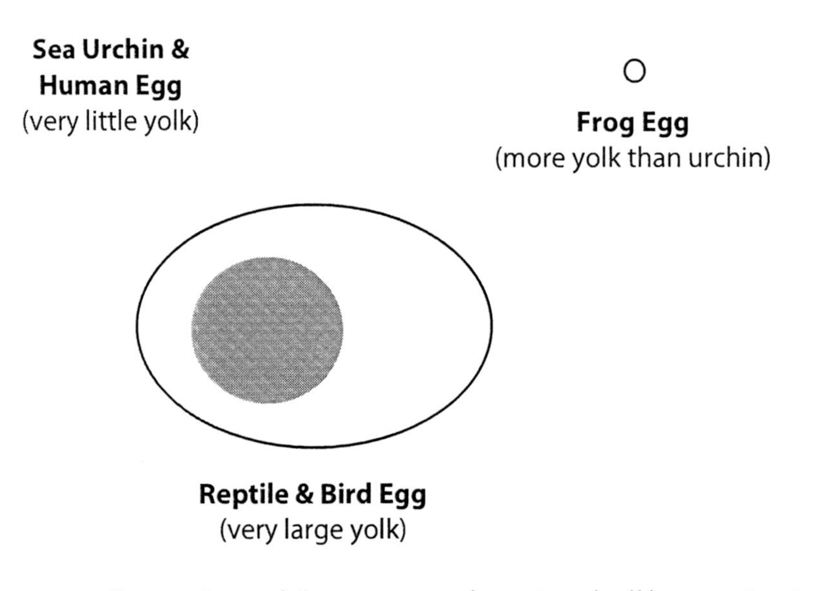

Figure 21.4. Comparison of the amounts of nutrient (yolk) occurring in the eggs of various organisms.

Placenta

The human embryo from the zygote to blastocyst stage must survive on the cytoplasm of the original egg cell. The egg divides but cells don't grow in size. In about one week the developing human embryo begins to implant into the uterine wall for the next stage of feeding (the placenta). The ***placenta*** is formed partly by the embryo and partly by the mother. Nutrients, wastes, and blood gases are exchanged between the embryo's blood system and the mother's blood system. The placenta begins as a few specialized cells on the outside of the implanting embryo. These special cells digest their way into the wall of the uterus. Soon after implantation, finger-like projections begin the formation of a more efficient structure. By the second month, the placenta is well developed and continues to grow with the fetus.

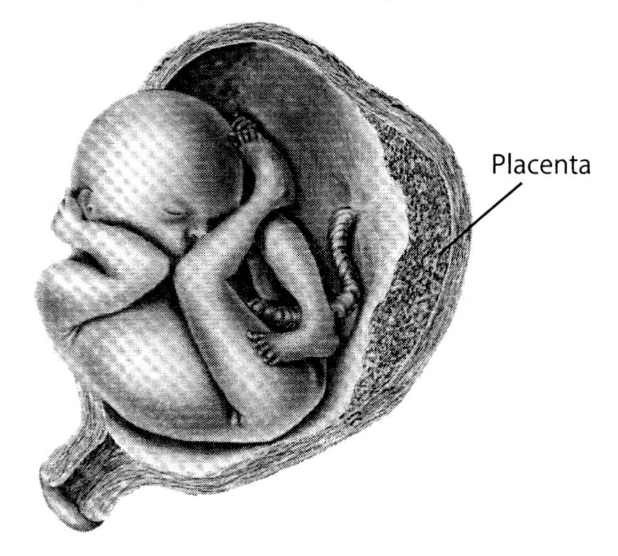

Figure 21.5. Structure of the human placenta.

? Question

1. List three methods for feeding early embryos by different organisms.

2. Why do complex animals require more specialized embryonic feeding mechanisms?

3. Why do most land animals require a more developed embryonic feeding mechanism than aquatic animals?

Exercise #4
Stages of Development

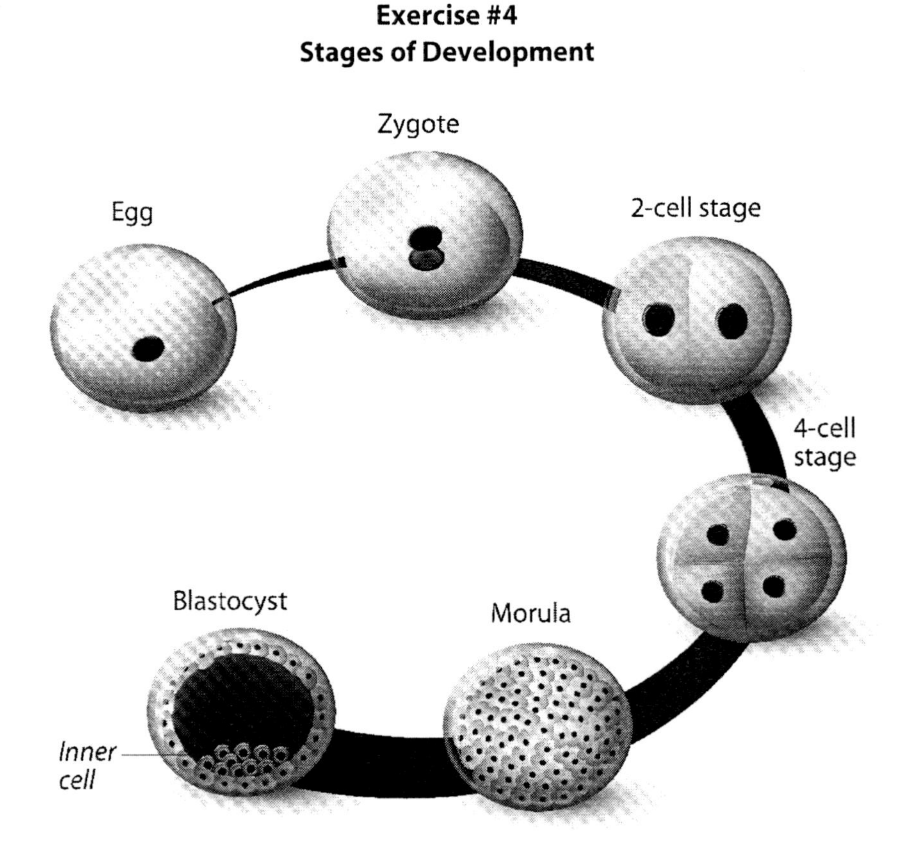

Figure 21.6. Early development of the human embryo.

During this Exercise, you will examine 10 stages of embryo development using various animals from sea urchin to pig embryos as illustrations of each stage. We will skip many of the details and focus only on a few structures as examples of the events.

Cleavage, Morula, and Blastocyst

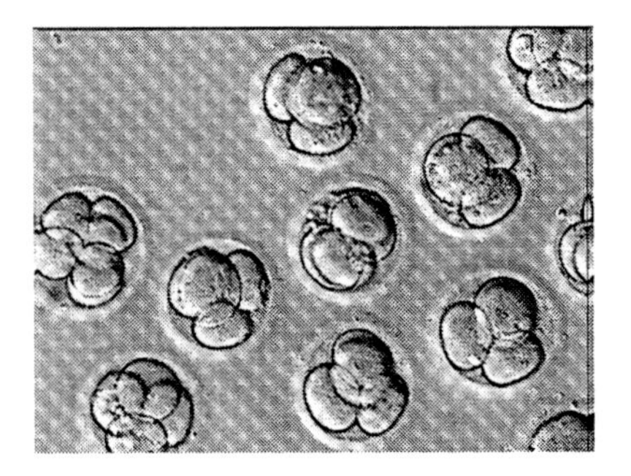

Early cleavage following in vitro fertilization (human)

The first divisions in development are called ***cleavage*** because the zygote divides (or cleaves) into two cells, then four, then eight, etc. Eventually, a small solid ball of cells forms called the ***morula***. Cells of the morula continue to divide, producing a hollow ball of cells called the ***blastocyst***. The human blastocyst develops in about one week after fertilization. You can get an idea of what these early stages look like by observing early embryos of sea urchin and frog.

Materials

- ✓ two microscope slides: Early Sea Urchin Development, and Frog Blastula (Stage 8)

Procedure

- Examine the early developmental stages of the sea urchin.
- Compare the sizes of the earliest embryos undergoing cleavage. Are the stages from zygote to morula about the same size? _____
  Are these early cells growing or just dividing? _______________
- Draw pictures of these stages.

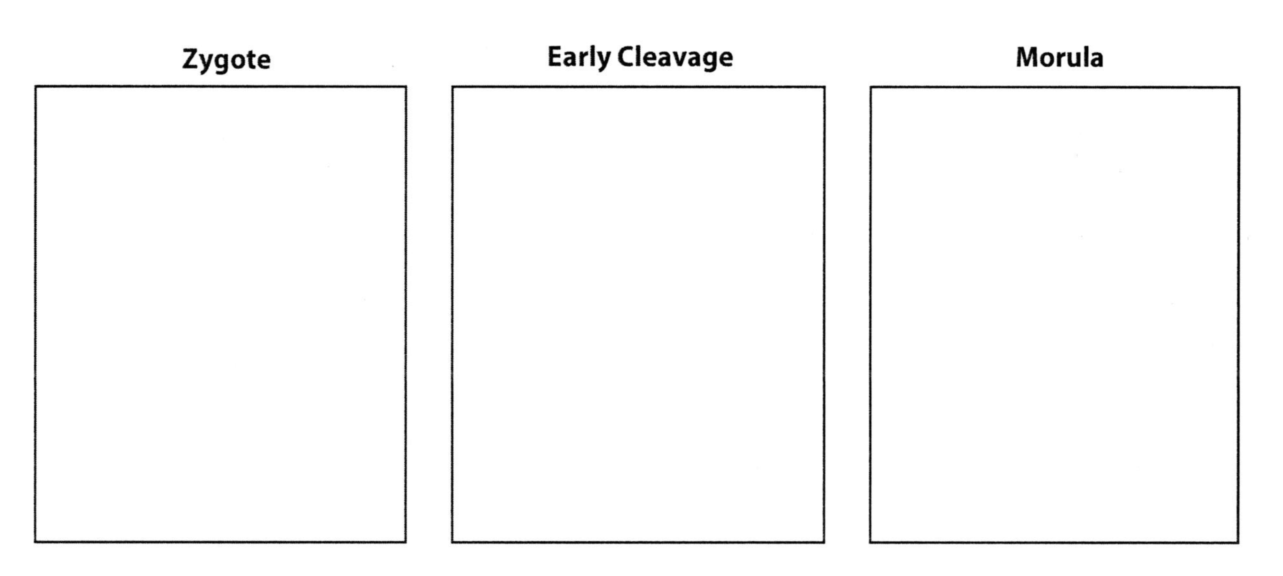

- Examine the early frog embryo (Stage 8). The frog blastocyst (also called blastula) is a partial hollow ball of cells. You should be able to see several different cell types. These cells are changing into new kinds of cells for the next stage of development.
- Draw a picture of the frog blastocyst, and show some different cell types.
- Return the slides.

Frog Blastocyst

Early Chick Embryos

Materials

✓ two microscope slides: 21-Hour Chick Embryo, and 28-Hour Chick Embryo

Procedure

- Work with another lab group, and put the two slides on different microscopes so that you can look back and forth between them. These are thick slide preparations. Use low power, and be very careful not to crush the coverslip when you swing the objective into place.
- The 21-hour embryo shows early development of the nervous system. The long ***neural fold*** becomes the spinal cord in a later stage. Perhaps you can see that the head end is slightly larger.
- The 28-hour embryo has a more developed neural fold, and you can see that the brain end has enlarged. Also, there are several paired blocks of tissue along the ***neural tube***. These blocks are called ***somites***. The somites develop into the internal organs, skeleton, and muscles of the adult body.
- Draw pictures of the two embryonic stages, and label the neural fold, neural tube, head end, and somites.
- Return the slides.

21-Hour Chick Embryo

28-Hour Chick Embryo

Later Chick Embryos

Materials

✓ two microscope slides: 56-Hour Chick Embryo, and 72-Hour Chick Embryo

Procedure

These chick embryos are comparable in development to three and four-week-old human embryos.

- Again, use low power magnification, and be very careful not to crush the coverslip.
- Put the two slides on different microscopes for comparison. These are side views of the embryos, which are shaped like a question mark (either forward or reversed depending on how the slide was made).
- You should be able to see changes in the brain development between the two embryonic stages. The brain consists of several lobes. In the 72-hour stage there is a lobe in front of the lobe with the eye in it. That front lobe is the ***cerebrum***.
- The somites are easy to see. How many are there? _____
- There is a structure bulging from the middle of the embryo and positioned below the brain. This is the developing heart. In the older embryo, the bottom chamber of the heart is larger. That larger chamber is the ***ventricle***.
- Draw a picture of the 72-hour embryo showing brain and spinal cord, eye, heart, and somites. Label your drawing.
- Return the slides.

72-Hour Chick Embryo

Pig Embryo

Materials

- ✓ slide of the 10-mm pig embryo.
- ✓ a dissection microscope would be helpful

Procedure

The 10mm pig embryo is comparable in development to a human embryo at six to eight weeks.

- Use the dissection microscope to get a general orientation of the pig embryo structure.
- You should recognize backbones, developing nervous system including large brain, and smaller eyes than in the chick embryo.
- Can you find the ***limb buds***? In the human embryo these buds develop into arms and legs.
- Just below the heart is a large dark structure, the ***liver***.
- Draw a picture of the pig embryo showing brain, eye, vertebral column, heart, liver, and limb buds. Label your drawing.
- Return the slides.

10mm Pig Embryo

Exercise #5
Disruption of Normal Development

Considering how much can go wrong during development, it is amazing that anything can end up right.

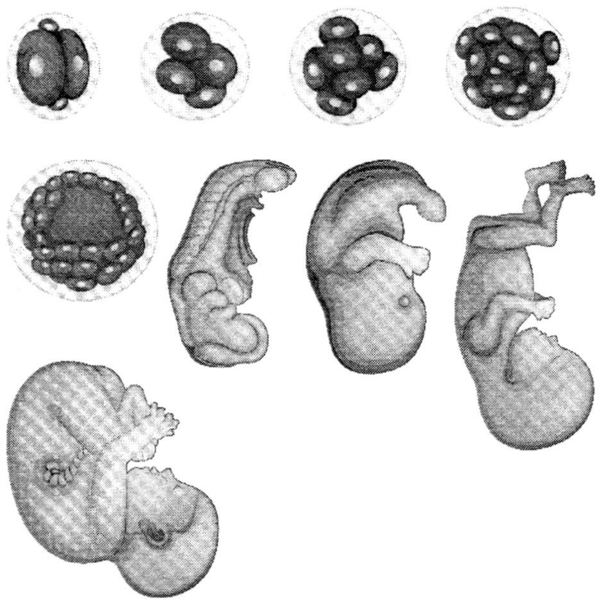

Figure 21.7. Normal Human Development. The earlier the disruption, the more serious are the consequences.

Considering all of the events that must go exactly right during embryonic development, it is easy to see that development can go wrong. A general principle for understanding developmental disorders is: The earlier the disruption, the more serious are the consequences. Some researchers estimate that one-half of all conceptions do not make it to birth. Women are usually unaware of the earliest embryonic failures which are also the most common. However, there are several disruptions during later development that are very serious and reveal themselves at birth (4–5% of births) or as late-term miscarriages.

Gene Mutations

Gene mutations can destroy the normal production of essential enzymes controlling basic metabolic processes in the fetus. Most states require the immediate testing of all babies for ***PKU*** (phenylketonuria) which is a deficiency of an enzyme responsible for the metabolism of the amino acid phenylalanine. This disorder is similar to other gene mutations — no product is formed, and there is a buildup of the substrate. The product is often essential for other steps in metabolism, and high concentrations of substrates are usually toxic or disruptive. Tay Sachs, galactosemia, and tyrosinosis are other metabolic disorders resulting from gene mutations.

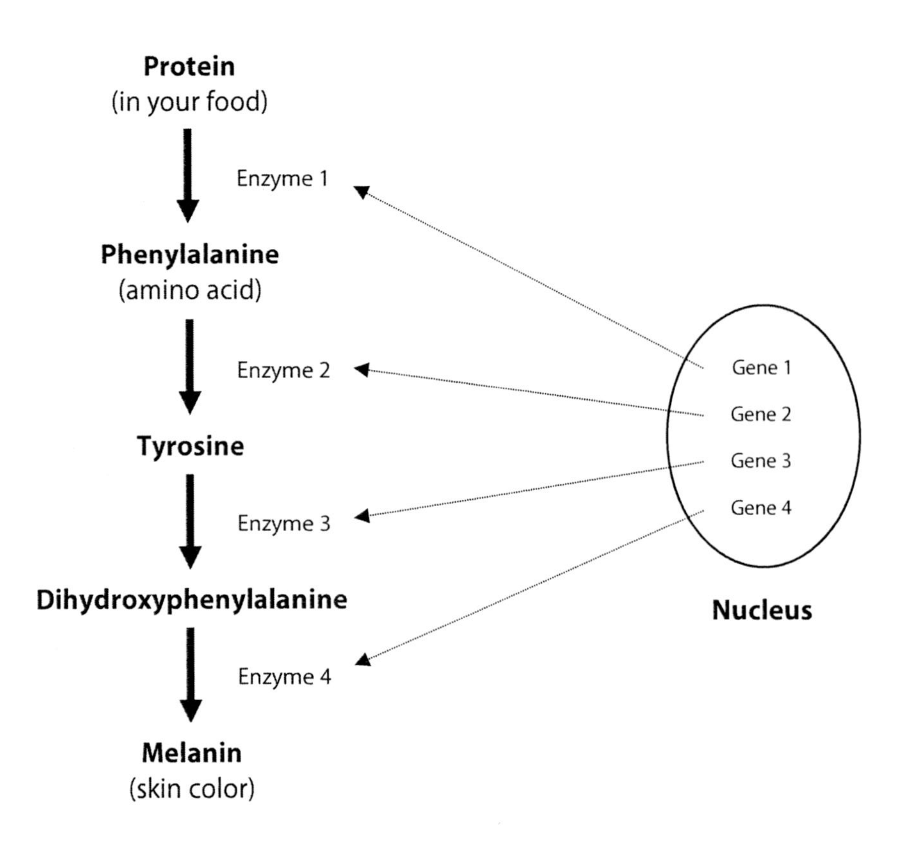

Figure 21.8. The role of genes and enzymes in metabolic pathways.

Chromosomal Aberrations

When mistakes occur during meiosis of sperm or egg production, abnormal numbers of chromosomes in the embryo are possible. These events are almost always lethal, but some chromosomal aberrations (X, XXY, and three #21 chromosomes) occur in 1% of live births. These babies have serious health problems. The trisomy of #21 chromosomes produces Down's syndrome, the most common form of mental retardation. The other chromosomal aberrations result in similar severity of health problems.

Ectopic Pregnancy

Occasionally, an embryo implants in the wall of a fallopian tube or on the outside of the uterus or intestine. There are two serious consequences of this abnormal event. First, the child will have to be delivered by Cesarean section. The second serious problem is that the placenta can attach to the wall of the wrong organ. That organ can be damaged or destroyed by placental development. This condition can be so dangerous to the mother's life, that termination of pregnancy may be her only chance of survival.

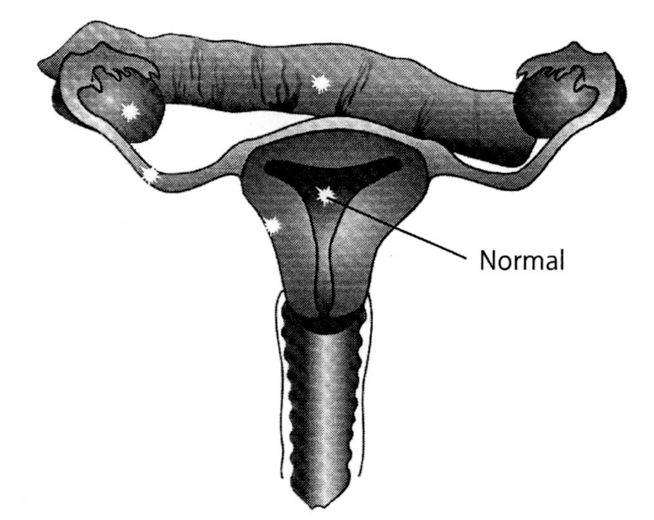

Figure 21.9. Ectopic pregnancy can be ovarian, tubal, intestinal, or outside wall of uterus.

Incomplete Development of Organ

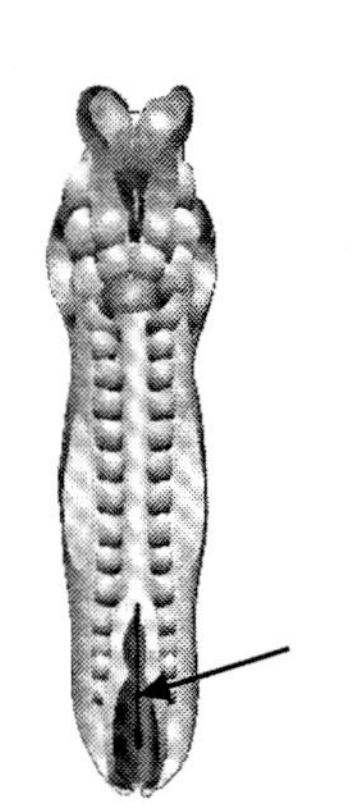

Spina bifida occurs if this part of the neural tube does not close during early development.

Sometimes an organ does not develop completely during early development. The cause can be a foreign chemical like thalidomide, or a disease like measles. Common examples are ***spina bifida*** (neural tube doesn't close), or holes between the heart chambers. The general principle of developmental disorders applies especially in these cases. The earlier the problem happens, the more serious are the consequences.

Bad Habits During Pregnancy

We have very little control over the previously described disruptions of normal development. In fact, many early miscarriages are the result of the mother's hormonal system "warming up" for future reproductive success. However, there are some disruptions in development that we can control—bad habits during pregnancy. Science makes a certain direct statement about bad habits during pregnancy:

"Drink alcohol, smoke cigarettes or marijuana, use drugs, or have poor nutrition, and your child will be below normal in mental abilities and may be physically deformed."

Fetal Alcohol/Drug Syndrome: This fetal damage is predictable and results from the mother's drinking or drug habit. Fetal Alcohol/Drug Syndrome includes characteristic facial deformations, a degree of mental retardation, and defects in internal organs (especially the heart and liver).

Cigarette Smoking: The majority of babies born by mothers who smoke have a smaller placenta, lower birth weight, and lower overall health. The effects on the fetus are not usually as extreme as with alcohol or drug use by the mother, but the birth results are 100% predictable.

Poor Nutrition: The usual problem with poor nutrition in this country is not a lack of Calories, but an unbalanced diet (not enough protein, vitamins, and minerals). The effects of poor nutrition are lower birth weight, lower overall health, and below-average mental ability. If the nutritional problems during pregnancy are severe, as seen in underfed parts of the world, all of the detrimental effects during embryonic and fetal development are greatly magnified.

? Question

1. What is the general principle for understanding developmental disorders?

2. Perhaps as many as __________ of all embryos don't survive to birth.

3. What is the approximate percent of births with developmental defects?

4. What is the most serious potential risk in an ectopic pregnancy?

5. List the two direct biochemical effects of most gene mutations.

6. What is the usual result of chromosomal aberrations?

7. What are the common effects on embryonic development caused by alcohol, drugs, cigarettes, and poor nutrition?

LAB

22

Biology and the Mind

This lab was written specifically for students interested in a health profession, and that ends up being all of us one way or the other. Our first two labs covered basic skills in scientific thinking and experimentation. This lab asks you to consider the value of using known psychological tools for applying scientific discoveries to health related fields.

The mind can be studied directly, but that would require another college course. We will consider only a few aspects of mind that influence the understanding of physiology, and then suggest how these might be used to understand other aspects of biological investigation. The topics in this lab warn us about how easily we are confused by scientific research related to health. We begin with a look at how our ways of thinking influence research into human physiology and our understanding of it. How does intuition sometimes confuse us? How strong is the bias when we think we have the right answer? And, what are some very specific methods for training people to improve their health and physical performance?

Exercise #1	**Intuition and Counter-Intuition**
Exercise #2	**Knowing That You Are Right**
Exercise #3	**Use of Reinforcement Schedules in the Health Professions**

Exercise #1
Intuition and Counter-Intuition

It is important for you to answer all of the following questions by yourself before starting this exercise. The answers to these questions will give clues about your own thinking processes, and they will be used as part of the class data.

"Questionnaire"

Each student answers these questions.

1. Define what you mean by the word "intuition."

2. What is the source of intuition?

3. Do you use intuition to make decisions?

4. If so, in what kinds of decisions?

5. Can intuition be wrong?

6. If so, when is it wrong?

7. Don't analyze the following theoretical problem. Make a quick decision after reading the setup information.

   There are two buttons: A and B. You must push one of them. (circle your choice)

   Button A—If you push this button, you will be given $3000.

   Button B—If you push this button, you have an 80% chance of getting $4000 and a 20% chance of getting nothing.

8. Again, don't analyze the problem. Make a quick decision after reading the setup information for another theoretical problem.

   There are two buttons: C and D. You must push one of them. (circle your choice)

   Button C—If you push this button, you lose $3000.

   Button D—If you push this button, you have an 80% chance of losing $4000 and a 20% chance of losing nothing.

9. How many people do you usually date before finding a more serious relationship?

10. How many different college majors are you considering as definite possibilities?

11. In your opinion, do most people stay in bad relationships longer than they should?

12. In your opinion, do most people stay in unsatisfying jobs longer than they should?

13. Your brain makes some decisions without your conscious awareness. What could you do to discover when this happens?

14. If you observed that all people make the same kind of mistake, what would you do?

An answer that comes quickly to your mind is most likely a reflection of intuition.

Intuitive and Counter-Intuitive Thinking?

Several concepts must be understood before we can start our investigation of intuition. First, we need clear definitions. The dictionary describes ***intuition*** as (1) the act or faculty of knowing without the use of rational process; (2) a capacity for guessing accurately; and (3) a sense of something not deducible. These definitions suggest that intuition does not depend on the analysis of a situation, and that it is often correct. In addition, the definitions imply that certain truths might only be discovered through intuition.

The primary process in science is called empirical thinking, and it is different from intuition. ***Empirical*** is defined as (1) relying upon or derived from observation or experiment; and (2) not guided by theory or intuition but by verifiable measurements or observations. Science is used to test ideas, and when it tests intuitive ideas, there are two possible results:

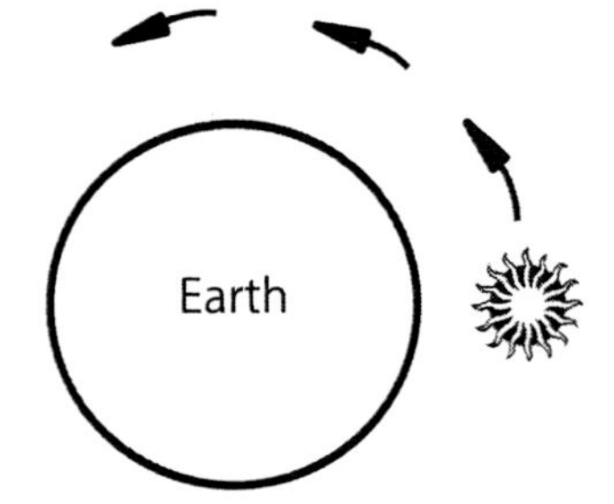

Counter-Intuitive Process- The sun appears to go around the earth.

- The intuitive idea agrees with the empirical evidence, or
- The intuitive idea does not agree with the empirical evidence.

When science explains a process differently than intuition does, that process is called ***counter-intuitive.*** One example of a counter-intuitive process is the rotation of the Earth. Our intuition tells us that the sun revolves around us. We see it rise in the east and set in the west. However, science has shown that our intuition is wrong in this case. The planet spins on its axis from west to east, and one complete rotation takes 24 hours. When you use your intuition to explain a counter-intuitive process, expect to be wrong! But how would we know whether a process is intuitive or counter-intuitive?

What are the Consequences When Intuition is Wrong?

Help!

There are two useful ways of describing the consequences when some aspect of the world operates counter to your intuition.

- It really doesn't matter that the intuitive idea is wrong; it has value anyway. (It's a great day!)
- The consequences of intuition being wrong might harm us.

While doing the following activities, consider if serious consequences might occur when intuition is wrong. Why would it be that we don't recognize and easily solve certain problems? Are some of these problems counter-intuitive? Finally, consider how we might use scientific thinking when our intuition fails. Check with your instructor now if you are confused about our starting definitions.

? Question

1. Define intuition.

2. Define empirical.

3. Define counter-intuitive.

Decisions

Can we agree that the ability to make a decision is an important human act? Can we agree that your relationships, your choice of a college major, and how you spend your money are important decisions? If you answered yes, then let's look at the discoveries made by two scientists, Daniel Kahneman and Amos Tversky, who have studied human decision-making.

Class Data

Refer to the "Button Problems" in the Questionnaire. Record your answers to questions #7 and #8 on the whiteboard, and write the class totals below.

Table 22.1. Record the Class Data.

Total # of Students In Your Lab Class	
# Choosing Button A	
# Choosing Button B	
# Choosing Button C	
# Choosing Button D	

Kahneman and Tversky found that most people choose Button A and Button D.

How does the class data compare with research findings?

Do their findings agree with the class data?______

Do their findings agree with your choices?______

Low Risk and High Risk

- A ***low-risk situation*** is one in which you are not going to lose something that you already have.
- A ***high-risk situation*** is one in which you are going to lose something whatever choice you make.

? Question

1. Which situation is low risk? (circle your choice)

   A and B or C and D

2. Which Button choice in the low-risk situation is "playing it safe"?
3. Which Button choice in the low-risk situation is "taking a chance"?
4. So, how do most people act when they are in a low-risk situation?

   They play it safe. or They take a chance.

5. Which Button choice in the high-risk situation is "playing it safe"?
6. Which Button choice in the high-risk situation is "taking a chance"?
7. So, how do people usually act when they are in a high-risk situation?

   They play it safe. or They take a chance.

Which is the best choice? A or B

Let's Examine the Facts

The button decision problems were designed to offer you a choice and determine whether you have an intuitive bias. The results reveal that intuition directs us to choose A and D, both of which are losing strategies in life. The winning choices are actually B and C.

In a low-risk situation there is no real personal loss to you if you take a chance. You are being offered money. The odds favor more money if you choose Button B. Your choice of Button A means that you stayed with the safe bet. It's just like staying with one major in college or dating only one person as you learn about relationships. You are limiting your options for better success. After all, who would insist that the first person we date is the best choice for a successful relationship? Likewise, our economy proves that

Was that C or D?

there are many successful choices for college majors. But, in money, love, and vocation we find ourselves playing it safe in a world of opportunity.

In the world of intuition, there are low-risk losers and high-risk losers.

By the way, if you happened to choose Button B and you're feeling pretty smug right now, then forget it. Research has shown that when the problem is restated in terms other than money, you tend to make the same mistake as everyone else (Button A). For example: Imagine that you are hungry. Button B offers you a free seven-course dinner at a posh restaurant with a 20% chance that the restaurant is closed. Button A offers you a free meal at your favorite fast-food emporium that is open 24 hours every day.

A more serious problem is revealed by the selection of Button D. Both Buttons C and D involve loss. However, the intuitive bias towards Button D means that you are willing to expose yourself to even more risk in an effort to avoid any pain. Choosing Button D compounds the risk. Choice C, while still a loss, represents the best strategy—take the loss and move on. Moving on to a new relationship has a greater chance of success than waiting for a bad relationship to improve. Getting a new job has more possibility of encouraging the development of your talents than staying at an unsatisfying job.

Procedure

Let's use the Personal Relationship example as a review of the concepts discussed so far. Divide the class into discussion groups. Assign one person the responsibility of coordinating the discussion. This person is not supposed to come up with all of the ideas—just keep the rest of the group on task. Assign another person to write down the group's answers. This discussion should reveal the influence of intuition bias on Personal Relationship.

Are you hoping to pull it out of the hat?

- Determine how using strategy A or D to solve the problem could actually make the situation worse.

  A = Playing it safe when there is opportunity to take a chance without loss.
  D = Taking even more risk in an attempt to avoid loss.

- Determine how using strategy B or C to solve the problem might improve the situation.

  B = Taking a chance for a new opportunity when there is no real risk at stake.
  C = Taking the loss and moving on to new opportunities.

- Discuss the following questions for 10 minutes. There will be a class discussion following your group discussions.

Group Discussion

Personal Relationships

1. Assume you are young (perhaps in high school or early college) and have been dating someone for a year or more. This relationship is OK, but neither of you thinks that the relationship would be the ideal kind for a lifetime marriage.

   What would be a **Strategy A** approach to this relationship?

2. What are possible negative consequences for using this **Strategy A**?

3. What would be a **Strategy B** approach to this relationship?

4. What are possible benefits of using this **Strategy B**?

5. Now, assume that you have been in a pretty bad relationship for some time. It's the kind of relationship that neither your friends nor relatives think is very good.

   What is a **Stategy C** approach to this relationship?

6. What are possible benefits of using this **Strategy C**?

7. What is a **Stategy D** approach to this bad relationship?

8. What are possible negative consequences of using this **Strategy D**?

Group Discussion

Procedure

Now, before leaving this topic, see if your group can list five important clinical, nutritional, or conditioning concepts that you think are counter-intuitive. Remember, counter-intuitive means that most all people will have the wrong understanding when using their natural intuition.

Can you list some Counter-Intuitive health problems?

1.

2.

3.

4.

5.

Exercise #2
"Knowing That You Are Right"

This exercise examines the intuitive feeling of "knowing that you are right" in a particular situation. Your group will solve a very common problem with city redevelopment.

Problem: ***An area of the city does not have enough jobs, and there are not enough cheap houses for the lower-income people who live there.***

Your group is to quickly write down your solutions for city redevelopment and predict the positive results of your plan. After you have done this, read the information on the next page. Don't read the Empirical Findings until after your group has devised a solution to redevelopment. If you cheat, you won't get as much value from testing your intuition (an opportunity foregone).

Write your ideas about city redevelopment here:

1.

2.

3.

It's obvious that my plan is best! Why would anyone question that?

Empirical Findings

The most common solution to the city redevelopment question is to build more cheap housing. This solution always fails. More lower-income people are attracted to the area and the local unemployment actually increases. Another common solution is to build factories in the area. This approach works only when building the new factories also reduces some of the existing housing and number of residents at the same time. An unexpected finding is that destroying some of the existing low-income housing by itself increases the percentage of employed people remaining in the area. These results have been known for several decades, yet the same failed approaches are continually recommended by politicians and planners.

Review your group's solution again. Do you still believe that your plan would work if only it was given a little time and a fair chance?______

Most people answer "yes" to this question. It is almost universal that when people are presented with empirical evidence that disproves their intuitive idea, they still feel that they are correct. The intuitive feeling of "knowing that you are right" apparently does not change even when the facts oppose you. This is a major challenge to solving some of the modern societal problems that are counter-intuitive. Furthermore, medicine, nutrition, conditioning and training, and other important health issues are also affected by the counter-intuitive bias. What are your solutions to this challenge?

Well, there you have it. Were we correct to include these first two topics for biology lab students? We hope so. And now, on to methods of changing health related behavior.

You just know you're right!

Exercise #3
Use of Reinforcement Schedules in the Health Professions

When thinking about advancements in medicine and health, most people imagine new drugs, medical devices, and surgical techniques. These technical advances will be part of future health professions, but another contributing therapy is behavior modification. The first step in using behavior modification is to define the health problem very specifically. Perhaps the person needs a stronger heart, or to lose weight, or strengthen muscles, or perform at a higher competitive level. Maybe it's just to "feel better". Once the basic challenge has been defined, we need to prescribe a set of behaviors to achieve a solution to the problem.

Dare you use training methods on your clients?

Primary reinforcers are "wired in" to be wanted.

Choosing Effective Reinforcers

How do you train people to get rid of bad behaviors and add the good behaviors? This requires help from the behavior experts. ***Reinforcement*** is any process that increases the frequency of a ***targeted behavior***. Do you want your patient or client to eat less or exercise more or take their medicine regularly? These are targeted behaviors, and reinforcement will increase their frequency. There are two basic categories of ***reinforcers*** (factors that increase reinforcement). ***Primary reinforcers*** operate directly because of something basic in our biology or because something has been so conditioned by our upbringing that it acts "wired in" to our actions. Primary reinforcers include pleasure, food, attention, and the like.

Secondary reinforcers operate indirectly because they are associated with (conditioned to) something that increases the behavior. Secondary

We desire secondary reinforcers because we can get something else with them.

reinforcers can be connected to a primary reinforcer or with another secondary reinforcer. It can get a bit complicated to figure out, but behavior experts have defined all combinations. Money is a secondary reinforcer, as is giving "points" to kids in the classroom. It is important to emphasize that a reinforcer is not necessarily something "pleasant". It is merely something that increases the frequency of a targeted behavior. A "slap in the face" could be a reinforcer in certain circumstances.

There are four ways to approach reinforcement.

- **Add a reward** to increase a good behavior. (Give a piece of candy to the child after she cleans the bedroom.)
- **Add a punishment** to decrease a bad behavior. (Yell at the child when she runs into the street.)
- **Take away something punishing** to increase a good behavior. (Perhaps the good behavior of taking an aspirin relieves the headache.)
- **Take away something rewarding** to decrease a bad behavior. (Driving privileges are removed because of bad grades.)

What is the perfect reinforcer?

A crucial part of helping people to change their behavior is deciding which reinforcers should be used. Most professionals shy away from primary reinforcers because they are considered too personal or controlling. But why couldn't you suggest that a person eat a favorite food after they perform the targeted behavior? That would be using a primary reinforcer. Or the health professional might have the client create a journal and use that for determining whether the client is accidently using primary reinforcers to increase bad behaviors. Regardless of the hesitation to use primary reinforcers, they can be used. And there are many secondary reinforcement programs that also can be devised to accomplish success for the patient.

? Question

1. What is the difference between a primary and secondary reinforcer?

2. Describe a possible "reward to increase a good behavior" that would help a person to improve health.

3. Describe a possible "punishment of a bad behavior" that would improve the situation.

4. Describe a possible "take away something punishing" to increase a good behavior.

5. Describe a possible "take away something rewarding" to decrease a bad behavior.

Reinforcement Schedules

A ***reinforcement schedule*** is defined by the way you deliver reinforcers to the subject in order to change their behavior. There are two basic approaches to reinforcement. The first approach is to change the required number of times the behavior must happen before the reinforcer is delivered to the subject. This is called the ***ratio method***. The second approach is to change the amount of time between the behavior and the reinforcer. This is called a ***time interval method***. Both of these approaches have many variations and combinations.

Ratio Methods:
How many times must I do it before I get what I want?

Ratio Methods

Behavior research on ratio methods is often started by setting up a situation where the subject gets a reinforcer every time the behavior happens. Then this response rate is compared with different ratios – perhaps five behaviors are required before the reinforcer is given. (e.g. Five chores have to be done before the child can go outside to play or watch TV.) The ratio can be ***fixed*** (every certain # of behaviors), or can be ***variable*** (random).

Time Interval Methods

Time Interval Methods: How long do I have to wait before I get what I want.

Time interval methods can also be fixed or variable. They range from immediate (a smile) to longer fixed intervals (Paycheck every Friday). The interval can be variable (random) like a slot machine. By the way, the slot machine is both variable ratio and variable interval, plus a lot more. More about that later.

What Works Best?

There are many useful findings about reinforcement schedules. ***Continuous reinforcement*** (behavior is reinforced every time) is the quickest way to start associating a behavior with the reinforcer. After accomplishing that link other schedules can be implemented to strengthen or maintain the behavior. One of the next steps could be to vary the ratio. A ***fixed-ratio schedule*** (reinforcer is delivered after a certain # of behavior repeats) produces a high and steady rate of responding by the subject. There is sometimes a bit of "lag" immediately following the reinforcer, but the desired trend continues. The ***variable-ratio schedule*** (subject can't predict the # of behaviors that are required before the reinforcer is given) is a bit slower at the start compared to continuous or fixed-ratio, but it also produces a high and steady rate of responding by the subject. And it is more resistant to ***extinction*** than fixed-ratio. Extinction is the process of a decrease in behavior when the "training" is over.

Continuous reinforcement is the best way to get a new behavior started. Once the new behavior is started, switch to either a fixed-ratio or variable-ratio schedule.

The time interval between the behavior and the reinforcer is another part of a health training program. The immediate delivery is quick to get response from the subject, but it is hard to do in the "real world". It is almost impossible to reward immediately unless you have unrestricted access to the subject's everyday life. You could use a ***fixed-interval schedule***, but the length of time must not be too long. When the time for reinforcement is too long, there is a high rate of responding as the time for reinforcement approaches and an immediate decline in response after the reinforcer has been delivered. This makes common sense but we often ignore this obvious consequence when planning a fixed-ratio schedule. The ***variable-interval schedule*** (subject can't predict when a reinforcer is delivered) is slower to train the targeted behavior, but the response is more resistant to extinction than fixed-interval.

Paycheck: Fixed-Interval Reinforcement

Special Problems – Satiation and Extinction

There are two more challenging problems with behavior training - ***satiation*** and ***extinction***. People can become amazingly satisfied with the "already delivered reinforcer". Pay raises do not maintain the hard work required to get them in the first place. Sad but true. Many people are content with what they're getting, and increasing a reward does not change their behavior. This is one of the least understood and most controversial challenges to generalized reward systems.

The other problem with behavior training is extinction. Will your patient continue with good diet and exercise after the training? The best way to maintain a behavior is ***random and infrequent reward***. Gambling (especially the slot machine) is nearly perfect at getting the repeat behavior. The slot machine uses random and infrequent delivery of rewards, and it always gives certain "signals" to the customer if a big payoff might happen. Most of the time those "signals" are not followed by payoff, but the signal always happens before a payoff. Billions are made using these reinforcement schedules. But the success of random and infrequent depends on a powerful reward. You will have to decide if random and infrequent reinforcement is practical and will work for your patient.

Random and Infrequent Rewards

? Question

1. Define satiation.

2. Define extinction.

3. What are the two basic approaches for designing a reinforcement schedule?

4. What is the difference between fixed and variable reinforcement schedules?

5. Which reinforcement schedule is the quickest to start associating behavior with the reinforcer?

6. Which schedule is faster to reinforce the behavior (fixed-ratio or variable ratio)?

7. Which schedule is more resistant to extinction (fixed-ratio or variable- ratio)?

8. Which schedule is faster to reinforce the behavior (fixed-interval or variable-interval)?

9. Which schedule is more resistant to extinction (fixed-interval or variable-interval)?

10. If the reinforcer is very strong, which reinforcement schedule is the most effective for maintaining a behavior?

What about Punishment?

Reinforcement doesn't always involve providing a reward. It could be delivering a punishment. In the case of using punishment, continuous reinforcement is the definite best first choice. If you use a ***fixed-ratio with punishment***, expect the subject to push the limit right up to the required number of behaviors. "That's one. If you do three, you're on time out!" If ***random and infrequent punishment*** is used, there is no productive result. One predictable consequence of this method is that the subject develops a bad attitude towards authority and punishment. And if a random and infrequent schedule is used with both reward and punishment, the subject loses the ability to think clearly and choose successfully. It is the most self-destructive and compulsive training method known, and many people have health problems because they were accidently conditioned by it. Solve this problem and you will have a very successful health career.

Punishment

One final example of problems arising from random and infrequent schedules can happen as you are trying to get rid of a behavior. The quickest extinction of the behavior is to eliminate all reinforcers immediately. Beware of "tapering-off"! You can accidentally create the dreaded "random and infrequent" schedule as you do so.

? Question

1. What is the best reinforcement schedule for using punishment in a behavior training program?

2. What is one disadvantage of a fixed-ratio schedule of punishment?

3. What are two serious problems caused by a random and infrequent schedule of punishment?

Designing an Effective Reinforcement Schedule

Our purpose has been to point out that behavior training principles are an essential set of tools for anyone in the profession of medicine, conditioning and exercise, or nutrition. Now it's your turn to work with these ideas.

Group Discussion

Procedure

- Organize the class into three "topic" groups of 6-8 students: (1) a clinical health problem, (2) a nutrition topic, and (3) a conditioning challenge. Pick the group topic most interesting to you.
- Each "topic" group is to take a few minutes to pick a very specific problem that your group will try to remedy with a reinforcement plan.
- Choose reinforcers that could be used to modify the health behavior.
- Design a specific ratio method of reinforcement that you think would be most effective.
- Design a specific interval method of reinforcement that you think would be most effective.
- Include one example of how you might use punishment effectively to modify a behavior related to your topic.

Your Group Topic

" ____ "

Describe which reinforcers you would use?

Describe your Ratio Method of reinforcement schedule.
(Continuous, Fixed-Ratio, or Variable-Ratio)

Describe your Interval Method of reinforcement schedule.
(Continuous, Fixed-Ratio, or Variable-Ratio)

Describe one effective Punishment you might use.

Now take 15 minutes as a class to briefly review and compare the behavior modification plans for each of the three topic groups. Summarize the three most useful ideas that you heard during the class discussion.

Group Discussion

1.

2.

3.

How could you use escalating reinforcers in your profession?

Using Escalating Reinforcers

Satiation and extinction are major challenges in behavior training programs. One commonly used reinforcement schedule to help counter this problem with meth addicts is a "Voucher" system. These vouchers are usually for spending money, and they are earned depending on results from daily urine monitoring. The reinforcement starts with a moderate voucher for no drug use and then increases slowly with the length of abstinence time. This is called an ***escalating magnitude of reinforcer***. A failed urine test "resets" the voucher back to the starting level.

? Question

Does this give your group any ideas about how you could deal with satiation and extinction in your Group Topic reinforcement plan? (Be specific in describing what you might do.)

Role of Superstitious Behavior in Health Problems

Does superstition play any role in health problems?

The last example for your group to consider is ***superstitious behavior***. It happens when a very big reinforcer (reward or punishment) immediately follows a behavior. A famous example is the athelete who tips his hat immediately before hitting his first home run. Can you think of one example of superstitious behavior that may play a role in the behavior training problem you've been working on?

? Question

Summarize your group's ideas about retraining superstitious behaviors that contribute to health problems.

Have we sold you on the topic of Biology and the Mind?